Ultrasonics in Clinical Diagnosis

Ultrasonics in Clinical Diagnosis

EDITED BY

B. B. Goldberg

Division of Ultrasound and Radiologic
Imaging, Thomas Jefferson University
Hospital, Philadelphia, USA

P. N. T. Wells

Department of Medical Physics,
Bristol General Hospital, Bristol, UK

THIRD EDITION

CHURCHILL LIVINGSTONE
EDINBURGH LONDON MELBOURNE AND NEW YORK 1983

CHURCHILL LIVINGSTONE
Medical Division of Longman Group Limited

Distributed in the United States of America by Churchill
Livingstone Inc., 1560 Broadway, New York, N.Y. 10036, and
by associated companies, branches and representatives
throughout the world.

First Edition 1972
Second Edition 1977
Third Edition 1983

ISBN 0 443 02141 4

British Library Cataloguing in Publication Data
Ultrasonics in clinical diagnosis—3rd ed.
    1. Diagnosis, Ultrasonic
    I. Goldberg, B.B.   II. Wells, P.N.T.
    616.07'543   RC78.7.U4

Library of Congress Cataloging in Publication Data
Main entry under title:

Ultrasonics in clinical diagnosis

    Includes index.
    1. Diagnosis, Ultrasonic. I. Goldberg, Barry B. II. Wells, P.
N. T. (Peter Neil Temple) [DNLM: 1. Ultrasonics—
Diagnostic use. WB 289 U47]
RC78.7.U4U445 1983   616.07'543   83-2094

Typeset by CCC, printed and bound in Great Britain by
William Clowes (Beccles) Limited, Beccles and London

Preface

Ultrasonic diagnosis has undergone significant advances from the time this book was first published in 1972. We have seen major innovations in the development of gray scale display, real-time scanning, improved resolution and Doppler techniques. These technical advances, accomplished largely through the application of modern solid-state electronics, have led to significant progress in the clinical utility of diagnostic ultrasound.

Patients will only benefit from this technology, however, if medical and paramedical professionals possess the level of skills necessary to maximise the imaging capabilities of an ultrasonic examination. *Ultrasonics in Clinical Diagnosis* forms an integrated review of all routine ultrasonic diagnostic procedures, their scientific basis and the provision and maintenance of an ultrasonic service.

In addition, there is a critical assessment of the biological effects of ultrasound.

Today there are numerous ultrasonic training programmes designed for clinicians, physicists, engineers and technicians, and no one text can possibly match all the different syllabuses. Nevertheless, this third edition provides the information necessary for all except the most specialised advanced courses. It is the hope of those who have contributed their expertise to this edition that *Ultrasonics in Clinical Diagnosis* will prove to be beneficial to the entire international ultrasonic community.

Philadelphia and   B.B.G.
Bristol 1983   P.N.T.W.

Contributors

R. N. Baird
Department of Surgery, Bristol Royal Infirmary, Bristol, UK

Catherine Cole-Beuglet
Division of Ultrasound and Radiologic Imaging, Thomas Jefferson University Hospital, Philadelphia, USA

D. O. Cosgrove
Nuclear Medicine and Ultrasound Department, The Royal Marsden Hospital, Sutton, Surrey, UK

R. L. Dallow
Department of Ophthalmology, Harvard Medical School, Boston, Massachusetts, USA

W. J. Garrett
Department of Diagnostic Ultrasound, Royal Hospital for Women, Sydney, Australia

B. B. Goldberg
Division of Ultrasound and Radiologic Imaging, Thomas Jefferson University Hospital, Philadelphia, USA

Albert Goldstein
Division of Medical Physics, Department of Diagnostic Radiology, Henry Ford Hospital, Detroit, Michigan, USA

David Graham
Department of Obstetrics and Gynecology and Diagnostic Radiology, The Johns Hopkins Hospital, Baltimore, Maryland, USA

C. R. Hill
Department of Physics, Institute of Cancer Research, Sutton, Surrey, UK

George Kossoff
Ultrasonics Institute, Sydney, Australia

H. B. Meire
Department of Radiology, King's College Hospital, London, UK

J. R. T. C. Roelandt
Thoracic Unit, Academic Hospital, Rotterdam, The Netherlands

R. C. Sanders
Department of Radiology, The Johns Hopkins Hospital, Baltimore, Maryland, USA

P. S. Warren
Department of Diagnostic Ultrasound, Royal Hospital for Women, Sydney, Australia

P. N. T. Wells
Department of Medical Physics, Bristol General Hospital, Bristol, UK

A. H. Wolson
Section of Ultrasound and Body Computed Tomography, Allentown and Sacred Heart Hospital Center, Allentown, Philadelphia, USA

J. P. Woodcock
Department of Medical Physics, Bristol General Hospital, Bristol, UK

Contents

Physics and instrumentation

P. N. T. Wells

FUNDAMENTAL PHYSICS

Ultrasonic diagnosis depends upon physical measurements of the interactions between ultrasonic waves and biological materials. An adequate knowledge of the basic physical processes involved in the generation, propagation and detection of ultrasonic waves is necessary for a proper understanding of ultrasonic techniques.

Wave motion

Ultrasound is a form of energy which consists of mechanical vibrations the frequencies of which are so high that they are above the range of human hearing. The lower frequency limit of the ultrasonic spectrum may generally be taken to be about 20 kHz*. Most diagnostic applications of ultrasound employ frequencies in the range 1–15 MHz.

Ultrasonic energy travels through a medium in the form of a *wave*. Although a number of different wave *modes* are possible, almost all diagnostic applications involve the use of *longitudinal* waves. The particles† of which the medium is composed vibrates backwards and forwards about their mean positions, so that energy is transferred through the medium in a direction parallel to that of the oscillations of the particles. The particles themselves do not move through the medium, but simply vibrate to and fro. Thus, the energy is transferred in the form of a disturbance in the

equilibrium arrangement of the medium, without any bodily transfer of matter.

Ultrasound and other mechanical waves are quite distinct from electromagnetic waves. Electromagnetic waves consist of electric and magnetic fields, one field supporting and generating the other.

In an ultrasonic field, cyclical oscillations occur both in space and in time—the simplest type is illustrated in Figure 1.1. The oscillations here are continuous at constant amplitude, and the particles move with *simple harmonic motion*—when a particle is displaced from its equilibrium position it experiences a restoring force which is proportional to its displacement. This *direct proportionality* is the characteristic which distinguishes simple motion from other, more complicated, disturbances. The *wavelength*, λ, is the distance in the medium between consecutive particles where the displacement amplitudes are identical; similarly, the wave *period*, T, is the time which is required for the wave to move forward through a distance λ in the medium. The *frequency*, f, of the wave is equal to the number of cycles which pass a given point in the medium in unit time (usually one second). Thus:

$$f = 1/T \qquad (1.1)$$

The wavelength and the frequency are related to the propagation speed, c, by the equation:

$$c = f\lambda \qquad (1.2)$$

For example, at a frequency of 1 MHz the wavelength in water ($c = 1500$ m s^{-1}) is 1.5 mm.

These relationships apply strictly only to continuous waves of constant amplitude. Other types of disturbance (for example, pulsed waves) are not

---

* 1 Hz (hertz) = 1 cycle per second. Thus, 1 kHz = 1000 cycles per second; 1 MHz = 1 000 000 cycles per second.

† A *particle* is a volume element which is large enough to contain many millions of molecules, so that it is continuous with its surroundings; but it is so small that quantities variable within the medium (such as pressure) are constant within the particle.

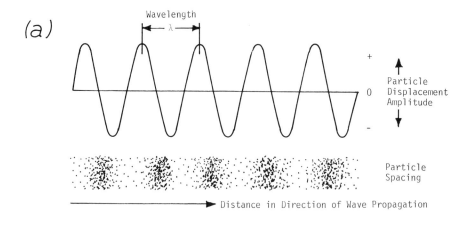

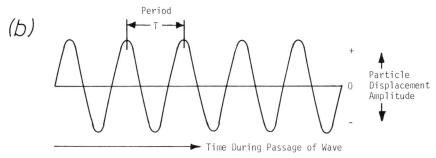

Fig. 1.1 Diagrams illustrating longitudinal wave motion (a) Particle displacement amplitude and particle spacing at a particular instant in time in the ultrasonic field: these diagrams represent the distribution of the wave in *space*. (b) Particle displacement amplitude at a particular point in space in the ultrasonic field: this diagram represents the distribution of the wave in *time*.

associated with a single frequency, and so λ and T are not constants (c is largely independent of frequency).

The speed at which the energy is transferred through the medium is determined by the delay which occurs between the movements of neighbouring particles. This depends upon the *elasticity*, K (because this controls the force for a given displacement in the medium) and the *density*, ρ (which controls the acceleration for a given force within the medium) according to the equation:

$$c = \sqrt{(K/\rho)} \qquad (1.3)$$

The speeds in different soft tissues are closely similar. The speed in bone is higher, whilst that in lung is lower. Speeds in various materials are given in Table 1.1.

Behaviour at boundaries

When a wave meets the boundary between two media at normal incidence, it is propagated without deviation into the second medium. At oblique incidence (Fig. 1.2), the wave is deviated by *refraction* unless the speeds in the two media are equal. The relationship is:

$$(\sin \theta_i)/(\sin \theta_t) = c_1/c_2 \qquad (1.4)$$

Sometimes a fraction of the incident wave is reflected at the boundary. In such cases, $\theta_i = \theta_r$, and the reflection is said to be *specular*.

In any given medium, the ratio of the instantaneous values of particle pressure and velocity is a constant. This constant is called the *characteristic impedance, Z,* of the medium, and it is related to the density and speed by the equation:

$$Z = \rho c \qquad (1.5)$$

Typical values of characteristic impedance are given in Table 1.1.

In a propagating wave, there are no sudden discontinuities in either particle velocity or particle pressure. Consequently, when a wave meets the boundary between two media, both the particle velocity and the pressure are continuous across the boundary. Physically, this ensures that the two media remain in contact. In each medium, however, the ratio of the particle pressure and velocity is fixed and equal to the corresponding characteristic impedance. If the characteristic impedances are equal, the wave travels across the boundary unaffected by the change in the supporting medium (apart from deviation by refraction, if the velocities differ, and the incidence is not normal). If the characteristic impedances are unequal, however, the incident energy is shared between waves reflected and transmitted at the boundary so as to satisfy the conditional requirements in the relationships between the particle pressures and the velocities. Because velocity is a directional quantity, whereas pressure is not, the calculation requires that account should be taken of the angle of incidence at the boundary. The most useful result is that corresponding to normal incidence — in this case, the fraction, R, of the incident energy which is reflected is given by the equation:

$$R = [(Z_2 - Z_1)/(Z_2 + Z_1)]^2 \qquad (1.6)$$

If $Z_1 = Z_2$, $R = 0$: thus, there is no reflection at a boundary between media of equal characteristic impedance. There is only a small reflection at the boundary between two soft tissues which have similar characteristic impedances. On the other hand, if $Z_2 \ll Z_1$ (e.g., at the interface between soft tissue and air), then $R = 1$, corresponding to almost complete reflection.

Plane waves, scattered waves and interference

The waves discussed so far in this chapter are known as *plane waves*. A plane wave is one in which, in any plane perpendicular to the direction of propagation of the wave and at any instant in time, every particle experiences a disturbance identical to that of every other particle.

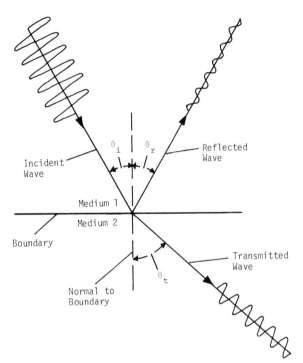

Fig 1.2 Diagram illustrating the behaviour of a wave incident on the boundary between two media.

Table 1.1. Ultrasonic properties of some common materials, including biological tissues.

	Propagation speed	Characteristic impedance	Attenuation coefficient at 1 MHz	Frequency dependence of attenuation coefficient
	$(m\ s^{-1})$	$(10^6\ kg\ m^{-2}\ s^{-1})$	$(dB\ cm^{-1})$	
Air	330	0.0004	10	f^2
Aluminium	6400	17	0.02	f
Bone	2700–4100	3.75–7.38	3–10	f–$f^{1.5}$
Castor oil	1500	1.4	1	f^2
Lung	650–1160	0.26–0.46	40	$f^{0.6}$
Muscle	1545–1630	1.65–1.74	1.5–2.5	f
Perspex	2680	3.2	2	f
Soft tissues (except muscle)	1460–1615	1.35–1.68	0.3–1.5	f
Water	1480	1.52	0.002	f^2

It is important to realise that the results of calculations of refraction and reflection conditions at a plane boundary may not apply to a similar characteristic impedance discontinuity in the form of a small obstacle, a rough surface, or an ensemble of small obstacles. (In this context, 'smallness' and 'roughness' imply dimensions of a few wavelengths or less.) The *specular* component of reflection is replaced, by an amount depending on the geometry, by components of *scattered* energy. As illustrated in Figure 1.3, the distribution of scattered ultrasound can be explained by considering the characteristic impedance discontinuity to be composed of very many tiny areas, each much smaller than the wavelength in size. Each of these areas scatters the incident plane wave as a spherical wavelet. These separate wavelets combine to form the re-radiated ultrasonic distribution.

In order to understand the way in which the wavelets combine, it is necessary to introduce the

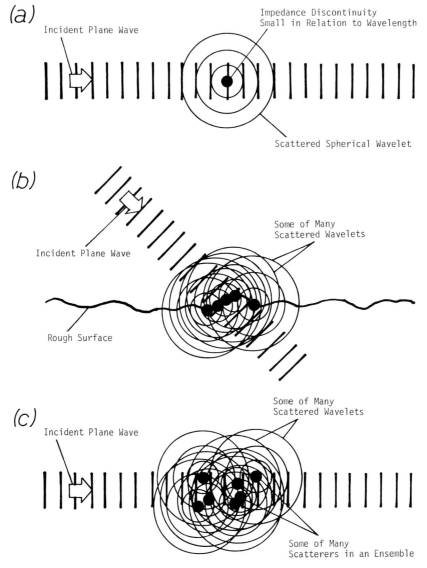

Page 1.3 Diagrams illustrating scattering. (a) Spherical scattering by a small isolated discontinuity. (b) Scattering by a rough surface; the scattered field is the resultant of every separate spherical wavelet. (c) Scattering by an ensemble of small discontinuities; again the scattered field is the resultant of the contributing wavelets.

concept of *interference*. First of all, consider the waves scattered by only two obstacles. If the positions in space of the maxima, minima and zeroes of the two waves are identical, the waves are said to be *in phase* and the resultant combination wave has a pressure amplitude equal to the sum of the separate pressure amplitudes of the two waves. This process is called *constructive interference*. On the other hand, if the maximum of one wave coincides spatially with the minimum of the other, and vice versa, the waves are said to be *in antiphase* and the combination wave has a pressure amplitude equal to the difference between the separate pressure amplitudes of the two waves. This is called *destructive interference*. The relationship between two waves can be expressed in terms of a *phase angle*. There are 360° in a complete wave cycle; waves are in phase if their relative phase angle is 0°, and in antiphase if it is 180°. Situations involving the combination of waves which are neither exactly in phase nor exactly in antiphase can be resolved by taking the phase angles into account.

Doppler effect

In the situation illustrated in Figure 1.4, the frequencies of the transmitted and the reflected waves are equal if the reflecting boundary is stationary. Movement of the reflecting boundary towards the source, however, results in a compression of the wavelength of the reflected wave, and vice versa. Since the velocity of propagation is constant, these changes in wavelength produce corresponding changes in frequency. The phenomenon is called the *Doppler effect*.

At normal incidence if f is the frequency of the incident wave, and v is the velocity of the reflecting boundary towards the source, the Doppler shift in frequency f_D which occurs in the reflected wave ($f_D = f^1 - f$, where f^1 is the received freqency) is given by:

$$f_D = 2vf/c \qquad (1.7)$$

provided that $v \ll c$, as is generally the case in diagnostic applications. In these applications, it often happens that the direction of the motion of

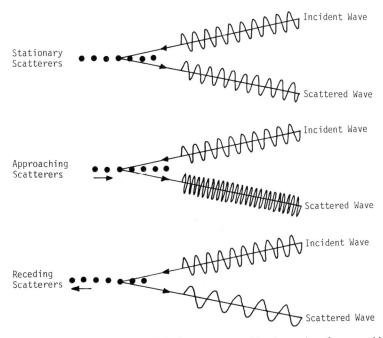

Fig. 1.4 Diagrams illustrating the Doppler shift in frequency caused by the motion of an ensemble of scatterers.

the reflecting boundary is at an angle γ with the incident wave, although the incident and reflected waves are effectively coincident. Then:

$$f_D = 2v\,(\cos\gamma)f/c \qquad (1.8)$$

Power, intensity and the decibel notation

Ultrasonic *power* may be expressed in any of several related units: one of the most useful is the *watt*. A power of 1 W is equivalent to a rate of flow of energy of 1 joule per second. The ultrasonic *intensity* is equal to the quantity of energy flowing through unit area in unit time — it may be expressed in terms of *watts per square centimetre.*★

The absolute value of ultrasonic intensity is an important consideration in relation to possible biological effects, and to the ability of a system to detect an ultrasonic wave in the presence of noise. It is frequently very convenient, however, to measure the *ratios* between pairs of intensities, or amplitudes, particularly if the *level* of one of these is taken as a reference for comparison with others. In this way, the need for absolute measurement is avoided and, because ultrasonic waves are generally both generated and detached electrically, relative wave amplitudes can be expressed as ratios of voltages.

Two advantages accrue if such ratios are expressed as logarithms. Firstly, this affords a simple method of expressing numbers which extend over many orders of magnitude. Secondly, the arithmetic product of two or more quantities is obtained by the addition of their logarithms (and similarly, by substraction, in the case of division). The logarithmic unit which is most commonly used in the *decibel*, defined as follows:

$$
\begin{aligned}
\text{(relative level in decibels)} &= 10\log_{10}(P_2/P_1)\\
&= 20\log_{10}(A_2/A_1),
\end{aligned}
$$
$$(1.9)$$

where P_1 and P_2 are the two powers, and A_1 and A_2 are the corresponding wave amplitudes.

★ The SI base unit is the [W m^{-2}], but this involves an inconveniently large area in relation to practical situations in biomedicine. The unit [W cm^{-2}] is permissible in the SI.

The intensity, I, is given by the relationship $I = \rho c v_0^2/2$, where v_0 is the peak particle velocity; and $v_0 = 2\pi f u_0$, where u_0 is the peak particle displacement.

The decibel levels corresponding to a wide range of power and amplitude ratios are shown in Table 1.2. It is important to appreciate that it is meaningless to express an absolute value of any quantity in terms of decibels, unless a reference level is also stated. Thus, for example, an intensity of 40 dB below 1 W cm^{-2} is equal to 0.0001 W cm^{-2} (i.e., 10^{-4} W cm^{-2}): note that *intensities* can be compared in the same way as *powers*. Similarly, the amplitude ratio of two waves, one of 20 dB and the other of 40 dB below a fixed reference, is equal to 10 (i.e., the first wave is 20 dB

Table 1.2 Power and amplitude ratios for various decibel levels.

dBs +ve		dB	dBs −ve	
Amplitude ratio	Power ratio		Amplitude ratio	Power ratio
1.000	1.000	0.0	1.000	1.000
0.989	0.977	0.1	1.012	1.022
0.977	0.955	0.2	1.023	1.047
0.944	0.891	0.5	1.059	1.122
0.891	0.794	1	1.122	1.295
0.794	0.631	2	1.259	1.585
0.708	0.501	3	1.413	1.995
0.631	0.398	4	1.585	2.512
0.562	0.316	5	1.778	3.162
0.501	0.251	6	1.995	3.981
0.447	0.200	7	2.239	5.012
0.398	0.159	8	2.512	6.310
0.355	0.126	9	2.818	7.943
0.316	0.100	10	3.162	10.000
0.282	0.0794	11	3.548	12.59
0.251	0.0631	12	3.981	15.85
0.224	0.0501	13	4.467	19.95
0.200	0.0398	14	5.012	25.12
0.178	0.0316	15	5.632	31.62
0.159	0.0251	16	6.310	39.81
0.141	0.0200	17	7.080	50.12
0.126	0.0159	18	7.943	63.10
0.112	0.0126	19	8.913	79.43
0.100	0.0100	20	10.000	100.00
0.0562	0.003 16	25	17.78	316
0.0316	0.001 00	30	31.62	1000
0.0178	0.000 32	35	56.23	3162
0.0100	0.000 10	40	100.00	10 000
0.0056	0.000 03	45	177.83	31 623
0.0032	0.000 01	50	316.23	100 000
0.001 00	10^{-6}	60	1000	10^6
0.000 32	10^{-7}	70	3162	10^7
0.000 10	10^{-8}	80	10 000	10^8
0.000 03	10^{-9}	90	31 623	10^9
0.000 01	10^{-10}	100	100 000	10^{10}

greater in amplitude than the second). The *half-power distance* (the distance for half the energy to be absorbed) is almost exactly equal to the length of path traversing which the intensity falls by 3 dB.

Attenuation

There are two processes by which the intensity of an ultrasonic wave may be *attenuated* during its propagation. Firstly, the wave may diverge from a parallel beam, or it may be scattered by small discontinuities in characteristic impedance, so that the ultrasonic power flows through an increased area. (Convergence of the beam results in an increase in intensity towards the focus.) Secondly, the wave may be *absorbed*, ultrasonic energy being converted into heat.

At any particular frequency, a constant fraction of the energy carried in an ultrasonic plane wave is lost by absorption in passing through a given thickness of a given material. For example, if the wave energy were to be reduced by a factor of 10 in passing through 100 mm of a certain material, then it follows that 200 mm of the material would reduce the wave energy by a factor of 100 (because the first 100 mm would absorb 90 per cent of the energy, the remaining 10 per cent being transmitted into the second 100 mm; in this, 90 per cent of the 10 per cent of the original energy would be absorbed, so that only 1 per cent would remain). In the absence of any other loss mechanism, the absorption represents the total attenuation which can be conveniently expressed in logarithmic units, such as decibels, because arithmetic products are obtained by the addition of logarithms. For instance, in the above example, 100 mm of material would reduce the wave energy by 10 dB, and 200 mm, by 20 dB — the *attenuation coefficient, α,* would be equal to 1 dB.cm^{-1}. Hence, 120 mm would reduce the wave energy by 12 dB (and it can be seen from Table 1.2 that this corresponds to a factor of 15.85 — it would be quite difficult to calculate this by linear arithmetic). Similarly, the half-power distance would be 30 mm.

The mechanisms by which ultrasound may be attenuated depend upon the properties of the material in which the wave is propagated. Some values for various typical materials are given in Table 1.1. In non-biological fluids, such as air and water, the attenuation coefficient is proportional to the square of the frequency; this is because absorption is due to viscosity, which itself is an aspect of friction between the particles of the fluid. In non-biological solids, absorption is mainly due to heat conduction; heat due to compression flows away so that energy is lost over the oscillation cycle, the attenuation coefficient being proportional to the frequency. In biological soft tissues, however, which are quasi-liquid, experimental measurements reveal that the attenuation coefficient is proportional to the frequency, at least over a limited frequency range. It seems that the most important contribution to absorption is due to *relaxation processes* in the protein constituents of the tissues. A relaxation process is one in which energy is first removed from the ultrasonic wave, and then returned at an appreciably later time in the wave cycle. Energy is transferred from the translational mode of motion (which is the ultrasonic wave) to other modes, such as vibration or rotation of the atoms within the molecules. Energy loss occurs when the translational motion of the ultrasonic wave is opposed by the translational motions returned to the wave from the transfer modes. The amount of energy loss depends upon the proportion of translational energy which is shared with other modes, and on the time constants of the relaxation processes. At very low frequencies, the time delay in energy transfer is negligible, and so the absorption is small. The absorption increases with increasing frequency, up to a maximum value when the shared energy is exactly opposed. Above this frequency, the absorption falls because there is less time available for transfer between the energy modes. The linear relationship between the absorption coefficient and the frequency which occurs in biological soft tissues over a limited range of frequency could be due to only a few distinct relaxation processes with different time constants.

The attenuation mechanism in lung is especially complicated; the behaviour of gas-filled cavities is not fully understood in this context. Likewise, the processes which occur in bone have not been satisfactorily explained, although it seems likely that scattering may have a significant role in the diploë.

GENERATION AND DETECTION OF ULTRASOUND

Piezoelectricity

In diagnostic applications, ultrasound is both generated and detected by the *piezoelectric effect*. Generation and detection are processes involving conversion between electrical and mechanical energies.

Piezoelectric materials are called *transducers* because they provide a coupling between electrical and mechanical energies. The electric charges bound within the crystalline lattice of the material are arranged in such a way that they can react with an applied electric field to produce a mechanical effect, and vice versa. In the undeformed state, the centre of symmetry of the positive charges coincides with that of the negative charges and, because the charges are equal in magnitude but opposite in sign, there is no effective charge difference across the *crystal*. If the transducer is compressed the centres of symmetry of the charges no longer coincide, so that a charge difference appears at the electrodes. The opposite charge difference appears if the transducer is extended. The converse piezo-electric effect occurs because the application of an electrical field tends to move the centres of symmetry of the crystalline charges in opposite directions, causing the transducer to deform.

Although there are many natural crystals which are piezoelectric (the best known is *quartz*), the most commonly used transducer material in ultra-sonic diagnosis is the synthetic ceramic *lead zirconate titanate*. This is a solid solution which can be polarised during manufacture so that it is strongly piezoelectric. It belongs to a group of materials called *ferroelectrics*. In ferroelectric material, there are many tiny electric charge domains which are preferentially orientated in a particular direction by the polarisation process. These domains are asymmetrical, as illustrated in Figure 1.5. The term *ferroelectric* is applied to this behaviour, because of its analogy to *ferromagnetism*.

Transducers for ultrasonic diagnosis

Narrow beams of ultrasound are almost always required in ultrasonic diagnosis. Such a beam is often generated by a disc of piezoelectric material electrically excited by means of two electrodes, one on each parallel surface. If an alternating voltage is applied between the electrodes, the piezoelectric effect causes a synchronous variation in the thickness of the transducer.

The movements of the surfaces of the transducer radiate energy into the media which are adjacent to them; the porportion of the energy which is radiated depends upon the corresponding charac-teristic impedances (as discussed on page 3). Therefore some of the vibrational energy is, in general, reflected into the transducer at each of its surfaces — this energy travels back within the transducer. Meanwhile, if the instantaneous value of the voltage applied to the transducer is varying, a new stress situation exists (due to the piezoelectric

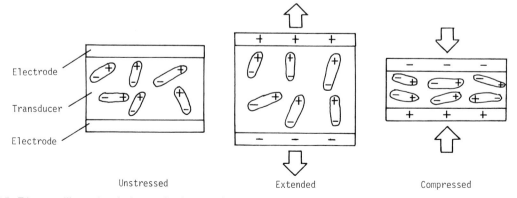

Unstressed   Extended   Compressed

Fig. 1.5. Diagrams illustrating the interaction between force and electric charge distribution in a piezoelectric transducer of the polarised ferroelectric variety. The arrows indicate the directions of the applied stresses, and the resultant surface charges are indicated at the electrodes. The converse effect leads to deformation of the transducer in response to an applied voltage.

effect) by the time that the energy reflected within the transducer arrives at the opposite surfaces. The net stress causing the tranducer to deform is equal to the sum of the piezoelectric stress and the stress due to the reflected wave. If the thickness of the transducer is equal to one half a wavelength at the frequency of the electrical excitation, these stresses reinforce each other. This condition is known as *resonance*. The displacement amplitudes of the surface of the transducer are greatest when the electrical driving frequency is equal to the mechanical resonant frequency of the transducer. Similarly, the transducer has maximum sensitivity as a receiver when it is driven by ultrasound at its resonant frequency. The ultrasonic velocity in lead zirconate titanate is about $4000\ m\ s^{-1}$; and, for example, at a frequency of 1 MHz, $\lambda/2 = 2$ mm.

In some applications, it is desirable for the transducer to operate at maximum efficiency. This requires that the electrical energy should be transferred with the minimum of loss from the transducer to the load or vice versa. In such a case,

the transducer is backed by air, to minimise the loss in its mounting, and some improvement can be gained by *matching* the characteristic impedance of the transducer to that of the loading medium by means of a matching layer attached to the transducer, of intermediate characteristic impedance and of thickness equal to one quarter of a wavelength. A more important requirement in many diagnostic applications, however, is that the transducer should be capable of responding to energy pulses of very short duration. High efficiency transducers are generally unsatisfactory in such applications. This is because transducers of this type rely upon the oscillation reinforcement which occurs at resonance, and this is greatest when the *damping* of the transducer is minimal. The efficiency is much reduced even at frequencies which deviate only by a small amount from the resonant frequency. The *response* of the transducer, however, becomes less dependent upon the frequency as the damping is increased. Thus, the efficiency (or sensitivity) of the transducer must

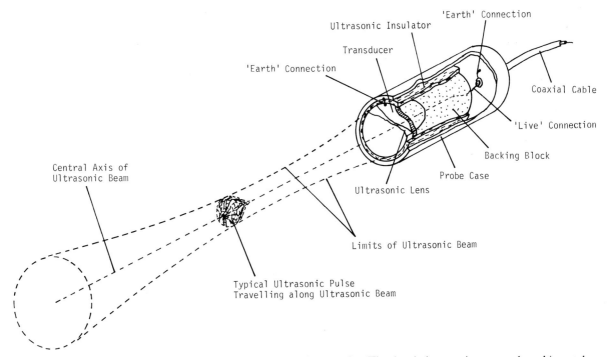

Fig. 1.6 Construction of a typical transducer probe for short pulse operation. The electrical connections are made to thin metal electrodes bonded to the flat surfaces of the transducer. Usually, the rear electrode is *live*, and the front electrode is connected to the metal case which is *earthed*. A plastic lens is used to focus the ultrasonic beam. The ultrasonic pulse is in the form of a teardrop with a pointed leading surface; it is, in fact, the volume formed by rotating the pulse shown in Figure 1.7 about the time-axis.

be reduced in order to achieve a response which is less critically dependent upon frequency.

An important property of a short-duration pulse of energy is that the energy is not confined to a single frequency. In general, the shortest pulse has the widest frequency *spectrum* of energy. Thus, in order for a transducer to respond to a short pulse of energy (which is spread over a wide frequency spectrum), it is necessary for the transducer to be damped so that its efficiency (or sensitivity) has a wide frequency spectrum.

Figure 1.6 shows a typical form of construction of a probe for the generation and detection of short ultrasonic pulses. The mechanical damping is provided by a block of highly absorbent material (e.g., fine particles of tungsten suspended in plastic) attached to the rear surface of the transducer. The front surface of the transducer is attached to a plastic matching layer. Ideally, this layer should be $\lambda/4$ in thickness, to achieve maximum efficiency; a compromise is necessary, however, because the wavelength is frequency-dependent, and the pulse energy extends over a wide frequency spectrum. The ultrasonic insulator (e.g., rubberised cork) between the case of the probe and the transducer-backing block assembly minimises the coupling of ultrasonic energy into and from the case. Such coupling is undesirable because the case may be made of a low-loss material (e.g., metal) and is likely to *ring* for some time in response to an ultrasonic transient. Ringing of the case would be detected by the transducer as an *artifact*.

The pulse response of this type of diagnostic ultrasonic probe is shown in Figure 1.7. The pulse

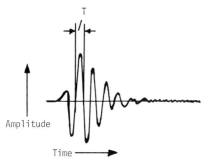

Fig. 1.7 Pulse response of a typical diagnostic ultrasonic probe. The zero-crossing frequency is equal to $1/T$. The ripple which follows the pulse is due to *ringing* of the probe case, and to radial-mode resonance of the transducer.

shape can be specified in terms of its *zero-crossing frequency*, and *duration* between specified *threshold* levels below the *peak amplitude*. These are important quantities in the estimation of the *resolution* of a pulse-echo diagnostic system (see page 23).

THE ULTRASONIC FIELD

Steady state conditions

The *ultrasonic field* of a transducer is the term used to describe the spatial distribution of its radiated energy. By reciprocity, this is identical to the sensitivity distribution of the transducer when acting as a receiver.

The analysis of the ultrasonic field is based on the application of *Huygen's construction*, in which the surface of the transducer is considered to be an array of separate elements each radiating spherical waves in the forward direction. The elements move in synchrony with equal amplitudes (i.e., a disc transducer is considered to be a piston the surface of which vibrates cophasally at constant amplitude). This is known as a *steady state* condition. The method of analysis is illustrated in Figure 1.8 by a simple example, in which the source consists of a six-element linear array. The waves *interfere* constructively to reinforce each other along the lines which touch the spherical waves due to every element. In other places, there is a tendency for maxima and minima to coincide, so that destructive interference occurs, and there is little net disturbance. Consequently, the ultrasonic field is concentrated within a *beam*, and this becomes more uniform with increasing distance from the array.

It is relatively easy to apply Huygen's construction to the analysis of one dimension of the ultrasonic field of a linear array. The situation is much more complicated, however, if the source is in the form of, for example, a circular piston. The problem is one of three-dimensional geometry — the theoretical field for such a source is shown diagrammatically in Figure 1.9. Moving along the central axis of the beam towards the source, the intensity increases until a maximum is reached at a distance $z|_{max}$ from the source given by:

$$z|_{max} = a^2/\lambda \qquad (1.10)$$

where a is the radius of the source and $a^2 \gg \lambda^2$ (as is usually the case in ultrasonic diagnosis). Increasingly closely spaced axial maxima and minima occur towards the source. At successive axial maxima and minima, starting at $z^{|}_{max}$ and moving towards the source, there are one, two, three, etc., principal maxima across the beam diameter. Thus,

the beam contains two distinct regions. The region between the source and the last axial maximum (at $z^{|}_{max}$) is known at the *near field* (or *Frésnel* zone); the region beyond this is the *far field* (or *Fraunhofer* zone). The beam is roughly cylindrical in the near field. Deep in the far field, the beam diverges at

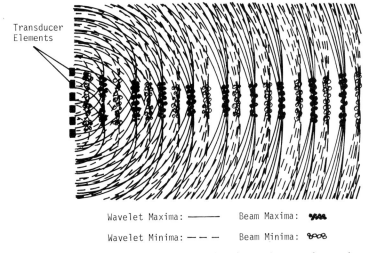

Wavelet Maxima: ———   Beam Maxima:
Wavelet Minima: — — —   Beam Minima:

Fig. 1.8 Diagram illustrating the formation of an ultrasonic beam by interference between the wavelets emitted from a six-element linear array.

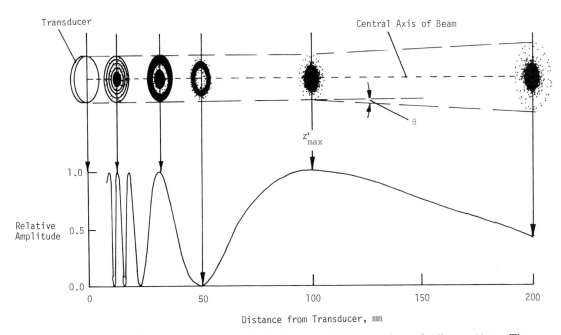

Fig. 1.9 The ultrasonic field. This example shows the distribution for a 1.5 MHz transducer of radius $a = 10$ mm. The ultrasonic beam normal to the central axis is circular in section, and the elliptical diagrams represent such sections, in the planes indicated on the graph. The graph shows the variation in the central axial amplitude with distance from the transducer.

angles $\pm \theta$ about the central axis, given by:

$$\sin \theta = 0.6\lambda/a \qquad (1.11)$$

Thus, the shape of the ultrasonic field depends upon the diameter of the transducer and the wavelength of the ultrasound. For example, a transducer of 20 mm diameter operating at 1.5 MHz in water has a near field length of 100 mm and a half-angle of divergence of 3.5° in the far field. The length of the near field increases with increasing diameter of the transducer and increasing frequency of the ultrasound (because $\lambda = c/f$); but the divergence in the far field decreases with increasing diameter and increasing frequency.

Transient conditions

In the steady state, the shape of the ultrasonic field is determined by interference between contributions from the entire surface of the source. If a short-duration pulse of ultrasound is involved, however, all the contributions which combine together to form the field in the steady state may not be present at any particular point in space during the passage of the transient. This is because, in general, the surface of the source is not equidistant from the point (except at the focus of a curved transducer). Consequently, the field shape changes with time. In the far field, steady state conditions apply within about half a cycle of the arrival of the transient; but in the near field, the transition between transient and steady state conditions may occupy from three to six half-cycles.

Focused fields

In the near field, an ultrasonic beam may be *focused* over a limited depth of field. This greatly improves the resolution in imaging systems (see page 23). The materials, such as plastics, from which ultrasonic lenses may be constructed generally have higher propagation speeds than water or soft tissues, so that converging lenses are concave.

In considering the phenomenon of focusing, a useful concept is that the focus occurs at the point in the field at which the contributions from the entire surface of the transducer all arrive together,

or *in phase*. Thus, lenses function by introducing appropriate thicknesses of material in which the speed differs from that in the medium, so that the transit times along ray paths of different length are all equal, as illustrated in Figure 1.10a. Focusing may also be achieved by the use of a concave transducer giving equal length ray paths from the transducer surface to the focus, as illustrated in Figure 1.10b. Figure 1.10c shows how the combined effects of convergence due to a concave transducer, and divergence due to a convex plastic lens, can make it possible for a flat-faced (or even slightly

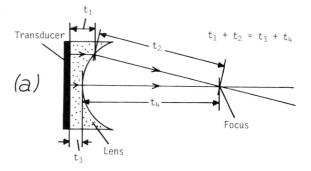

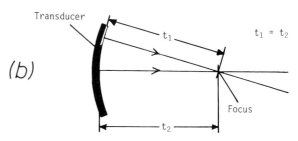

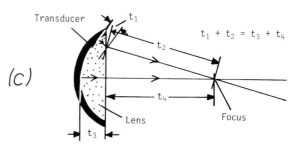

Fig. 1.10 Methods of focusing an ultrasonic beam. The distances indicated in these diagrams are expressed in terms of the propagation times, t, along each path. In all cases, the time delays are equal along all ray paths between the surface of the transducer and the position of the focus. (a) Plane transducer with concave plastic lens. (b) Concave transducer. (c) Concave transducer with convex plastic lens, producing a focused beam from a probe with a flat (or even slightly convex) surface.

convex) probe to produce a focused ultrasonic beam. This approach is often used in practice, because a concave probe face is difficult to use in skin contact scanning, as it tends to trap air which interferes with ultrasonic coupling.

The effective *depth-of-focus* is an important factor affecting the resolution in ultrasonic imaging (see page 23). In general, a large aperture transducer has a strong focusing action and a short depth-of focus; a small aperture has weak focusing and long depth-of-focus. Moreover, for a given transducer aperture, focusing close to the transducer is stronger and the depth-of-field is shorter than when focusing is at greater axial depth.

Lenses and curved transducers have fixed focal lengths. Another method by which focusing can be introduced is by means of appropriate time grading across an *array* consisting of small transducer elements. This makes it possible for the focal length to be controlled electronically (see page 22).

PULSE-ECHO DIAGNOSTIC METHODS

Ultrasonic diagnosis is based, with very few exceptions, upon the reflection of ultrasonic waves which occurs at the boundaries between different tissues within the body. A fraction of the incident ultrasonic energy is reflected if there is a change in characteristic impedance at such a boundary. Although the echoes which correspond to soft tissue boundaries have very small amplitudes,* they can be detected by a sensitive receiver. The energy which is not reflected travels beyond the boundary, and may be reflected at deeper boundaries. The maximum penetration is limited by the attenuation of the ultrasound in passing through the tissues.

Ultrasound is almost completely reflected at boundaries with gas. This prevents examination through gas, and is a serious restriction in investigations of and through gas-containing structures. In addition, it is necessary to exclude air from between the transducer and the patient — this is done by means of a liquid coupling medium.

The attenuation rate and the propagation speed

---

* For example, substitution of the appropriate values in Equation 1.6 shows that, at a flat interface between kidney and fat, about 0.6 per cent of the incident energy is reflected.

are both much higher in bone than in soft tissues. Consequently, examinations through bone are difficult, and they are only really satisfactory in a few specific applications.

The time delay which occurs between the transmission of the pulse and the reception of its echo depends upon the propagation speed and the path length. Range-measuring systems based on this principle are widely applied in medical diagnostics. The propagation speeds are similar in different soft tissues, and any error introduced by assuming a constant relationship between time and distance can usually be neglected. At a velocity of 1500 m s^{-1}, ultrasound travels 10 mm in about 6.7 μs.

The A-scope

The basis of the pulse-echo method is illustrated in Figure 1.11. The diagrams show how an ultrasonic pulse may be used to measure the depth of an echo-producing boundary. If the process is repeated sufficiently rapidly (at a rate greater than about 20 times per second), the persistence of vision of the observer gives the impression of a steady trace on the display. In soft tissues, the ultrasonic echoes are delayed in time from the transmission pulse by about 13.3 μs for each centimetre of penetration, during which time the ultrasonic pulse travels along a total go-and-return path length of 20 mm.

The kind of display illustrated in Figure 1.11 is called an *A-scan*. (This term, like several others used in ultrasonics, is the same as that which describes the equivalent display in radar.) The method can be extended to the examination of many interfaces lying along the ultrasonic path.

The basic elements of an A-scope instrument are shown in the block diagram in Figure 1.12. The *rate generator* (or *clock*) simultaneously triggers the *transmitter*, the *swept gain generator* (or *time gain control* generator) and the *timebase generator*. The voltages which appear at the probe are amplified by the receiver, and applied to the vertical deflection plates of the cathode ray tube. The output from the timebase generator is applied to the horizontal deflection plates.

A substantial improvement in the usefulness of the displayed information is obtained if the echo

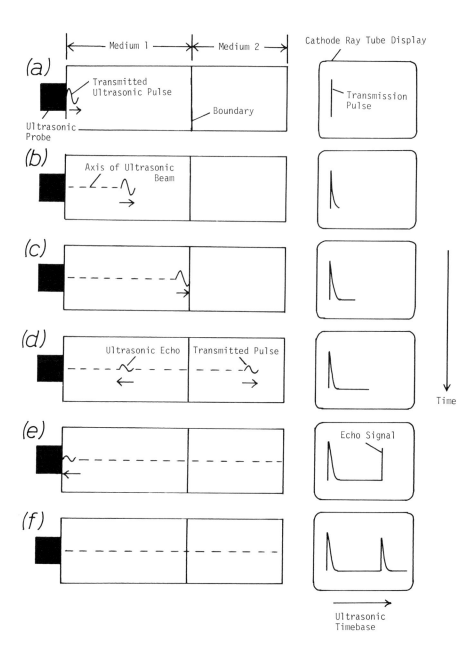

Fig. 1.11 Basic principles of the pulse-echo system. (a) The probe emits a short-duration ultrasonic pulse into medium 1, and simultaneously the spot on the cathode ray tube display begins to move at constant velocity from left to right. The vertical deflection plates of the cathode ray tube are connected to the output from an amplifier, the input of which is derived from the ultrasonic probe. The spot is deflected vertically at the instant that the ultrasonic pulse is emitted because the exciting voltage is also applied to the amplifier. (b) The ultrasonic pulse travels along a narrow beam at constant speed through medium 1, and the spot traces a horizontal line on the display. (c) The ultrasonic pulse reaches the boundary between media 1 and 2. (d) If there is a characteristic impedance discontinuity at the boundary, some of the ultrasonic energy is reflected back into medium 1. (e) The reflected wave reaches the probe, and generates a voltage which produces a second deflection on the display. (f) The distance between the two deflections on the display is proportional to the distance between the probe and the boundary between the two media.

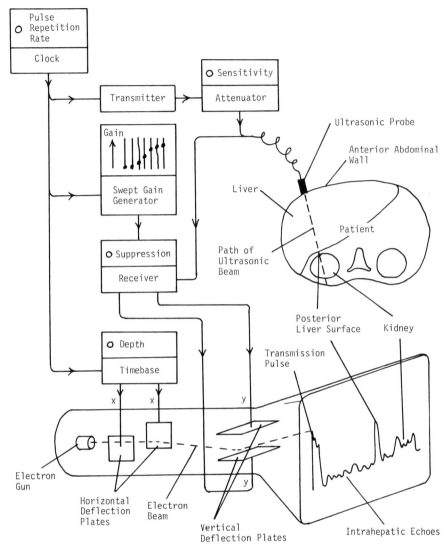

Fig. 1.12 Basic elements of the A-scope system. In this diagram, the ultrasonic pulse is illustrated in contact with the right side of the abdomen, and the display on the cathode ray tube shows echoes of the ultrasonic pulse as it travels through the anterior abdominal wall, the liver, the right kidney and the posterior abdominal wall.

signals from deeper structures are amplified more than those which originate closer to the probe. This function is performed by the swept gain generator, which is arranged to increase the gain of the receiver with time at an appropriate rate. Ideally, swept gain should lead to similar registrations of similar surfaces, irrespective of their distances from the probe. In practice, however, accurate swept gain is difficult to achieve, for two main reasons. Firstly, there is a variation in the attenuation rates of different tissues. Secondly, the

energy in the ultrasonic pulse is distributed over quite a wide frequency spectrum, and the higher frequency components of the pulse are increasingly attenuated with increasing penetration (the attenuation coefficients in soft tissues are approximately proportional to the frequency, see page 7).

In certain circumstances, diagnostic information may be obtained by studying the amplitudes of the displayed echoes. Such investigations are often made easier if swept gain is applied to the system — it is essential in grey-scale scanning, to obtain the potential advantages of the method (see page 19). It is important to remember that an extensive smooth surface gives rise to specular reflection, and so the amplitude of the echo received from such a surface is very directionally dependent. However, the echo from a small discontinuity, although smaller in amplitude, is less markedly directional (see page 4).

Although in a few systems there is a linear relationship between the input and output signals of the receiver, it is more usual for some form of *signal processing* to be used. Thus, the signal is generally *demodulated* so that the output is a undirectional representation of the alternating voltage which occurs at the probe. In addition, the *dynamic range* of the input signals may be compressed, so that a wide variation in amplitude at the input produces a smaller variation at the output — this may be achieved by an amplifier with a logarithmic characteristic.

Operation of the A-scope

A-scope systems are normally fitted with a number of manually operated controls. There is considerable variation in this respect between different instruments. A brief description of the more important of their functions is given here.

Rate generator

The *repetition rate* of the *clock* may be adjustable. A higher repetition rate gives a brighter display, but restricts the maximum penetration (because the interval between the ultrasonic pulses becomes closer to the time during which echoes are being received), and increases the possibility of biological hazard.

Transmitter

A variable *attenuator* may be fitted to control the amplitude of the transmitted ultrasonic pulse. This has the effect of controlling the overall sensitivity of the system.

Swept gain generator

The characteristics of the swept gain function, such as its *timing* and its *rate*, may be adjustable. It

may be possible to display simultaneously the A-scan and the swept gain function, so that the effect of the latter may be more easily appreciated.

Receiver

This may be fitted with a variable *gain* control, to adjust the sensitivity of the system (in view of the possibility of biological hazard, this is a less desirable method of sensitivity control than is provided by the transmitter attenuator), and a variable *suppression* control, to reject small echoes from the display. In most systems, the demodulated signal is displayed; but in some instruments, it is possible for the undemodulated signal to be displayed. In addition, the operating *frequency* may be adjustable, often within preset limits about a specified centre frequency.

Timebase generator

The *velocity* of the timebase may be adjustable. A *time-marker* circuit is sometimes included — this generates time-markers on the display at positions corresponding to known intervals of distance in soft tissues.

Display

This may have the usual controls for the adjustments of *focus, brightness* and *astigmatism*. It may be possible to *invert* the trace, and to *shift* it both horizontally and vertically.

Multiple reflection artifacts

A serious limitation of the pulse-echo method is due to the *multiple reflections*, or reverberations, that the ultrasonic pulse may suffer during its propagation. For example, in Figure 1.11 the pulse which returns to the probe from the boundary between the two media, and which causes the corresponding deflection of the spot, is not completely attenuated within the probe. This is because there is a characteristic impedance discontinuity at the probe surface, and quite a large proportion of the echo energy is reflected back into the first medium. This reflected pulse behaves as if it were a second transmitted pulse, of smaller amplitude than the first, delayed in time by an interval equal to the delay in the return of the first echo. Consequently, a second echo returns from the boundary between the first and second media, and this produces a registration on the display at a position corresponding to twice the range of the boundary. Similarly, third and subsequent reverberation *artifacts* appear, until the ultrasonic pulse becomes too small to be detected.

Multiple reflections causing artifacts are sometimes quite easily recognised by the regularity of their spacings. Those due to gas and bone are a fundamental limitation in ultrasonic diagnostics.

The B-scope

The information obtained with a pulse-echo system is a combination of range and amplitude data which can often be conveniently presented as an A-scan. The same information, however, may alternatively be displayed on a brightness-modulated timebase, in such a way that the brightness varies with the echo amplitude, as shown in Figure 1.13. This type of display is called a *B-scan*. The B-scope is the basis of two-dimensional scanning and time-position recording.

The two-dimensional B-scope

Static B-scanners

It is possible to mount the ultrasonic probe on a mechanical scanner which allows movements in two dimensions, and which links (either electrically or mechanically) the direction and position of a B-scope timebase on a cathode ray tube to those of the ultrasonic beam within the patient. Then, by continuously recording the display whilst the probe is moved around the patient, a cross-sectional picture in the plane of the scan (a *tomograph*) is constructed in two dimensions. The process is illustrated in Figure 1.13, and this type of two-dimensional B-scope is called a *static scanner*. The image usually stored in a *scan converter*, although

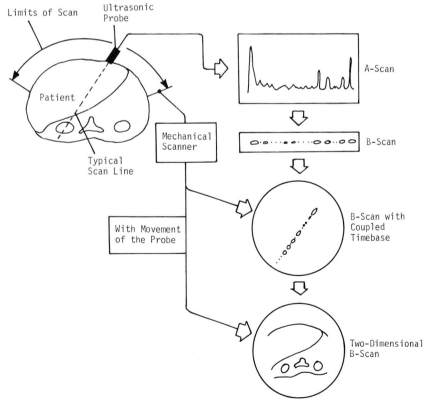

Fig. 1.13 Diagrams illustrating the relationship between the A-scan and B-scan display, and the basic principles of two-dimensional ultrasonic scanning.

it may be photographed directly. Two of the most important types of static scanner are illustrated in Figure 1.14, and details of the articulated arm type are shown in Figure 1.15.

Scan converters

The image obtained during ultrasonic scanning, e.g., as the probe is moved across the surface of the patient, may be stored by an electronic device called a *scan converter*. The name reflects the role of the device in converting from the line arrangement of the ultrasonic beam within the anatomical scan plane — which may be unpredictably determined by the scanning motion adopted by the operator, or in a rectangular, trapezoidal or sector format determined by the instrument design — to the raster line arrangement appropriate for display on a cathode ray tube TV monitor.

There are two main types of scan converter in ultrasonic systems. The first of these, now becoming obsolete, is the *analogue* scan converter. This device resembles a small cathode ray tube, but in which the viewing screen is replaced by a fine matrix of insulated charge storage elements. The image is built up as a charge pattern on this matrix, and is subsequently read out onto a TV monitor by raster scanning of the charge pattern. The analogue scan converter has the advantages of excellent

resolution, wide dynamic range, and the ability to store the peak echo amplitude detected from each element of the anatomical section. Its disadvantage is that, being an analogue device, it is prone to drift out of optimal adjustment.

Nowadays, image storage and scan format conversion is usually accomplished by means of a *digital* scan converter. This is a device which breaks the image up into a regular matrix of picture elements, called *pixels*, rather in the same way that a half-tone printed picture is made up of a matrix of dots. Each pixel memory element is able to store a digital signal related to the analogue voltage produced by the echo signal from the corresponding area in the ultrasonic scan plane. The data stored in the pixels are scanned sequentially in a TV raster format, and converted to analogue form for display on a cathode ray tube, which can be viewed directly, photographed or stored, for example on videotape or videodisk.

The dynamic range of the display is a measure of its ability to represent input amplitude variation in terms of spot brightness. It may be expressed as the decibel ratio between the minimum input amplitude at which a registration is just visible on the display, and the maximum input amplitude above which there is no further visible increase in brightness. A second, and equally important, consideration is the number of separate shades of

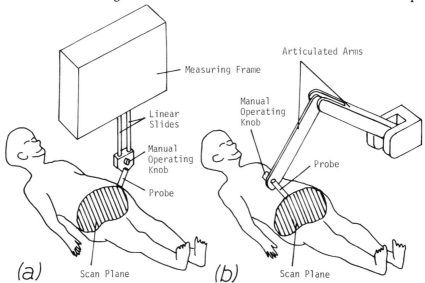

Fig. 1.14 Two types of commercially available two-dimensional static B-scanners in common use. (a) A manually-operated contact scanner with a rectilinear measuring frame. (b) A manually-operated contact scanner with articulated arms working in polar coordinates; more details of this arrangement are given in Figure 1.15.

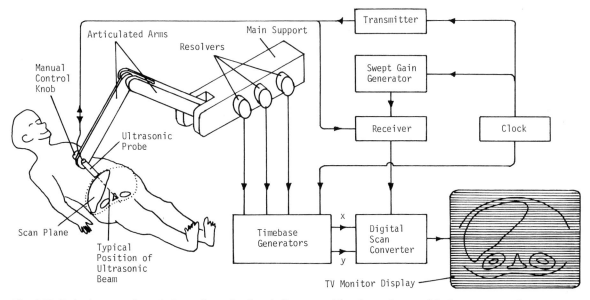

Fig. 1.15 Basic elements of a typical two-dimensional static B-scanner. The ultrasonic part of the instrument produces a brightness-modulated signal that is written into the digital memory of the scan converter in address locations determined by the x and y timebase generators. These timebase generators are controlled by electrical signals from three resolvers mounted on the main support of the scanner. The spatial signal input to the digital memory represents the direction and position of the ultrasonic beam across the patient. The information in the digital memory is continuously read out in TV raster format and displayed on a cathode ray tube monitor.

brightness which may be identified on the display over its dynamic range. This depends on a number of factors, including both the technical properties of the display, and the psychophysics of the observer. Some general principles are illustrated in Figure 1.16.

Nowadays, there is much emphasis on *grey scale* capability in two-dimensional scanning. The grey scale display allows deductions to be made from the scan concerning quite subtle characteristics of the echo amplitude arising from within organs and structures. Typical digital scan converters have a matrix of 512×512 pixels, and each pixel is represented by a digital word with a length of 4, 5 or 6 bits. A word length of 4 bits, for example, allows $2^4 = 16$ discrete grey levels to be stored, corresponding to a dynamic range of about 24 dB. Likewise, 5 bits is equivalent to 32 grey levels or 30 dB, and 6 bits, to 64 grey levels or 36 dB.

Digital scan converters have the important characteristic of being free from drifting in their performance. Moreover, they allow an image, once stored, to be viewed for an unlimited time; desired images can be *frozen* for unhurried study, calibrated scales can be overlaid on the display, and

the image can be refreshed line-by-line, so that new image information can replace old information during selective scanning of anatomical areas of interest.

Compound scanning

The angles of incidence and reflection of an ultrasonic beam are equal at a flat surface. The maximum echo amplitude is detected at normal incidence, since only then does the echo return directly to the transducer. Consequently, in traditional tissue-mapping applications at least, an improved image (a *compound B-scan*) is obtained if the probe is oscillated as it is moved around the patient, because this increases the likelihood of the occurrence of normal incidence with specular reflectors. Complete information cannot be obtained in this way, however, if the surface being examined is not normal to the plane of the scan, nor is it usually possible to scan through a complete circle. Another consideration is that inaccuracies in registration, and movement of the patient occurring during the scanning process, degrade compound scan images. For these reasons, com-

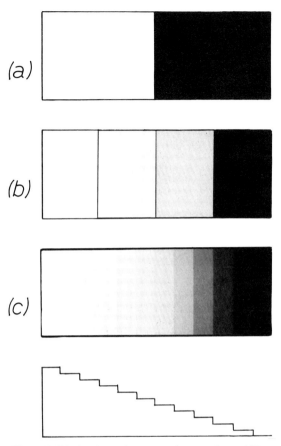

Fig. 1.16 Display dynamic range and grey scale capability. The graph at the bottom of the diagram represents the input amplitude. (a) This is a *bistable* display, in which there is a threshold input level above which the display is fully white, and below which it is fully black. (b) In this display, there are two discernible brightnesses between white and black, and the *grey scale* has 4 levels. (c) This grey scale has 12 discernible levels.

pound scanning is seldom used nowadays with static scanners. Moreover, digital scan converters operated in the usual way, so that new information is substituted for old information, do not provide the integration of echoes implicit in compound scanning.

Large water-bath B-scanners

Another class of two-dimensional scanners employs a large water-bath to provide coupling between the ultrasonic transducer and the patient. These instruments, typical examples of which are illustrated in Figure 1.17, have the advantages that

the movements of the transducer may be automatic, and that the patient may be examined without any local disturbance of body structures being caused by the movement of the probe. On the other hand, there may be considerable difficulty in achieving a satisfactory coupling between the patient and the water (unless immersion is possible), because air is likely to be trapped between the flexible membrane, which forms the wall of the water tank, and the patient's skin. This difficulty restricts the areas of the body which can be examined by the method.

Real-time scanners

A *real-time scanner* may be defined as being one which produced sequential image frames at a rate fast enough to follow changes in spatial relationships within the anatomical scan plane. Such changes may be due either to physiological movements within the defined spatial plane, or to movements of the scanner bringing different structures into the scan plane, or to a combination of these two types of motion. For example, in studying the movements of heart valves, a frame rate of 60 per second might be necessary to satisfy the definition of 'real-time', whereas a frame rate of four per second might be adequate to follow changes as the scanner is moved over the surface of the abdomen, provided that suitable image storage arrangements eliminate flicker.

Ultimately, the image line rate (and hence, the frame rate) is limited by the speed of ultrasound in tissue. If, for example, a penetration of 150 mm is required, the time which elapses between the transmission of the ultrasonic pulse and the reception of the echo from the maximum range is equal to 200 μs (taking the speed to be 1500 m s^{-1}). the corresponding maximum pulse repetition rate is about 5000 s^{-1} (although in practice 2000–3000 s^{-1} would be likely to be used). Thus the maximum image line rate is about 5000 s^{-1}. This is equal to the product of the number of lines per frame and the number of frames per second, e.g., at 40 frames per second, there would be 125 lines per frame, and so on.

Some of the many methods of rapid scanning to produce real-time images are illustrated in Figure 1.18.

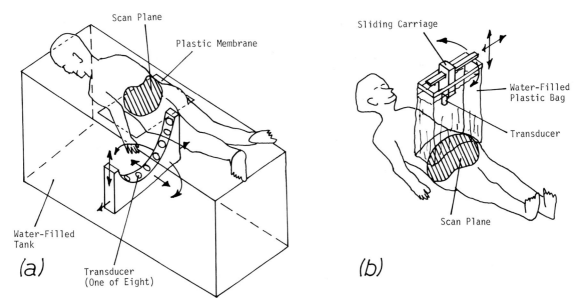

Fig. 1.17 Two types of commercially-available two-dimensional large water-bath scanners. The solid arrows indicate available adjustments in scan plane orientation. (a) An advanced scanner employing several (eight) transducers, each of which can be oscillated through any chosen sector to produce a scan with or without compounding. (b) A simple scanner producing scans with rectangular format in vertical planes.

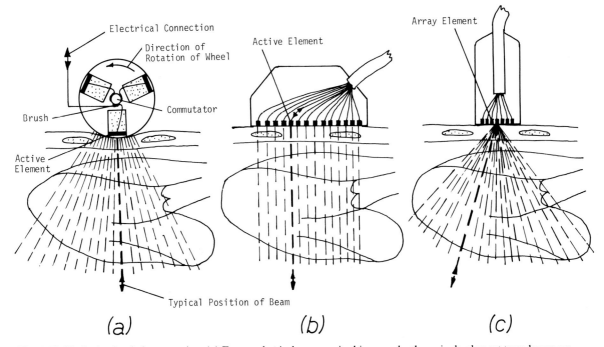

Fig. 1.18 Methods of real-time scanning. (a) Fast mechanical scanner: in this example, three single-element transducers are mounted on a wheel which rotates continuously, and as each transducer in turn comes into contact with the patient (or with a thin membrane in contact with the patient), it sweeps out a new frame made up of an image sector. (b) Electronically scanned linear array transducer; the transducer elements are addressed sequentially, to sweep out image frames with a rectangular format (see Figure 1.20 for more detailed information). (c) Transducer array with electronic beam steering; the signal paths associated with the individual elements in the array are delayed appropriately to sweep out image frames with a sector format (see Figure 1.22 for more detailed information).

Mechanical real-time scanners. Two types of this class of scanner are illustrated in Figure 1.19. Figure 1.19a shows a rapidly-oscillated annular array (see page 13), resembling the large water-bath scanner (Fig 1.17b) but operating fast enough to produce trapezoidal-format real-time images. This is suitable for abdominal studies. Figure 1.19b has a fixed transducer and an oscillating mirror, and produces sector scans. Different versions of this are suitable for abdominal, cardiological, and small parts scanning. There are many other possible arrangements.

Electronically switched linear array real-time scanners. The principles of this class of scanner, sometimes called a *linear array*, are illustrated in Figure 1.20. In practice, the transducer elements are not addressed one at a time, but typically in groups of four stepping one element at a time along the array. This is done to combine the advantages of a larger aperture (to give good resolution, see page 23) and high image line density. By operating with groups of four elements, a 64-element array gives 61 image lines, and so on for other elements groupings and numbers.

The thickness of the scan plane can be reduced for a fixed depth-of-field by attaching a cylindrical lens to the face of the array. Dynamic focusing can be applied within the scan plane, as illustrated in Figure 1.21. This type of focusing is the electronic analogue of the lens focusing (page 12), in that the required time delays are introduced on the electronic side of the transducer, rather than on the ultrasonic side. Fixed focusing only can be applied on transmission, but on reception *dynamic* focusing can be introduced by sweeping the time delay circuits continuously to coincide with the echo-producing targets along the ultrasonic beam axis.

Electronically steered array real-time scanners. This class of scanner, sometimes called a *phased array*, is a refinement of the electronically switched

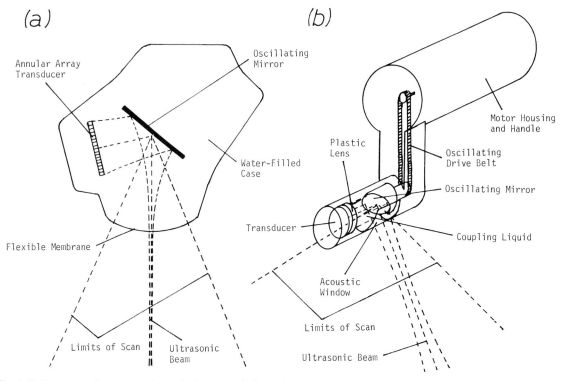

Fig. 1.19 Two types of commercially-available mechanical two-dimensional real-time scanners. (a) A large instrument suitable for abdominal scanning. The instrument casing is supported by an easily-manoevrable gantry. In this example, the transducer is an array of annular elements, and electronic time grading across the array sweeps the focus on reception to give good lateral resolution over a long depth-of-focus. (b) Small, hand-held, scanner head using an oscillating mirror to produce sector scans. This ingenious design allows lens focusing to be used, and the beam shape is unaffected by the mirror angle since the oscillation of the mirror takes place around the ultrasonic beam axis.

linear array. It employs an array typically composed of 20 small strip elements mounted side-by-side in a probe similar in overall size to the single-transducer-element used in a static B-scanner. The ultrasonic beam can be dynamically focused in just the same way as a linear array. The beam can also be steered through a sector, as illustrated in Figure 1.22.

Electronically steered arrays have the advantage, in competition with mechanical real-time sector scanners, of allowing the beam direction to be changed instantaneously, and of being steered for any desired time in any desired direction. This is valuable when simultaneous real-time scanning and M-mode recording (page 26) are carried out. Steered arrays have the disadvantage of relatively high cost, and the tendency to suffer from ultrasonic beam side-lobes which may result in strongly-reflecting structures being artifactually misregistered on the display.

Resolution in pulse-echo systems

The *resolution* of a pulse-echo system is defined as the reciprocal of the distance which appears on the display to be occupied by a small reflector in the ultrasonic field. As in the definitions of display dynamic range and grey scale (page 19), considerations of the psychophysics of vision are involved in this definition.

Two different resolutions are of importance: these are the *lateral resolution*, which describes the resolution along the beam diameter normal to the axis and the *range resolution*, which is the resolution along the axis.

The range resolution depends upon the shape of the ultrasonic echo, and the characteristics of the receiver. The waveform of a typical echo pulse at the output from the transducer is shown in Figure 1.7. Given such an echo, the resolution corresponds to the time interval during which the echo

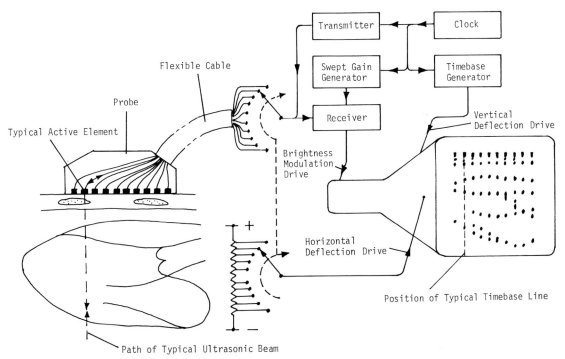

Fig. 1.20 Principles of the linear array real-time scanning system. In this diagram, the probe contains 10 separate transducer elements. The clock triggers the transmitter, and, in this simple example, the transmitter pulse is applied, through a sequencing switch, to one of the transducer elements. Simultaneously, the clock triggers the timebase generator connected to the vertical deflection plates of the cathode ray tube display. Echoes returning from within the patient are detected by the transducer element which emitted the original pulse, fed through the sequencing switch, and amplified (under swept gain control) to brightness-modulate the display. Each element is rapidly addressed in sequence, and a two-dimensional image is built up by a second sequencing switch applying appropriate horizontal deflection voltages to the cathode ray tube.

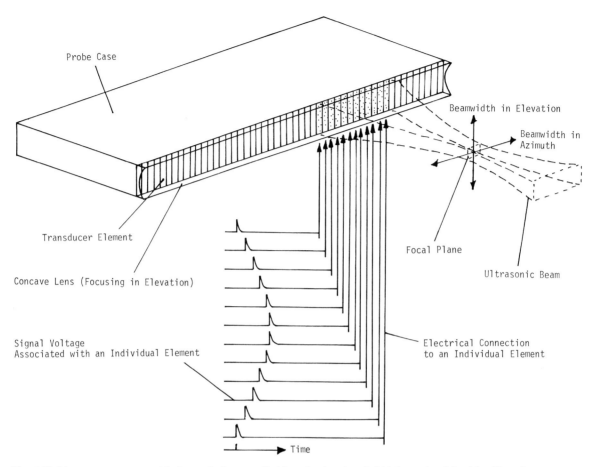

Fig. 1.21 Linear array scanner with electronically controlled focusing in azimuth (this has to be of fixed focal length on transmission, but can be swept on reception), and with lens focusing in elevation. In this example, the elements are used in groups of 12, and there are 51 elements in the array, giving 42 separate lines in the image.

amplitude is large enough to produce a registration on the display. Some systems respond only to the leading edge of the echo, and so the 'resolution' seems to be better than pulse length considerations would indicate — such a display is said to be *differentiated*.

The lateral resolution depends upon the effective width of the ultrasonic beam and the characteristics of the receiver. For example, the beam shape of a typical probe is shown in Figure 1.6. At any given range, the resolution corresponds to the width of the beam from within which the echo amplitude from a point reflector is large enough to produce a registration on the display.

These definitions neglect the complication of grey scale in discussing resolution. It may be more meaningful to define the resolution as the reciprocal of the minimum distance between two point objects in the ultrasonic field, which produces separable registrations on the display. Whilst this is a useful concept, it does involve two difficulties. The first is the psychophysical problem of defining perceptible brightness differences. The second is a physical problem which arises because the echoes from adjacent targets may interfere, destructively or constructively according to their phase relationship. Nevertheless, some excellent *phantoms* have been developed, and these are very useful for checking scanner performance. In addition to simple wire phantoms, such as the *AIUM test object*, which give insight into resolution capability, phantoms containing controlled distributions of scatterers are available for assessing grey scale, dynamic range, and the detectability of subtle

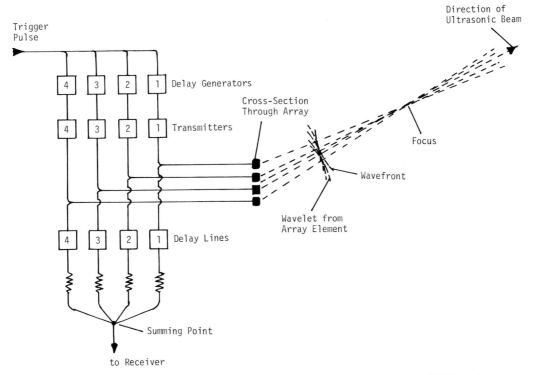

Fig. 1.22 Principles of electronic beam steering. The trigger pulse starts the delay generators 1–4, which introduce progressively longer delays in the excitation of the elements along the array. Consequently the wavefront, which is the surface joining the cylindrical wavelets emitted by the transducer elements, has its normal (the direction of the ultrasonic beam) lying along a bearing which can be changed by changing the delay times. Similarly on reception, the delay lines introduce time delays in the individual signal paths associated with each element in the array so that echoes from the focus arrive simultaneously at the summing point.

characteristics such as small cystic areas within homogeneous tissue.

Elementary considerations indicate that the resolution should improve with increasing frequency. The attenuation coefficient of soft tissues increases with frequency, however, and this leads to a reduction in the maximum penetration (because there is a limitation to the maximum amplification of the receiver). In addition, the errors in the swept gain compensation increase as the range of the swept gain control is increased to compensate for the greater attenuation at higher frequencies. These errors lead to a deterioration in the resolution which may offset any theoretical advantage of operating at a higher frequency. Optimum results are obtained in abdominal and neurological investigations at frequencies of 1–3 MHz, in cardiovascular work at 2–5 MHz, in small parts scanning at 4–6 MHz, and in ophthalmology at 8–20 MHz. It is helpful to consider the

wavelength to be the factor which controls the dimensions of the ultrasonic field. In most applications, a transducer diameter of 20λ, and a penetration of 200λ, give satisfactory results.

The C-scope

The *C-scope* is a brightness-modulated display of a scan plane in which deflection along one axis corresponds to the angular position (the *azimuth*) of the echo-producing structure, and that along the other axis to the corresponding distance from a reference datum (the *elevation*). The display is somewhat similar to a conventional X-ray picture, because it is in the form of a plane 'view'; but unlike the X-ray, the depth and thickness of the section can be controlled by electronic time-gating of the echo signals.

Because C-scanning is inherently slow — only one data point per plane is collected for each transmitted ultrasonic pulse — its place in diagnosis has hardly been explored. Results obtained in a few applications (notably in breast scanning) indicate its potential value.

Time-position (M-mode) recording

Pulse-echo information from moving structures can be recorded to generate time-position waveforms. The simplest technique is to photograph a B-scan display with a camera in which the film is moving at constant velocity in a direction at right angles to the ultrasonic timebase. In modern instruments, the recording is made in this way by means of a fibre-optic cathode ray tube on self-processing paper sensitive to ultraviolet radiation, or on dry-silver photographic paper, as shown in Figure 1.23.

Time-position recordings are usually referred to an *M-mode* recordings. The principal application of the technique is in cardiac studies, and it is usual to record the electrocardiogram simultaneously;

the phonocardiogram is sometimes recorded as well. Time and distance markers are normally included on the recording, so that measurements may be made.

In echocardiography (see page 103), the positioning of the ultrasonic beam for M-mode recording requires much skill. The operator can often be greatly helped if there is a method of examining the heart with a real-time scanner to assist in structure identification and M-mode beam placement. This can be done with a mechanical scanner (page 22), with a scan converter to provide a 'frame freeze' display (page 18), but it is easier with an electronically steered array which permits simultaneous real-time imaging (page 22) and M-mode recording.

DOPPLER DIAGNOSTIC METHODS

Ultrasonic Doppler methods are nowadays both widely used and of established value in the study

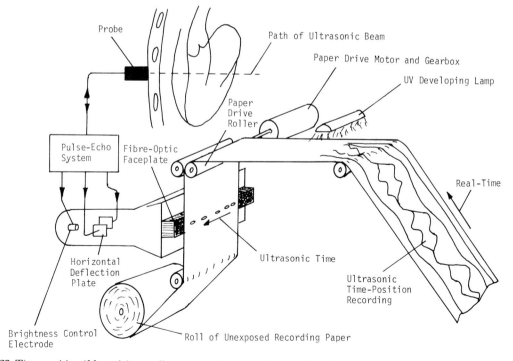

Fig. 1.23 Time-position (M-mode) recording system using a continuous strip of photographic paper sensitive to ultraviolet light. The B-scan is displayed on a cathode ray tube with a fibre-optic faceplate. This display is extremely bright, and a continuous recording of the time-position trace is produced (and developed within a few seconds) as the paper is driven at constant speed past the cathode ray tube.

of moving structures in clinical diagnosis. In most applications, the Doppler shift in frequency of a continuous wave ultrasonic beam reflected from a moving structure is used to provide information about the velocity of the structure, either for interpretation by ear, or for analysis by instrument.

The same restrictions and limitations (such as the necessity to maintain good ultrasonic coupling, and the inability to operate successfully through gas) which apply to ultrasonic pulse-echo methods, also apply to ultrasonic Doppler methods.

The choice of the ultrasonic frequency depends upon the clinical application. A compromise is necessary between the penetration, the variation of Doppler shift frequency for a given variation in target velocity, the sensitivity to small reflectors and the size and shape of the ultrasonic field. In both obstetrics and cardiology, the frequency is generally 2–3 MHz, but, in blood flow studies, it may be as high as 10 MHz.

The relationship between ultrasonic frequency, Doppler shift frequency, reflector velocity and angulation, is given by Equation 1.8 (page 6). For example, in the case where no correction for angulation is required (i.e., $\cos \gamma = 1$), a 2 MHz ultrasonic beam is shifted in frequency by about 260 Hz after reflection from a surface moving at a velocity of 100 mm.s^{-1}. For practical purposes, the Doppler shift frequency may be taken to be proportional both to the ultrasonic frequency and to the reflector velocity.

Continuous wave Doppler systems

A block diagram of a continuous wave ultrasonic Doppler frequency shift detector is shown in Figure 1.24. The transmitter operates continuously, providing an output of constant amplitude and frequency. The ultrasonic probe contains separate transmitting and receiving transducers. These are necessary because it is important to minimise the direct transfer of energy from the transmitter to the receiver, in order to avoid overloading the receiver amplifier. The output from the receiver consists of a collection of signals, some of a frequency equal to that of the transmitter (these are due to reflections from stationary structures in the ultrasonic field, and electrical leakage), and some of the frequencies shifted by the Doppler effect (these are due to reflections from moving structures). These signals are combined in the *mixer*, the output from which contains the difference frequencies between the transmitted ultrasonic wave and the Doppler shifted recieved waves. The output from the mixer is filtered to remove signals of unwanted frequencies, and the remaining Doppler shift signals are further amplified.

In clinical applications, the Doppler shifted signals generally do not consist of a single frequency, but they extend over a frequency spectrum. This is because the system simultaneously detects the movements of several different struc-

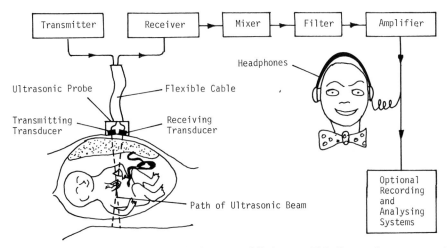

Fig. 1.24 Basic elements of the continuous wave Doppler frequency shift detector. This diagram shows a system designed to detect the movements of the fetal heart.

tures, e.g., blood cells move at differing velocities in a blood vessel according to the flow profile. For this reason, measurements of Doppler shift signals made by *ratemeters* (which can only indicate some kind of frequency average) need to be interpreted with caution. The ear is extraordinarily good at recognising sound patterns in complex spectra, and it is often most convenient, in clinical practice, simply to listen to the Doppler shift signals. Used in this way, the Doppler system can be considered to be an active stethoscope of great sensitivity and excellent directivity. In some applications, such as in studies of blood flow, additional information can be obtained by frequency analysis of the Doppler shifted signals. (A *frequency spectrum analyser* is an instrument which generates a chart in which time and frequency are plotted on orthogonal axes, and the corresponding sound intensity is represented according to the density of the recording.)

Directionally sensitive Doppler systems

Simple continuous wave Doppler systems of the type just described do not provide information concerning the *direction* of movement of a reflector — they only indicate its *velocity*, which may be either towards or away from the probe. The directional information is defined by the arithmetic sign of f_D as given by Equations 1.7 and 1.8, but this is rather difficult to determine because $f_D \ll f$. There are two chief ways in which this directional information may be obtained. The first method, which is the method most commonly used in commercial instruments, simply indicates *forward* or *reverse* flow — it is clearly misleading when both forward and reverse flows exist simultaneously! The second method really depends on frequency spectrum analysis, and it is capable of displaying a spectrum of velocities, simultaneously both forward and reverse.

Pulsed (range-gated) Doppler systems

Continuous wave Doppler systems do not provide information concerning the distances from the probe to the moving structures which give rise to the shifted frequencies. It is possible, however, to separate the Doppler signals by a range-gating technique in which the frequencies of the echoes

selected from a sample of distance lying between chosen limits are compared with that of a continuously running oscillator which itself is gated to provide the transmitted pulse. Several systems capable of extracting this information from the Doppler signals have been devised (including some which also provide directional information), and it has already become clear that they have significant roles in cardiology (page 108), vascular studies (page 145) and, as discussed in the next sections, in vessel imaging and duplex scanning.

Two-dimensional Doppler scanning

The Doppler shifted signals from flowing blood are sufficiently characteristic to allow them to be identified by electronic logic circuitry. This ability is exploited in a two-dimensional scanner designed to map out blood vessels. The position of the Doppler probe in space in a two-dimensional plane is measured by means of resolvers mounted on a scanning frame, and this information is used to control the horizontal and vertical deflection circuits of a cathode ray tube display, as shown in Figure 1.25. The output from the Doppler detector controls the brightness of the display. The probe is scanned regularly over the entire surface of skin beneath which it is desired to map the blood vessels, and only when the ultrasonic beam passes through moving blood does a registration appear on the display. Arterial and venous blood are distinguished by their different flow directions. The simpler two-dimensional Doppler scanners use continuous waves, but the more complicated scanners use range-finding Dopplers and are also capable of producing cross-sectional images of blood vessels (page 146).

DUPLEX DIAGNOSTIC METHODS

Although two-dimensional Doppler imaging, as discussed in the previous section, is clinically useful because of its ability to display regions and patterns of flowing blood, it has two important limitations. Firstly, the method does not display neighbouring (stationary) anatomical structures. Secondly, scanning to form an image is a lengthy process,

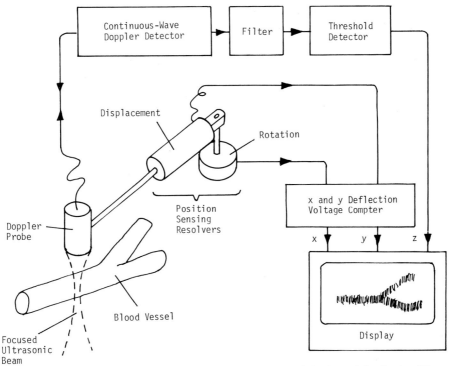

Fig. 1.25 Continuous wave Doppler system for two-dimensional visualisation of blood vessel distribution. The position of the probe is measured by a polar coordinate scanner, the resolvers of which control the x (horizontal) and y (vertical) deflection voltage computer driving the cathode ray tube display. The output from the Doppler detector is fed through a filter to remove low-frequency signals, due mainly to skin contact changes, and then to a threshold detector so that the z (brightness-controlling) signal is stored in the display system to form the two-dimensional image.

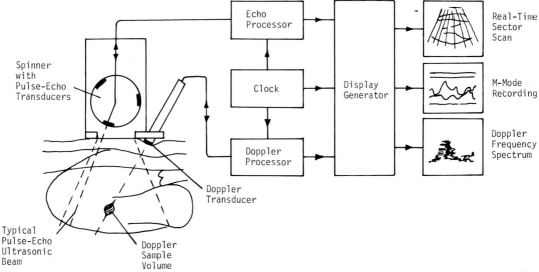

Fig. 1.26 A duplex scanning system, combining real-time two-dimensional pulse-echo imaging (using a rotating wheel, like that illustrated in Figure 1.18a) with pulsed Doppler capability. Stopping the rotation of the spinner allows M-mode recordings to be made. This particular instrument employs a separate transducer for Doppler measurements, arranged so that the sample volume can be positioned at any desired region in the image plane. A popular, but less versatile, variant of this arrangement requires the spinner to stop rotating during Doppler studies — one of the spinner transducers is then operated in pulsed Doppler mode and directed at the region of interest identified on a frozen frame two-dimensional image.

since at least one complete cardiac pulse is necessary to form each element of picture information. Because of these restrictions, the potential of combining two-dimensional pulse-echo real-time imaging with a limited amount of pulsed Doppler information (so-called *duplex scanning*) is extremely attractive. As with real-time scanning and M-mode recording (page 26), the pulse-echo image (either obtained continuously, or with a frozen frame display) allows regions of interest to be located, and the pulsed Doppler system allows structure or blood flow within them to be studied.

The basic elements of a typical duplex scanner are illustrated in Figure 1.26.

FURTHER READING

Kremkau F W 1980 Diagnostic ultrasound: physical principles and exercises. Grune and Stratton, New York
McDicken W N 1981 Diagnostic ultrasonics: principles and use of instruments. 2nd edn. Wiley, Chichester
Rose J. L, Goldberg B B 1979 Basic physics in diagnostic ultrasound. Wiley, New York
Wells P N T 1977 Biomedical ultrasonics. Academic Press, London

Wells P N T 1982 Ultrasonic imaging. In: Wells PNT, (ed) Scientific basis of medical imaging. Churchill Livingstone, Edinburgh, p. 138–193
Wells P N T, Ziskin M C (eds) 1980 New techniques and instrumentation in ultrasonography. Churchill Livingstone, New York

2

Obstetrics and gynaecology

David Graham and R. C. Sanders

ULTRASOUND IN OBSTETRICS

Since its introduction into widespread clinical use by Donald in the 1950s and '60s, ultrasound has developed to the extent that it is now an invaluable tool in the assessment and management of pregnancy and its complications. While it has been proposed that every patient should have at least one sonogram during her pregnancy, this at present is not a cost effective procedure. The major indications for sonographic examination in pregnancy are:

Diagnosis of pregnancy and assessment of
    gestational age
Localisation and assessment of anomalies of the
    placenta
Prenatal diagnosis and management of fetal
anomalies
Diagnosis and management of multiple
    pregnancy
Intrauterine growth retardation
Assessment of vaginal bleeding during pregnancy
Determination of fetal death
Aid in amniocentesis
Characterisation of a coexistent adnexal mass
Localisation of an intrauterine device
Diagnosis of extrauterine pregnancy
Fetal biophysical assessment
Management of induced abortion and its
    complications
Postpartum complications
Gestational trophoblastic disease

Diagnosis of pregnancy

The earliest sonographic evidence of an intrauterine pregnancy is the gestational sac — a small ring of echoes from the trophoblast surrounding a lucent chorionic cavity: the sac may be seen 5–6 weeks after the last menstrual period. At about 6–7 weeks of gestation a small collection of echoes, the fetal pole, may be seen within the sac. The use of real-time ultrasound allows visualisation of fetal movement or fetal cardiac movement or both, even at this early stage of pregnancy.[28] At this time the gestational sac has an asymmetric shape, the trophoblastic shell appearing thicker at one pole where the maternal component of the placenta, the decidua basalis, and the fetal component, the chorion frondosum, are forming the placenta. As pregnancy advances the placenta becomes more recognisable as a distinct structure.

With careful scanning of the gestational sac a small additional circular collection of echoes may be seen adjacent to the embryo representing the yolk sac. At 10–11 weeks the fetal head is recognised as a distinct circle of echoes and from this time onwards increasingly greater fetal detail may be recognised.

Gestational age assessment

Several methods are available for estimation of fetal size and hence, indirectly, fetal age.

Gestational sac measurements

In the first trimester, the gestational sac diameter bears a linear relationship to gestational age with an accuracy of ± 2 weeks.[17] Sometimes, especially when distorted by an overdistended bladder, the gestational sac is not a circular structure, and in such cases the average of the diameters obtained in three planes can usually be used with equal accuracy. When measuring sac diameter the inner margin of the sac is used. Gestational sac area and volume have also been utilised and are only slightly more accurate but a good deal more cumbersome.

Crown-rump length

The embryo may be recognised as early as six weeks after the LMP, measurement of the largest axis of this fetal pole (excluding the limbs) representing the 'crown-rump length' (Fig. 2.1).

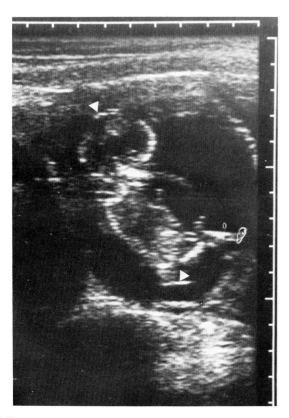

Fig. 2.1 Longitudinal section through the uterus in a 13-week gestation showing the measurement used for crown-rump length (arrows). The fetal head, body and limbs are readily visualised.

Real-time ultrasound allows rapid visualisation of the crown-rump length, which is the single most accurate method of assessing gestational age because of its minimal biological variation in size in the first trimester and the rapid growth of this measurement at this stage of pregnancy. A statistical study has shown that a prediction of gesta-

tional age can be made to within ±4.7 days in 95 per cent of cases if a single measurement is made and even more accurately than this with several measurements.[29]

General appearances of the sac

The descriptive approach was first used for assessment of gestational age, assigning an age depending on the relative size of the sac in relation to the uterine cavity, whether a fetal pole was present, whether a placenta was seen, and so on. Whilst this approach is useful to give an estimate of gestational age, more formal measurements described are to be preferred.

Biparietal diameter (BPD)

The fetal head may be first seen at 10–11 weeks of gestation. From this time until term, measurement of the widest transverse diameter of the head is used as a method of dating the pregnancy or of observing fetal growth. In the period 10–14 weeks both the BPD and crown-rump length (CRL) may be used to assess gestational age but after 12 weeks differing degrees of flexion of the fetus may affect the CRL and make it less accurate. A number of large studies has shown that the BPD grows in a linear fashion until approximately 30 weeks of gestation.[4,35]

After this time incremental growth slows and the standard deviation widens from approximately ±7–11 days at approximately 20 weeks up to ±3–4 weeks near term. Several composite charts are available which attempt to correct for possible differences in different populations and to give large enough samples to narrow the variation of given measurements.[22,23]

The growth-adjusted sonographic age (GASA) attempts to correct for variations in growth in different fetuses.[30] Using this method two BPD measurements are obtained, the first at 20–24 weeks and the second at least six weeks later. At the first examination a provisional gestational age is assigned. At the following examination the fetus is placed into an average, above average, or below average growth rate group. If the second examination places the fetus in the average growth group the initial assigned age is accepted as correct, while if the growth is < 25 per cent or > 75 per cent then the original gestational age is corrected by several days. Using this method it is claimed that the accuracy of gestational age assessment is greatly improved.

With grey-scale techniques allowing visualisation of intracranial anatomy it is now easier to obtain reproducible accurate representations of the BPD. This should be a section at the level of the thalamus and cavum septum pellucidum (Fig. 2.2).

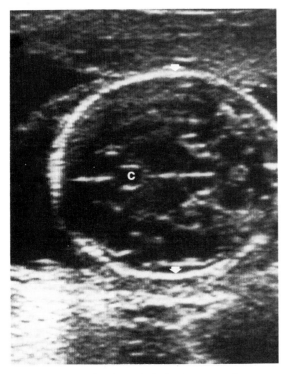

Fig. 2.2 Cross-section through the head showing the measurement used for the biparietal diameter (arrows); the cavum septum pellucidum is also seen (c) as are the thalami.

Use of real-time makes it possible to obtain a representative section rapidly with an accuracy comparable to that of static scans.

Femoral length

Measurement of long bone length is a relatively uncomplicated procedure especially with real-time equipment. Charts of femoral length and gestational age show a linear relationship with a fairly narrow standard deviation comparable to the BPD.[10,27] Measurement of femoral length is of value where there is a problem in visualising the fetal head or in cranial anomalies such as anencephaly, hydrocephalus or iniencephaly.

Placenta and placental localisation

The placenta is seen from 8–9 weeks of gestation as a localised thickening of the trophoblastic ring and after 9–10 weeks it is usually consistently seen as a homogeneous structure. Ultrasound is of value in the assessment of placental location, 'maturation', masses, morphology and volume.

Placental location

Use of ultrasound has revealed several phenomena which were previously unsuspected clinically. One of these is the concept of placental 'migration' — the change in the relative position of the placenta in the uterus as gestation advances. The effect of this is that a placenta which might be partially or totally praevia in mid-pregnancy is likely to have a location away from the cervix as term approaches. This was originally suggested as being due to an actual change of the placental attachment.[20] It seems, however, more likely to be due to a combination of technical factors associated with the degree of bladder distension and also to differential growth of the lower uterine segment in the last weeks of pregnancy. The over-distended bladder may markedly distort the appearance of the lower uterine segment and produce an appearance mimicking placenta praevia.[36] Where placenta praevia is diagnosed in the presence of a full bladder, the scan should be repeated with the bladder partially empty to see whether the finding is persistent. The concept of differential uterine growth with development of the lower uterine segment in the last weeks of pregnancy is well known clinically. Because of this, when diagnosis of placenta praevia is made prior to term a repeat examination is required before any decision is made regarding mode of delivery.

Placental maturation

As pregnancy progresses, the placenta may show echogenic changes thought to be due to calcification within its substance, this echogenicity evolving in a specific manner. The appearance may be classified according to a system of grading, dependent on the internal appearance of the placenta:[13]

Grade 0: Homogeneous texture, no echogenicities
Grade I: Scattered echogenicities throughout the placenta
Grade II: Definite evidence of basal echogenicity
Grade III: Evidence of cotyledons with intercotyledenous echogenicity

The time of appearance of echogenicity is variable and in any particular pregnancy the placenta may show any of the above changes at term, e.g., less than 10 per cent of pregnancies at term show Grade III changes. The importance of the grading system is that some prediction of fetal lung maturity can be based on the appearances — Grade I placentas have a 60 per cent correlation with a mature L/S ratio. Grade II changes have an 80 per cent correlation and Grade III changes, a 100 per cent correlation. Average time of appearance of Grade I changes (if they occur) is 30–32 weeks, Grade II 33–34 weeks, and Grade III 36–37 weeks.

Placental anatomy

The normal placenta may, in addition to the changes of calcification as described already, show regions of lucency in the subchorionic area or intraplacental area, or masses within or projecting from the surface of the placenta. Areas of lucency in the immediate subchorionic area are often seen with good resolution equipment and represent vascular structures on or near the fetal surface of the placenta. Occasionally areas of lucency are seen within the placenta — these have been related to areas of fibrin deposition and vascular lakes and unless extensive have little clinical significance.

Tumours of the placenta are rare, chorioangiomata being the most common. Chorioangiomata appear as areas of increased echogenicity within the placental substance or as irregular masses protruding from the fetal surface of the placenta.

In certain conditions such as Rhesus disease and infections such as toxoplasmosis or cytomegalovirus, changes may been seen in the placenta — the placenta appearing enlarged and oedematous.

Fetal anomalies

Since bistable techniques allowed only visualisation of major fetal structures, the first anomalies to be described using that technique were major, such as anencephaly and gross hydrocephalus. With grey-scale techniques, however, an increasing number of subtle fetal anomolies may now be visualised.[18] The normal anatomy and most common anomalies are described in this section for each organ system.

Central nervous system

The fetal cranium can be visualised well enough to obtain a biparietal diameter measurement from approximately 10–11 weeks. As well as BPD measurements, a good deal of intracranial anatomy can be seen. In the early second trimester the choroid plexuses are quite prominent as echogenic structures. Up to 17 weeks of gestation, the lateral ventricles are relatively large and may occupy more than 50 per cent of the diameter of the head.[8] At the level at which the BPD is obtained, the thalami may usually be seen, together with the cavum septum pellucidum, the tips of the anterior horns of the lateral ventricles and occasionally the third ventricle. At a slightly higher level the bodies of the lateral ventricles may be seen as two parallel lines adjacent to the midline. Other intracranial structures which may be visualised are the cerebral peduncles, cerebellum, orbits, petrous ridges and, using real-time equipment, several of the major arterial structures.

The spine may be examined in both longitudinal and in serial transverse sections. In longitudinal section the normal spine appears as two adjacent rows of dense echoes which in their major portion are parallel, but which diverge in the cervical area and to a lesser extent in the lumbar area. In transverse sections a ring of echoes representing an intact neural arch should be visualised.

The most common anomalies of the CNS which are diagnosable are:

Anencephaly. When the neural tube fails to close in its lower portion, spina bifida is the result. When there is failure of closure in the upper portion anencephaly may result. Anencephaly may be diagnosed readily on ultrasound by failure to visualise the normal cranial echoes, with occasionally a small 'nubbin' of tissue representing the base of the skull and remnants of brain tissue being seen. In approximately 50 per cent of cases associated polyhydramnios is visualised. Caution should be exercised in evaluating the head which is deeply engaged in the pelvis, since normal cranial structures may not be visualised.

Hydrocephalus. The lateral ventricles may be visualised on transverse sections of the head, and the ventricle/head ratio measured. An abnormal ventricular size allows diagnosis of hydrocephalus at an early stage, usually before the BPD is abnormal. Unfortunately, in some cases hydrocephalus is first reflected in dilatation of the posterior horns of the lateral ventricles which are less often visualised with ultrasound. As well as the degree of ventricular dilatation the width of the remaining cortical tissue may be measured, allowing prediction of neonatal outcome. Serial assessment allows determination of progressively worsening hydrocephalus which may necessitate early delivery and immediate neonatal treatment. Hydrocephalus is often secondary to an X-linked recessive condition giving a 25 per cent chance of occurrence in subsequent pregnancies and 50 per cent chance of a male being affected.

Encephalocele. Encephaloceles usually occur in the occipital area and are visualised sonographically as a cystic mass adjacent to the occipital bone. There is often solid tissue, representing cerebral tissue in the cystic mass. Careful scanning may show the defect in the occipital bone. Anterior encephaloceles are usually small but again are readily recognisable.

Hydranencephaly. In hydranencephaly the cerebral tissue supplied by the anterior and middle cerebral arteries is entirely replaced with fluid and therefore not visualised. The basal ganglia and cerebellar tissue are usually spared. It is important to distinguish hydranencephaly from hydrocephaly because of the poor prognosis of the former.

Iniencephaly. Iniencephaly is characterised by defective formation of the occipital bone in association with an open neural tube defect involving the cervical and upper thoracic spine to a varying degree. Sonographic appearances are marked hyperextension of the fetal head, a spina bifida of the cervical spine and a solid mass representing cerebral and spinal tissue adjacent to the occipital bone and spine. There is usually associated polyhydramnios.

Microcephaly. Microcephaly may be difficult to diagnose prior to the third trimester when it is most often recognised as a consistently smaller BPD than expected, with a discrepancy between the head size and the trunk circumference. A BPD > 1.5 cm less than the trunk diameter is highly suspicious of microcephaly.

Spina bifida. Spina bifida has an incidence of 1 in 1000 pregnancies in the United States and 4–7 per 1000 in parts of the United Kingdom and Europe. After one affected pregnancy, a woman has a 5 per cent chance of recurrence of a neural tube defect (NTD) which is not necessarily of the same type. After two affected pregnancies the risk is approximately 12 per cent. NTDs may be screened for in two manners — by maternal alpha feto-protein (AFP) measurement and by ultrasound examination.[3,25.]

Although AFP screening may miss closed defects which account for about 10–15 per cent of the total, these closed defects may potentially be detected with ultrasound screening. A spina bifida may be seen as a localised widening of the parallel echoes in longitudinal views or by a V-shaped or U-shaped appearance of the neural tube on transverse sections. A cystic or solid mass representing the meningocele or myelomeningocele may be seen adjacent to the spinal defect. With small defects the longitudinal views may appear relatively normal necessitating serial transverse sections. Care must also be taken in the sacral area since the normal sacral dip between the iliac bones may mimic a spina bifida. Small lesions in the lower lumbar and sacral area may be missed even with a careful examination. It has been noted that the degree of limb movement in utero has no correlation with the neurological defect postpartum — a severe degree of paralysis after birth may have normal limb movement in utero.[3]

Genitourinary system

From the early second trimester the fetal kidneys and the bladder may be seen. In a transverse section of the trunk the kidneys are seen adjacent to the spine as oval structures. The renal pelvis appears echogenic when collapsed but often, with physiological dilatation, has a sonolucent centre. The corticomedullary junction and arcuate arteries may be seen in normal subjects. Standards are available for renal size throughout gestation.[14] The bladder appears as a cystic structure in the lower abdomen which when scanned serially over some hours may be seen gradually to fill and then to empty.

In genital tract anomalies associated with lack of renal function such as bilateral ureteric or bladder outlet obstruction there is usually oligohydramnios which is readily apparent sonographically. The commonest genitourinary anomalies to be recognised with ultrasound are:

Renal agenesis. Renal agenesis may be inherited as an autosomal recessive condition. In association with severe oligohydramnios it is impossible to visualise the normal renal outlines or bladder. The normal adrenals, however, may be mistaken for kidneys.

Hydronephrosis. It is not uncommon to see slight dilatation of the renal pelvis in utero as a normal finding. Unilateral or bilateral hydronephrosis may be diagnosed prenatally by a greater dilatation of the renal pelvis and is most often due to UPJ obstruction. The degree of hydronephrosis may be followed on serial examinations allowing decisions to be made concerning early delivery. Bilateral hydronephrosis may also be due to bladder outlet obstruction, secondary to urethral valves, urethral atresia, or the Eagle-Barrett syndrome. In such instances, the bladder is markedly enlarged and on sequential examinations is not seen to empty. Greatly dilated ureters may be visualised sonographically.

Polycystic kidney. The infantile form of polycystic kidney, inherited as an autosomal recessive condition, may be diagnosed prenatally as bilaterally enlarged echogenic kidneys without the normal corticomedullary differentiation. The cysts are usually in the order of 1–2 mm in diameter and therefore are not usually visualised sonographically. The adult form of polycystic kidney, inher-

ited as an autosomal dominant, rarely presents in the neonatal period and most would be expected to have a normal renal appearance prenatally. Theoretically in an infantile presentation evidence of cyst formation in the kidneys and liver would be expected to be seen.

Multicystic kidney. Multicystic kidney involves replacement of the normal renal parenchyma by cysts of varying sizes, with absence of renal pelvis and calyces and atresia of the ureter. The affected kidney is enlarged and involved with numerous cysts. Occasionally one cyst predominates to such an extent that it may mimic the appearance of a marked hydronephrosis. Even with bilateral multicystic kidneys there may be enough urine excretion to allow visualisation of the bladder.

Eagle-Barrett syndrome (prune belly syndrome). This syndrome is characterised by absent abdominal wall musculature, renal hypoplasia, hydronephrosis, megacystis and megaureter. The collecting system changes may be readily visualised. Free peritoneal fluid, due to urine ascites may also be seen.

Ovarian cysts. Ovarian cysts may be diagnosed prenatally. They may, however, be on a long pedicle and located at some distance from the pelvis. Such cysts pathologically are usually benign cysts of germinal or graafian epithelial origin.

Hydrocele. Occasionally in an otherwise routine scan, a male fetus may be seen to have a hydrocele — the scrotum is seen to contain a cystic collection adjacent to the testis. Hydroceles may be associated with fetal ascites.

Meckel-Grüber syndrome. This is an autosomal recessive condition associated with infantile polycystic kidneys, encephalocele and polydactyly, all of which are diagnosable sonographically. Most die in the neonatal period.

Gastrointestinal system

During fetal life there is continued swallowing of amniotic fluid. Therefore the stomach is usually visualised as a fluid-filled structure in the left upper abdomen. Similarly, loops of bowels may be visualised as small cystic structures in the abdomen which, with real-time scanning, may be seen to have peristalsis. It is not, however, possible to differentiate small from large bowel prenatally.

The liver is readily visualised as a homogeneous structure in the upper abdomen — branches of the hepatic and portal systems may be seen.

Intestinal atresia. Where large fluid filled structures are visualised in the abdomen, without change of appearance over time, intestinal atresia should be suspected. Intestinal atresia, especially proximal atresia, is often associated with polyhydramnios. Duodenal atresia has a sonographic appearance analogous to the radiographic 'double-bubble' sign — two fluid filled structures are seen in the upper abdomen but no bowel loops are seen elsewhere in the abdomen.

Tracheo-oesophageal fistula. One type of tracheo-oesophageal fistula has associated proximal oesophageal atresia. In such a case, the stomach is not visualised and the presence of the stomach therefore rules out this type of tracheo-oesophageal fistula. In most instances there is associated polyhydramnios.

Abdominal wall defects. Omphalocele and gastroschisis are the two major types of abdominal wall defects.[33] Omphalocele involves an umbilical defect in which bowel and intra-abdominal viscera, contained in a peritoneal sac, protrude from the anterior abdomen. Ultrasound examination shows an abnormal contour to the abdomen with loops of bowel or the liver (or both) present outside the abdominal cavity. Other congenital defects such as congenital heart disease are more common in association with omphalocele. Gastroschisis is a skin-covered paraumbilical defect in the abdominal wall, through which intestine eviscerates. It may be distinguished from omphalocele by its paraumbilical location, normal insertion of the umbilical cord and lack of abdomino-peritoneal sac containing abdominal viscera. Loops of bowel anterior to the abdominal wall are seen. Associated congenital anomalies are not as common as with omphalocele.

Skeletal system

Ultrasound allows visualisation of the soft tissues and bones of the limbs at a stage before radiographic diagnosis is practical, giving the potential for prenatal diagnosis of dwarfism syndromes and anomalies of long bones.[10,18]

Achondroplasia. Although most often a result of a sporadic chromosomal anomaly, achondroplasia, being an autosomal dominant condition, has a 50 per cent chance of being transmitted to the child of an affected parent. Visualisation of the long bones makes it possible to determine whether the limbs are within the normal range for length. Unfortunately, it has been noted that examination prior to mid-pregnancy may demonstrate normal length limbs in an infant which subsequently proves to be affected.[10]

Thanatophoric dwarfism. This form of short-limbed dwarfism, fatal in the neonatal period, may be recognised prenatally by short limbs with soft tissue thickening (due to a 'telescoping' of the soft tissues on the shortening bones), together with a small chest and relatively large head. Occasionally a clover-leaf deformity of the skull or Kleeblattschädel may be seen. Bowing of the limbs (campomelia) may be a feature.

Osteogenesis imperfecta. If the fetus is severely affected in utero by the recessive form of osteogenesis imperfecta, fractures of the long bones may occur. Since bone is well visualised on sonography these long bone changes may be seen with ultrasound (before they can be detected with radiography) as irregularity, bowing and shortening of the bone. A thin outline to the skull, due to poor mineralisation, may also be seen.

Hypophosphatasia. This rare, recessively inherited metabolic disorder of alkaline phosphatase formation produces bony changes histologically similar to rickets. From early in the second trimester, there is defective bone mineralisation of the skull outline and suboptimal visualisation of long bones. If a radiograph is performed the visualisation of bony structures is even poorer.

TAR syndrome. The thrombocytopenia-absent radius syndrome is one of the syndromes associated with absence of the radius (other syndromes include Fanconi's anaemia, deLange syndrome). Although absent radius is not a constant feature of this syndrome, it does tend to recur in any subsequently affected pregnancies of a particular individual. Scanning the forearm readily demonstrates whether one or two limb bones are present.

Ellis van Creveld syndrome. Cardiac malformations are present in association with mesomelia and polydactyly, all of which may be visualised with ultrasound.

The chest

Ultrasound may be used in evaluation of the anatomy and function of the heart and major intrathoracic problems such as diaphragmatic hernia, cystic malformations of the lungs, pulmonary hypoplasia and fluid collections within the chest.

Cardiac evaluation. The heart and its chambers and valves, the vena cava and the aorta are readily recognised with real-time ultrasound (Fig. 2.3).[21] The heart may be investigated by M-mode echocardiography, which allows assessment of chamber size, valve structure and function, pericardial effusions and also some physiological parameters such as ejection fraction and end-diastolic volume. Rhythm and conduction disorders may be diagnosed. Two-dimensional studies of the heart allow diagnosis of structural anomalies.

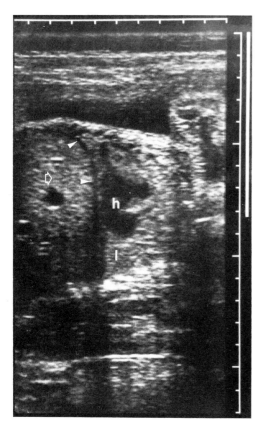

Fig. 2.3 Longitudinal section through the fetal thorax and upper abdomen to show the heart (h), lungs (l) and liver (open arrow). Diaphragm is also well visualised (closed arrows).

Pulmonary and thoracic anomalies. Little detail of the lungs may be seen in utero. It may be possible, however, to detect pleural effusions, as in fetal hydrops; cystic adenomatoid malformations of the lung, which usually presents as non-immune hydrops (cystic changes may be seen replacing normal pulmonary echoes); and diaphragmatic hernia, in which fluid filled bowel loops may be seen in the chest adjacent to the heart (there may also be displacement of the heart).

Soft tissue anomalies

Cystic hygroma. This is a congenital malformation of the lymphatic system. Approximately 50 per cent of these lesions are present at birth and so are potentially diagnosable in the antenatal period. They appear as a mass, usually cystic in appearance, in the region of the neck or axillae. Other soft tissue lesions in this area are potentially diagnosable such as branchial cleft cysts and haemangiomas.

Sacrococcygeal teratoma. Prenatally a large mass in the sacral area is seen protruding from the fetal rump. There is a variable intrapelvic component.

Miscellaneous conditions

Fetal hydrops. The combination of fetal pleural effusion, ascites and skin thickening may be secondary to Rhesus or other iso-immunisation or to nonimmune causes (Fig. 2.4).[16] Nonimmune causes include infections such as CMV or toxoplasmosis, renal vein thrombosis, cystic adenomatoid malformations of the lungs, thalassaemia and sickle cell anaemia. The sonographic appearances are quite characteristic with free fluid in the pleural and peritoneal cavities. In the abdomen free fluid may be distinguished from a large cystic mass by the fact that it is seen around the liver and spleen. Skin and scalp thickening may be seen to a varying degree.

Where hydrops is due to an iso-immunisation, it indicates a severely affected infant, although not all severely affected infants necessarily show hydrops. The diameter of the umbilical vein in the liver and in the fluid may be measured sonographically and it has been suggested that an increase in the diameter of the umbilical vein precedes changes in the amniotic fluid bilirubin concentration.[9] In iso-immunisation the placenta may show characteristic changes being enlarged, thickened and oedematous. Ultrasound may be utilised to direct amniocentesis to assess the Δ OD450 level and also may be used instead of fluoroscopy to direct intrauterine transfusion.[6] Ultrasound has the advantages of not involving radiation and allowing two-dimensional location of the tip of the needle. The catheter tip can be seen and determined to lie in a correct position. Real-time examination allows direct visualisation of the injection of blood.

With nonimmune hydrops, transabdominal decompression of the fetal ascites may be used to decrease the abdominal girth sufficiently to allow abdominal delivery.[19]

Diabetes. The infant of the diabetic mother is more prone to growth problems in utero and also to an increased incidence of congenital anomalies. Ultrasound may be used to detect macrosomia, in which the growth rate is increased with an above average estimated fetal weight and thickening of the skin and scalp, shown as a double outline to the fetus; intrauterine growth retardation, which is commoner in the more severe insulin-dependent

classes of diabetes (IUGR is discussed in detail in a later section); and congenital anomalies such as cardiac defects of the atrioventricular type and spinal anomalies such as sacral regression which are more common in the infants of diabetic mothers.

Multiple pregnancy

First trimester

Two or more gestational sacs may be seen from the time that the sac first becomes visible on ultrasonic scanning. A large proportion of pregnancies in which twin gestation is diagnosed in the first trimester eventually deliver a single fetus.[23] Thus, there is either a large 'wastage' in multiple pregnancies or there are a number of diagnostic errors. Certainly there are a number of potential pitfalls in diagnosis. Occasionally a septum may be seen across a sac, with only one 'sac' containing fetal echoes. This may be a normal finding, perhaps due to visualisation of both the chorionic and amniotic cavities or perhaps due to an extra-amniotic collection of blood or fluid. To make a firm diagnosis of twins in the first trimester, it is necessary to see two gestational sacs with a viable fetus in each (Fig. 2.5). Even in such cases a

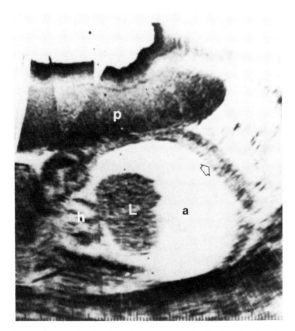

Fig. 2.4 Longitudinal section through the fetal body showing marked pleural and peritoneal effusions (a), the heart (h), liver (l) and abdominal wall (arrow). The placenta is anterior and appears slightly oedematous (p).

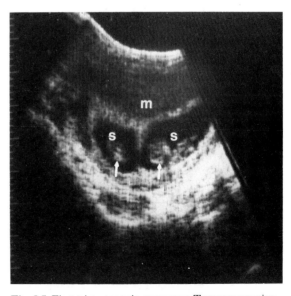

Fig. 2.5 First trimester twin pregnancy. Transverse section of a uterus in the first trimester showing gestational sacs (s) each of which contained a viable fetal pole (arrows). Myocardium is also visualised (m).

number subsequently show death and resorption of one of the embryos. Occasionally fetal death occurs at a later stage of gestation, with the dead twin being retained as fetus papyraceus which may be visualised sonographically.

Second and third trimesters

When diagnosed after the first trimester, the majority of multiple gestations do continue as such. Ultrasound is of value in the management of twin gestation in:

1. Dating the pregnancy, since clinical estimation of the fetal size is unreliable.

2. Evaluation of fetal growth and diagnosis of the twin-to-twin transfusion syndrome. There is some controversy concerning the growth pattern of normal twin pregnancies compared to singletons, usually because most centres have a relatively small number of twin pregnancies serially examined with ultrasound. It appears that twins have a growth which parallels that of a singleton up to 28–30 weeks and then show an incremental BPD growth and growth rate less than that of a singleton pregnancy. With dizygous twins a discrepancy in fetal sizes and slight discrepancy in BPD is not unusual. Twin-to-twin transfusion syndrome is a condition which may be seen in monozygous twins, when there is arterio-venous crossing of the two placental circulations. In such a case one twin receives a disproportionately large proportion of the placental blood flow and becomes plethoric while the second, 'donor' twin becomes anaemic, growth retarded and may die in utero.

3. Presence of polyhydramnios. When excessive, polyhydramnios may predispose to premature labour and require transabdominal removal of fluid.

4. Diagnosis of fetal anomalies which are more common in twins, especially monochorionic twins.

5. Placental position. With the increased size of the placenta, placenta praevia is more common.

It is not possible definitively to determine prenatally by ultrasound examination whether a twin pregnancy is monozygous or dizygous. A septum dividing the two amniotic sacs may be seen in all dizygous twins and in all but a few monozygous twins. The exceptions are monoamniotic twins which have a high incidence of

congenital anomalies, cord accidents and locking during delivery.

Intrauterine growth retardation (IUGR)

Using ultrasound two patterns of IUGR have been described:[2]

1. 'Low profile', symmetric type in which the fetus is below average size with a small biparietal diameter throughout the pregnancy and which is more commonly seen in association with fetal anomalies and intrauterine infections.

2. 'Late onset', asymmetric type in which the BPD measurements and estimated fetal weights are normal until approximately 28–30 weeks of pregnancy when serial examinations show slowing of the incremental growth of the BPD and a lower than average estimated fetal weight. This type of IUGR is most commonly seen in placental insufficiency as in pre-eclampsia, diabetes or for unknown cause. The slowing of the incremental BPD increase is a delayed phenomenon since there is a 'brain-sparing-effect' in severe IUGR.

Several other parameters may be used in screening for IUGR:

Total intrauterine volume (TIUV). Measurement of the axis of the uterus in three dimensions allows measurement of the TIUV using the formula:

$$\text{TIUV} = 0.523 \times \text{length} \times \text{width} \times \text{AP diameter} \tag{2.1}$$

This measurement reflects the volume of the fetus, the placenta, and the amniotic fluid, and has been proposed as a screening tool for IUGR, a decreased TIUV being the first associated change.[12]

Head/trunk (H/T) ratio. This is defined as:

$$\text{H/T ratio} = \text{head circumference/abdominal circumference} \tag{2.2}$$

The H/T ratio changes according to gestational age, the head circumference being measured at the plane of the BPD and the abdominal circumference being measured at the umbilical vein level. Up to 36–37 weeks the H/T ratio is $\geqslant 1.0$, but it subsequently becomes $\leqslant 1.0$. In asymmetric growth retardation decreased glycogen deposition in the liver results in decreased liver size, and is reflected in a decreased abdominal circumference. Since in asymmetrical growth retardation brain growth is spared until the insult is severe, the head circumference remains relatively normal for some time and the H/T ratio becomes abnormal.[5]

Estimated fetal weight. Various methods have been proposed for estimation of fetal weight, one of the most reliable being that which uses a combination of BPD and abdominal circumference.[34] Using this method a standard deviation of ± 30 g kg^{-1} has been obtained.

Placental grading. In IUGR the placenta may be seen to show a placental grade more advanced than expected for the gestational age.

Ultrasound may also be used in the assessment of fetal well-being in the fetus affected by IUGR. As well as use of Doppler recording of the fetal heart rate for the non-stress test, physiological parameters of breathing, movement and tone may be assessed and prediction made as to the fetal status.[24]

Bleeding in pregnancy

Early pregnancy

Vaginal bleeding occurs in about 25 per cent of all early pregnancies. After clinical examination has excluded local causes and shown the cervix to be closed, ultrasound should be used to determine whether the pregnancy is viable. In such a patient the ultrasound appearance may be:

1. Normal, showing an intrauterine pregnancy appropriate to the expected gestation, with evidence of a viable embryo. Such pregnancies have a high probability of proceeding to term.

2. Blighted ovum (anembryonic pregnancy), in which the gestational sac has a poorly defined trophoblastic reaction, is smaller than expected and does not contain a fetal pole (Fig. 2.6). In the appropriate clinical situation the ultrasound appearances are diagnostic. If there is some doubt, however, a repeat examination after 7–10 days helps to distinguish a blighted ovum from a very early normal pregnancy. The repeat examination in the blighted ovum patient shows little or no interval growth.

3. Missed abortion, where there has been fetal demise some weeks prior to the ultrasound examination. The uterus may be smaller than expected, containing no recognisable fetal structures but instead a collection of amorphous echoes representing retained products of conception. Alternatively a small gestational sac with an immobile fetal pole, without cardiac activity, may be seen.

Late pregnancy

In late pregnancy the major causes of uterine bleeding are placenta praevia and placental abruption. Clinically, because of the danger of profuse haemorrhage if pelvic examination is performed in the patient with placenta praevia, ultrasound is used as the initial method to determine placental location (Fig. 2.7). Although the relationship of the edge of the placenta to the internal os is usually apparent on serial longitudinal scans, with a posteriorly located placenta the presenting fetal part may shadow the placental edge making diagnosis of marginal praevia difficult. Scanning the patient in a Trendelenberg position or with gentle abdominal manipulation of the presenting part (or both) may allow visualisation of the lower segment. In some patients with active bleeding a small collection of echogenic material due to blood may be seen adjacent to the internal os. Because of changes in the appearance of the lower uterine segment produced by a distended bladder, where

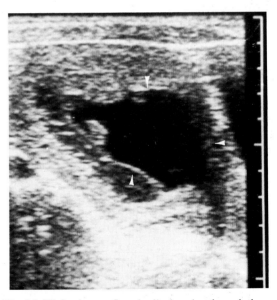

Fig. 2.6 Blighted ovum. Longitudinal section through the uterus showing an irregular gestational sac (arrows) without a demonstrated fetal pole.

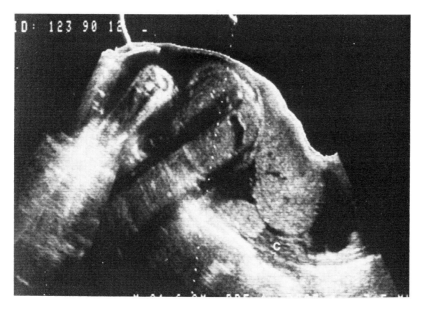

Fig. 2.7 Placenta praevia. Longitudinal section through the uterus showing the edge of the placenta completely covering the cervix. The area of the internal os is marked (c).

placenta praevia is diagnosed with a full bladder, the bladder should be partially emptied and the uterus rescanned to confirm the placental position.[36]

With placental abruption, the sonographic appearances are quite likely to be normal. Where the patient has experienced a major abruption the clinical picture is usually so characteristic that the patient does not come to ultrasound examination. In those patients with abruption who are examined, the sonographic appearances may vary from normal to those of a retroplacental blood collection with separation of the placenta from the uterine wall. Occasionally blood from an abruption may collect in an extrauterine location adjacent to the internal os and mimic a placenta praevia.

Fetal death in utero

Real-time ultrasound provides a rapid and reliable method of assessing fetal viability. Fetal cardiac motion is apparent from approximately seven weeks of gestation and can usually be readily seen with real-time scans. If only static scanners are available, fetal cardiac activity may be demonstrated by scanning very slowly over the heart; motion of the heart is shown by a wavy line.

Where fetal death has been present for some days, changes analogous to those seen on radiography may be demonstrated with overlapping of skull bones, thickening of the scalp echoes, abnormal fetal position and a double contour to the skull.

Amniocentesis

Ultrasound is of great value in performance of amniocentesis, both in midtrimester for genetic diagnosis and in late pregnancy for assessment of lung maturity.

Midtrimester amniocentesis may be performed to obtain fetal cells for cell culture to diagnose chromosomal anomalies, for determination of enzyme deficiencies or for measurement of AFP. Ultrasound is of value in:

1. Dating the pregnancy to ensure optimal time for amniocentesis (16–17 weeks) and also as a standard for AFP, since AFP changes markedly with gestational age.

2. Diagnosis of multiple pregnancy, again to

explain abnormally high AFP levels. Ultrasound is essential to aid in the aspiration of both sacs. After the fluid from the first sac is removed, 0.5 ml of indigo carmine is injected. When clear fluid is withdrawn on the second aspiration, it is certain that the other sac is being sampled.

3. Determination of the presence of fibroids or adnexal masses which might make amniocentesis more difficult.

4. Placental localisation to avoid the placenta and thus to minimise risks of fetomaternal transfusion or induced placental bleeding.

5. Fetal localisation to minimise chances of producing fetal injury.

6. Determination of fetal anomalies (as already discussed).

7. Determination of fetal death. Where fetal death is diagnosed, amniocentesis is avoided along with potential medicolegal problems. Since FDIU and certain fetal anomalies increase AFP levels their diagnosis is essential.

Three large studies have been undertaken to investigate the value of ultrasound in midtrimester amniocentesis. Multicentre North American studies revealed no increase in subsequent neonatal or maternal morbidity or mortality, while a less well-controlled British study did show a slight but significant increase in fetal deaths, neonatal abnormalities, and obstetric complications. Further analysis of the British study by the NICHD has suggested that their findings were no different from American and Canadian studies.[26] It is certainly apparent that routine use of ultrasound in amniocentesis does decrease the number of bloody taps. When ultrasound is utilised, it is essential that both the ultrasound examination and the amniocentesis be performed in the same area with the scan performed immediately before amniocentesis to avoid changes in placental position due to bladder emptying and to recognise transient myometrial contractions.

In the third trimester ultrasound may be used to localise a suitable pocket of amniotic fluid and to avoid placenta, fetus and umbilical cord. If it is in fact necessary to traverse the placenta, an area near the periphery of the placenta is chosen to minimise chances of puncturing a placental vessel. After the amniocentesis is performed, real-time scanning allows a rapid check of the fetal heart.

Adnexal mass in association with pregnancy

In association with a first trimester pregnancy, an adnexal mass is most likely to represent a corpus luteum cyst. Since these are hCG dependent, as hCG levels fall after the first trimester, corpus luteum cysts should regress. A persistent mass is less likely to be a corpus luteum cyst and may be evaluated sonographically. A corpus luteum cyst appears as a cystic mass, usually with a smooth wall and without internal echoes; the presence of internal septa or solid material makes the mass unlikely to be a corpus luteum cyst. When there has been bleeding into the cyst, low level echoes or a fluid-fluid level may be seen.

Other nonfunctional masses, the most common being a dermoid, may occur in association with pregnancy. Nonfunctional masses require operation in midtrimester to prevent possible complications and to exclude malignancy.

Intrauterine devices (IUDs) and pregnancy

Depending on their type, IUDs have a failure rate of two to three per 100 woman-years. Because of the widespread use of IUDs, a significant number of women become pregnant with an IUD in place or thought to be in place. Where the string is not seen on pelvic examination the device has either been expelled, is in an extrauterine location or the string has been drawn up into the uterus as it has grown in size. When in an extrauterine location, the IUD may be impossible to see sonographically. When intrauterine, the device and its relationship to the gestational sac can be assessed. Where the device is difficult to see in an early pregnancy, scanning with decreased gain may allow the device to be seen more readily. After mid pregnancy, an IUD, even when known to be intrauterine, may be impossible to visualise sonographically.

Extrauterine pregnancy

There is a great deal of controversy concerning the role of ultrasound in the management of the patient with suspected ectopic pregnancy. Sonographic appearances in the patient with subsequently proven tubal pregnancy may be normal, may show an adnexal mass of varying complexity or, with a ruptured tubal pregnancy, may show evidence of a haematoperitoneum.[11] Probably the greatest value of ultrasound in the patient with a positive immunological pregnancy test and suspected ectopic is in diagnosis of an intrauterine pregnancy. The presence of an intrauterine pregnancy makes a coexistent extrauterine pregnancy extremely unlikely.

The finding of an adnexal gestational ring with a viable fetal pole is quite unusual and the 'adnexal ring' itself may be mimicked by other lesions such as PID and corpus luteum cyst (Fig. 2.8). With

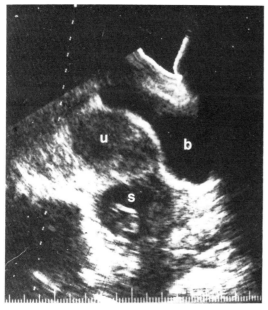

Fig. 2.8 Ectopic pregnancy. Longitudinal view of the pelvis showing the bladder (b) and a gestational sac (s) with viable fetal pole, posterior to the uterus (u). The pregnancy was in the ampullary portion of the Fallopian tube.

ectopic pregnancy a 'decidual cast' may be seen in the uterus and this may mimic an early gestational sac, without a fetal pole. If no decidual cast is seen there may be increased endometrial echoes. The uterus is usually slightly enlarged.

A pregnancy in the cornual area may continue to grow until about 10–12 weeks of gestation. Clinically such patients are found to have an enlarged uterus with a localised swelling. Sonographically the gestational sac is asymmetric and

it is not possible to demonstrate a rim of myometrium going around the sac. It should be noted, however, that in the first trimester implantation of a normal gestational sac may be quite asymmetric; when scanned 1–2 weeks later a normal appearance is seen.

Abdominal pregnancy is quite a rare occurrence and is usually diagnosed only in retrospect. Sonographic features are the presence of a solid mass adjacent to the bladder, representing the uterus with no myometrium surrounding the gestation.

Fetal biophysical assessment

At present, methods of investigation of the antepartum wellbeing of the fetus are limited and consist primarily of the nonstress test (NST), the oxytocin challenge test (OCT) and measurement of urinary oestriol excretion. These methods have several disadvantages:

1. NSTs have a high false-negative rate.

2. OCTs are invasive with a high false-positive rate.

3. Oestriol collections are subject to wide variations in excretions and reflect the status during the time of collection and not at the current time.

Since deterioration in fetal wellbeing may occur very rapidly, particularly in the pregnancy affected by diabetes, more accurate information is needed of the status of the fetus at any given moment.

Observation of several physiological parameters of the fetus has been proposed as a means of assessing fetal wellbeing.[24] The physiological parameters which may be evaluated include fetal breathing, muscle tone, movement and heart rate, together with the amount of amniotic fluid present. Using such parameters, a planning score analagous to the postnatal Apgar score may be derived and has value in predicting subsequent outcome. A low score (on a 10 point scale) is associated with intrauterine fetal compromise and subsequent poor neonatal outcome in a high percentage of cases.

Induced abortions and associated problems

In the management of the patient presenting for pregnancy termination ultrasound may:

1. Give gestational age, therefore allowing choice of the safest method.

2. Reveal the presence of uterine fibroids, coexistent mass or uterine anomalies which might make the procedure, especially an amnioinfusion, more difficult.

3. Reveal the presence of multiple gestation.

After commencement of the procedure there may be problems such as failure to abort or failure to obtain tissue in a suction procedure. The pregnancy may be reassessed and a second procedure may be performed under direct ultrasound guidance in the operating room.

Other complications in which ultrasound evaluation may be helpful include:

Retained products of conception. Immediately after a termination the uterus is enlarged with increased echoes due to blood within the cavity, the appearance returning to normal over the next few weeks. Since the appearance of blood and those of tissues in the uterine cavity are similar, finding such increased echoes is not helpful in patient management. Where the endometrial cavity appears normal, however, it is unlikely that there are significant products of conception remaining.

Haematocrit drop. Haematocrit drop following pregnancy termination may be secondary to uterine perforation. If this results in a broad ligament haematoma, a mass containing echoes may be seen adjacent to the uterus. If bleeding occurs directly into the peritoneal cavity, an appearance similar to ascites may be produced or there may be masking of the normal pelvic anatomy.

Postoperative fever. Postoperative endometritis is a not uncommon sequel to induced abortion. If not promptly treated, infection may spread to the parametrium and lead to a pelvic abscess. The sonographic features of pelvic infection are discussed later in the next section.

Postpartum problems

After a term delivery it takes about six weeks for the uterus to return to normal size. Postpartum problems such as haematocrit drop and postpartum fever are more likely to occur in the patient who has had an operative delivery.

Post-caesarean section changes. After a lower segment transverse incision, careful scanning may show a localised bulge into the bladder representing the area of the surgical repair. Where haemostasis

has been incomplete, a localised haematoma may collect and be visualised as a cystic collection.

Pelvic infection. Pelvic abcess following caesarean section may be visualised as a cystic mass with irregular walls often showing internal debris or a fluid-fluid level.

Wound problems. The Pfannensteil, transverse skin incision often used for caesarean sections is prone to development of a subfascial haematoma or abscess. Longitudinal incisions are prone to subcutaneous collections, both of which can be recognised on ultrasound. Since the fascia may be visualised the location of the collection may be identified.

Retained products of conception. In an uncomplicated pregnancy the postpartum endometrial cavity appears as a thin or a slightly thickened line. Retained products of conception appear as a collection of dense echoes within the central area or as homogeneous placenta-like echoes. As already mentioned, blood in the uterus may also give a collection of dense central echoes.

Trophoblastic disease

Trophoblastic disease is a spectrum of conditions ranging from hydatidiform mole through the invasive mole to choriocarcinoma.

Hydatidiform mole occurs approximately one in 1500 pregnancies in the United States and is considerably more common than this in the Orient. Clinically, vaginal bleeding in association with a uterus which is larger than expected is the common presentation. In a significant percentage, however, the uterus is of a size equal to or less than expected by dates. Sonographically, the uterus is seen to be enlarged with a characteristic vesicular pattern. Focal sonolucencies may be seen within or at the periphery of the uterus and these represent areas of necrosis or haemorrhage. Similar uterine appearances may be seen sonographically in missed abortion, degenerated fibroids and necrotic ovarian tumours. Theca lutein cysts may be seen in approximately 50 per cent of cases as bilateral multilocular cystic masses (Fig. 2.9).

Choriocarcinoma histologically does not have the hydropic villi seen in hydatidiform mole and therefore has a different sonographic appearance. The uterus may be enlarged with a nonhomoge-

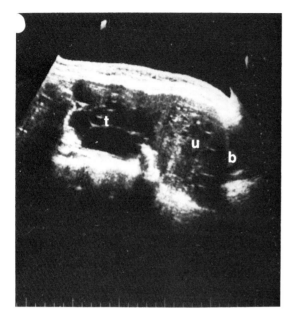

Fig. 2.9 Hydatidiform mole. Longitudinal section of the pelvis and lower abdomen showing the bladder (b) and the uterus (u) to be enlarged with nonhomogenous echoes, compatible with hydatidiform mole. Superior to the uterus are large theca-lutein cysts (t).

neous collection of echoes in its centre or in the myometrium. Where there has been extension to the serosal surface there is irregularity of the uterine contour.

ULTRASOUND IN GYNAECOLOGY

Ultrasound is now widely used as a primary or adjunct diagnostic procedure in the patient with symptoms referrable to the pelvis. Whilst ultrasound may show an abnormality with a limited differential, in many instances the sonographic appearances together with the clinical history allow a specific diagnosis to be made.

The following are the major indications for pelvic sonography:
Diagnosis and characterisation of a pelvic mass.
Assessment and management of the patient with pelvic inflammatory disease (PID).
Diagnosis and management of endometriosis.
Localisation of intrauterine devices.
Diagnosis and follow-up of gynaecological malignancy.

Assessment of the infertility patient.
Evaluation of postoperative fever.
Evaluation of certain gynaecological and urological problems.

Normal anatomy

A full bladder is a prerequisite to a satisfactory pelvic sonogram — the bladder acts as a sonic window to the uterus and adnexae and also displaces bowel away from the pelvis.[32]

On longitudinal section of the pelvis of a patient in the reproductive years, the uterus may be visualised as an oblong structure in the centre of which is a thin echogenic line representing the endometrial cavity (Fig. 2.10). In the premenstrual phase this thin line may become thickened representing endometrial thickening and secretions. The cervix may be seen, as may the vagina. In longitudinal section, the vagina appears as two zones of decreased echogenicity (the walls of the vagina), with a central echogenic line representing the cavity lying immediately posterior to the bladder (Fig. 2.10).

On transverse section, the uterus is seen in cross section as a round or slightly ovoid structure

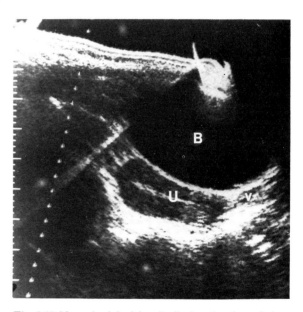

Fig. 2.10 Normal pelvis. A longitudinal section through the normal pelvis showing the uterus (u), vagina (v) and bladder (b). The endometrial cavity is seen as a thin line in the centre of the uterus.

containing a central echogenicity, the endometrial cavity. The ovary is seen usually lateral to the upper portion of the uterus as an ovoid solid structure. With grey-scale equipment, the ovaries may be visualised in the majority of instances. Occasionally the Fallopian tubes may be seen passing from the uterine fundus to the ovaries.

On transverse sections the musculature of the pelvis may also be recognised. The iliopsoas muscles are seen anterior to the iliac wings. Each contains a central echogenicity due to connective tissue and nerves (Fig. 2.11). Other muscles which

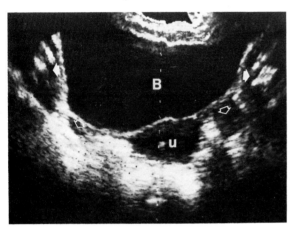

Fig. 2.11 Transverse view of the normal pelvis showing the bladder (b), uterus (u) and ovaries (open arrows) and the iliopsoas muscles (closed arrows). The bright echo at the centre of the uterus is the stem of a Copper 7 intrauterine device.

may be recognised are the obturators lying immediately adjacent to the lateral aspect of the bladder and the piriformis muscles lying more posteriorly adjacent to the midline. Oblique scanning of the pelvis can be used to demonstrate the iliac vessels on the pelvic sidewall and these vessels may indeed be used as a guide to ovarian localisation since the ovaries lie immediately medially.

Bowel may be visualised in the pelvis either as areas of shadowing or as cystic collections. Where liquid masses, perhaps due to bowel, are seen in the pelvis, real-time examination may demonstrate peristalsis. Where there is still doubt a water enema should help to distinguish between bowel and a pelvic mass.

Characterisation of pelvic masses

Uterine masses

Fibroids. Fibroids are the commonest benign tumours of the uterus occurring in more than 20 per cent of women in the later reproductive years. Fibroids are usually multiple and since they are smooth muscle tumours they affect the uterine body more commonly than the cervix. Minimal involvement of the uterus may be seen only as a slight enlargement of the uterus with maintenance of the normal outline. Often this produces a characteristic impression on the urinary bladder. Interstitial or subserous fibroids often produce an obvious, irregular outline to the uterus. Careful adjustment of technique usually demonstrates a texture difference between the fibroid and the normal myometrium (Fig. 2.12). Pedunculated fibroids may appear to lie adjacent to the uterus and on occasions may be difficult to distinguish from a solid adnexal mass.

When uncomplicated these tumours usually have a homogeneous texture, different from that of normal myometrium. Where degeneration occurs, however, the texture usually changes and may become nonhomogeneous. Various types of

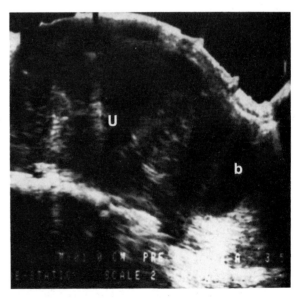

Fig. 2.12 Fibroid uterus. Longitudinal view of the bladder (b) and the uterus showing a uterus (u) markedly enlarged by fibroids. The texture is nonhomogenous and there is decreased through-transmission.

degeneration may be seen. Cystic degeneration appears as decreased echogenicity of the tumour with a concomitant decrease in attenuation by the tumour. Calcareous degeneration is often demonstrated sonographically before it is apparent radiographically, and is visualised as areas of increased echogenicity within the tumour together with acoustic shadowing.

Malignant tumours of the uterus. The main malignancies involving the uterus are endometrial carcinoma, sarcoma of the uterus and carcinoma of the cervix.

Endometrial carcinoma is primarily a disease of the postmenopausal woman presenting clinically as postmenopausal bleeding and uterine enlargement. Sonographically it is difficult to distinguish endometrial carcinoma from uterine enlargement due to fibroids, but where uterine carcinoma has involved the myometrium and extended through the serosal surface the edge of the lesion is irregular and ill-defined in contrast to the smooth, well-defined border of the leiomyoma. Early endometrial cancer may be visualised as increased endometrial echoes, again a nonspecific finding.

Sarcomas, the commonest of which is the leiomyosarcoma, are uncommon malignancies of the uterus. The uterus may be very large and nonhomogeneous, due to areas of necrosis within the tumour. Appearances are often indistinguishable from large fibroids.

Ultrasound is of limited value in the initial diagnosis of carcinoma of the cervix since clinical examination together with cytology or biopsy diagnose the majority. Staging of cervical carcinoma is primarily performed clinically and is based on the degree to which the tumour extends beyond the cervix into the parametrium and towards the pelvic sidewall. Although not an accepted investigation in staging of cervical cancer, ultrasound may be of benefit in correlation with physical findings and in assessment of metastatic spread.

Ovarian masses

Non-neoplastic ovarian cysts. Several types of non-neoplastic or functional cysts may occur in the ovary. In the preovulatory phase of the normal menstrual cycle the developing follicle may be visualised and its growth to ovulation may be followed. The follicle is first visualised as a cystic structure in the ovary, usually with well defined borders and without internal echoes, and usually several days prior to the LH surge. At the time of ovulation the follicle has an average diameter of 20 mm, but after ovulation the follicle decreases immediately to approximately 50 per cent of this size.[7]

After ovulation the corpus luteum may become cystic due to blood or fluid and be visible sonographically as a cystic mass. Whilst a corpus luteum cyst may be indistinguishable from the follicular or retention cyst of the ovary, occasionally the wall may be slightly thickened and irregular, whilst that of a follicle cyst is usually thin and well defined.

The follicular or retention cyst has the typical sonographic features of a cystic mass, i.e., well-defined border, enhanced through transmission and no internal echoes. Where there has been bleeding into the cyst, however, internal echoes may appear. Follicular cysts are common in the reproductive years and are usually followed clinically through several menstrual cycles, when they should regress. It is important to remember that adnexal cysts in the prepubertal or postmenopausal patient or in the patient on oral contraception are unlikely to be functional cysts.

Functional cysts are usually sonographically simple and the presence of solid tissue within a mass makes the diagnosis of a functional cyst less likely, although septa may be seen in theca lutein cysts and uncommonly in corpus luteum cysts. The corpus luteum cyst in association with pregnancy is hCG dependent and therefore should markedly decrease in size or disappear as pregnancy continues.

Neoplastic tumours of the ovary. Several classifications have been proposed for neoplastic lesions of the ovary. They are generally based on the cell type from which the tumour is derived. In one scheme, the groups are:

1. Epithelial tumours, which include serous cystadenoma and cystadenocarcinoma, mucinous cystadenoma and cystadenocarcinoma, endometroid tumours, clear cell neoplasms and Brenner tumour.

2. Germ cell tumours including dysgerminoma, teratoma and choriocarcinoma.

3. Stromal tumours including granulosa cell tumours, Sertoli-Leydig tumours and thecoma-fibroma.

4. Gonadoblastoma.

5. Metastatic tumours.

Serous epithelial tumours. The serous tumour is the commonest tumour of the ovary, accounting for 60 per cent of cases. A large proportion, approximately 30 per cent, are bilateral. Serous tumours sonographically appear as cysts of varying size, usually containing one or more septa. The presence of a significant amount of solid tissue within the cyst or the presence of ascites, or both, are suggestive of malignancy.

Mucinous tumours. Mucinous tumours are less common than the serous type and are more likely to be benign. The mucinous cystadenoma usually has the typical sonographic appearance of a cystic mass with a number of internal septations (Fig. 2.13). Although these septa are usually thin, a considerable amount of solid tissue may be present even though the tumour is benign. Low-level echoes representing the mucin content may be seen within the cyst. Mucinous cystadenomas

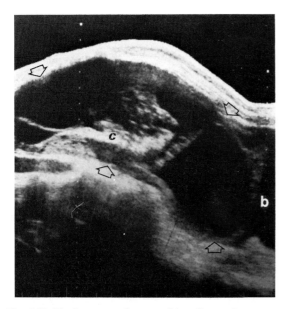

Fig. 2.13 Mucinous cystadenoma with malignant focus. Longitudinal section through the pelvis and lower abdomen showing a large septated, cystic mass (arrows) superior to the bladder (b). There is an area of solid tissue (c) within the mass, more than usually seen in purely benign lesions and compatible with an area of malignancy.

are less likely to be bilateral than the serous variety. The benign tumour may rupture causing pseudo-myxoma peritonei which is visualised sonographically as free fluid in the peritoneal cavity.

Mucinous cystadenocarcinomas contain larger amounts of solid tissue, are more likely to be bilateral and are often associated with ascites. Capsular infiltration of the tumour may be represented by poor wall definition.

Endometroid, clear cell and mesonephric tumours may have sonographic appearances similar to those of a complex or solid ovarian mass.

Germ cell tumours. Cystic teratomas are benign tumours derived from all three germinal layers. They are bilateral in 10–15 per cent of cases. Being derived from all three layers a number of differing tissues are found in the tumours, including hair, bone, cartilage and sebaceous material. The sonographic features are variable and reflect the relative amount of the different tissues.[15] Sonographic appearances may be predominantly cystic or complex with cystic areas, dense echogenic areas and areas of shadowing. The hair component when present is characteristically echogenic. Areas of calcification or ossification may cause acoustic shadowing. Fluid-fluid levels or areas of solid material 'floating' within cysts are very suggestive of dermoids. There may also be areas of acoustic shadowing. Occasionally densely echogenic areas near the surface of a dermoid may cause shadowing of the rest of the mass. This is called the 'iceberg sign'.

The remainder of the germ cells tumours and the stromal mesenchymal tumours are predominantly solid and have a similar sonographic appearance.

Metastatic ovarian tumours. Approximately 10 per cent of ovarian cancers are metastatic, the commonest primary sites being breast and bowel. There is also an increased incidence of primary ovarian malignancy in patients who have a breast neoplasm.

Pelvic inflammatory disease

The sonographic appearances of pelvic inflammatory disease vary depending on the severity and extent of the disease.[31] In the early stages of the disease the endometrium is predominantly affected, as an endometritis. Sonographically an area of decreased lucency may be seen surrounding the endometrial echoes, which may be prominent, either from debris within the cavity or because of the stage of the cycle. When the inflammation spreads to the Fallopian tubes to produce an endosalpingitis, the pus produced may cause tubal distension resulting in a pyosalpinx or an abscess involving both Fallopian tube and ovary, the tubovarian abscess (Fig. 2.14). A cystic mass

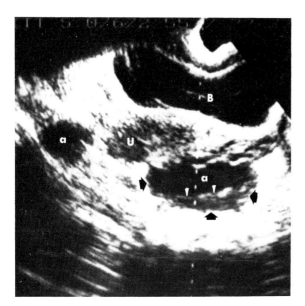

Fig. 2.14 Pelvic inflammatory disease with tubo-ovarian abscess. Longitudinal view through the pelvis showing the bladder (b) and the uterus (u) posterior and superior to which are irregular cystic collections (a), the larger one containing debris with a fluid-fluid level (arrows).

usually with thickened, slightly irregular walls and containing internal echoes or a fluid-fluid level may be seen sonographically. Involvement of adjacent pelvic organs may produce several such abscess collections. In addition, localised reactive ileus of small bowel may contribute to the sonographic appearance of cystic collections within the pelvis. Acute PID often causes blurring of sonographic borders, and on occasion the inflammatory reaction may be severe enough to cause the boundaries of the pelvic organs to be obscured making identification of uterus and ovaries impossible. A similar appearance may be caused by blood in the pelvis as in a ruptured ectopic.

After the acute disease the residua of PID may also be recognised sonographically. A cystic dilatation of the Fallopian tube, a hydrosalpinx, appears as a tubular fluid-filled structure, often folded upon itself. Scarring after PID may result in distortion of normal pelvic anatomy. The uterus may be in a fixed retroverted position or may distort the posterior aspect of the bladder. The ovaries may also be drawn into a more medial location.

Endometriosis

Endometriosis is a condition where endometrial tissue is found in locations other than the endometrial cavity. Most often such deposits are found in the true pelvis, in the cul-de-sac and on the ovaries. Characteristically, endometriosis produces pelvic pain, dysmenorrhea, and infertility. Often the symptoms produced are disproportionate to the amount of disease seen at laparoscopy or laparotomy. Sonographically, endometriosis most often produces the appearance of one or more cystic masses, often with thick walls in the pelvis.[32] Occasionally the cysts are seen to contain numerous low-level echoes, representing the blood. Alternatively solid material, usually representing clotted blood, may be seen. Other appearances produced are adnexal thickening and fixed retroversion of the uterus. The sonographic appearance may be indistinguishable from that of pelvic inflammatory disease.

Adenomyosis is the presence of endometrial tissue in the muscular wall of the uterus. Although the uterus may be seen to be enlarged, this is a nonspecific finding and the internal appearance of the uterus is usually indistinguishable from normal.

Localisation of IUDs

When intrauterine or intramyometrial in location, IUDs are readily assessed by ultrasound. Inability to locate an IUD string after insertion of the device suggests that the device has been expelled, the device has perforated (this is thought to occur most often at the time of insertion), the string has retracted into the uterus or the string is too short, the patient is pregnant and the device has been drawn up into the uterine cavity, or there has been growth of the uterus from some other cause (e.g. fibroids) resulting in retraction of the device.

Various methods are available for IUD localisation including plain radiography, radiography after insertion of a probe or a second IUD, sonography and hysterosalpingography. Since pregnancy may be a cause of a lost string, ultrasound should be considered as a first investigation. Where the device is indeed intrauterine it can be visualised almost invariably with ultrasound. Where the device is in an extrauterine location, it may lie between loops of bowel and therefore cannot be visualised.

Different types of IUD give different sonographic appearances:

Lippes loop. In a longitudinal scan of the uterus, the Lippes loop appears as a series of 4–5 adjacent dots in the endometrial cavity (Fig. 2.15). In a

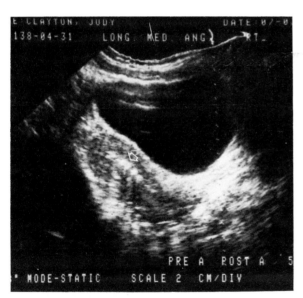

Fig. 2.15 Lippes loop. Anterior section through the pelvis showing an anteverted uterus containing a Lippes loop. Characteristically, the limbs of the loop are shown as four to five echogenic dots within the uterus (arrow). There may be subtle shadowing from the limbs, as seen in this study.

properly situated device the most cephalad limb should be in the uterine fundus. On transverse sections the appearances vary, depending on whether a horizontal or vertical limb is scanned.

Copper 7 and Copper-T devices. The stem of these devices, being wound with copper wire, is very readily seen as an echogenic line within the endometrial cavity. On transverse section the stem appears as a bright dot. The upper limb of the '7' or T, although less echogenic than the stem, may be seen on transverse views.

Progestasert. This device also has a T-shape, but having no copper winding is somewhat less echogenic than the Copper-T.

Dalkon shield. Although this device has been withdrawn from the market, it is still in use by a number of patients. Since the device is a flat oval it is seen as two lines of echoes in both longitudinal and transverse scans.

Saf-T-coil. On longitudinal scan the stem appears as a single line, whilst on transverse scan the two coils appear as a series of echoes, somewhat similar to the Lippes loop.

Ultrasound may be used as a routine procedure after IUD insertion to confirm that the device is suitably located. In certain circumstances a device may be more difficult to visualise when there are increased central echoes premenstrually or where there is intrauterine bleeding or a missed abortion. Similarly in an early intrauterine pregnancy, trophoblastic echoes may obscure a device. In such instances, rescanning with low gain usually results in loss of these obscuring low level echoes and better visualisation of the device.

When pregnancy coexists, the relationship of the device to the developing sac may be seen. In the first trimester a coexistent IUD is unusually easily seen. In more advanced pregnancy, however, the device may be undetectable.

Ultrasound in gynaecological malignancy

Ultrasound potentially has a significant role in the assessment and management of the patient with suspected or proven pelvic malignancy.[32]

The main indications for ultrasound in such patients are:

1. Characterisation of a pelvic mass as ovarian or uterine and as cystic, solid or complex.

2. Determination of involvement of adjacent structures such as bladder and rectum.

3. Presence of intraperitoneal and extra-abdominal spread of pelvic malignancies.

4. Ultrasound guided biopsy of adenopathy or pelvic masses.

5. Postoperative follow-up and evaluation of recurrent pelvic diseases.

6. Radiotherapy planning, external and internal.

7. Assessment of postoperative complications, such as fever and haematocrit drop.

Characterisation of pelvic masses

The characterisation of masses of the cervix, uterine body and ovary are described in this chapter.

Involvement of adjacent organs

Involvement of the bladder or rectal mucosa greatly worsens the prognosis and affects the treatment of gynaecological malignancies. Involvement of the bladder mucosa causes an irregular ill-defined border of the wall and may be associated with a mass protruding into the bladder lumen. Rectal involvement is more difficult to demonstrate but it has been suggested that similar sonographic findings may be seen if fluid is introduced into the rectum.

Spread

Adenopathy. Pelvic malignancies metastasise to lymph nodes in a pattern determined by the primary site, the principal groups potentially involved being those of the obturator, internal, external and common iliacs, presacral nodes and para-aortic groups. Where metastatic involvement has caused enlargement of lymph nodes they may be visualised by sonography. The nodes of the iliac chain may be seen as small solid masses adjacent to the pelvic sidewall. Para-aortic adenopathy results in lobulated masses adjacent to or surrounding the aorta.

Omental involvement. The omentum may become involved with secondary extension of pelvic tumour, particularly from an ovarian primary. Minimal omental involvement is usually not detectable by sonography. When there is involvement to the extent that an omental 'cake' is produced, perhaps with matting of adjacent bowel, a mass lesion may be produced.

Ascites. Ascites is a common accompaniment of malignant ovarian tumours and of omental and peritoneal spread of tumour. Minimal ascites is first shown by collection of fluid in the cul-de-sac; as the amount of fluid increases, there are collections in the flanks and around the liver. Ascites of malignant origin may have different sonographic features from that due to a benign cause such as hepatic cirrhosis. In malignancy, ascites is more likely to be loculated and to show evidence of bowel matting. It may also contain low-level echoes. Since with malignant disease there may be loculation of ascites and areas of adhesed bowel, sonographic examination is helpful to aid in paracentesis by locating an appropriate site.

Peritoneal implants. Ovarian malignancy characteristically spreads transperitoneally and produces studding of the peritoneal surfaces with metastases. These may be recognised sonographically as small masses on the peritoneal surface, more readily visible when there is associated ascites. The subdiaphragmatic peritoneal surfaces, particularly on the right side, are also common sites for metastatic spread. Such metastases are usually subtle sonographically but are more obvious if there is adjacent ascitic fluid.

Hepatic metastases. Although intrahepatic spread of pelvic malignancies is readily assessed with ultrasound the sonographic appearances are variable and may differ even with the same primary histology. Intrahepatic metastases may produce echodense lesions, sonolucent solid lesions, complex lesions or 'bull's eye' type lesions. Sonographic examination of the liver should be included in each instance where a pelvic mass appears to be malignant and in all instances of a pelvic mass in the postmenopausal years.

Ureteric involvement. A pelvic mass may produce ureteric obstruction by extrinsic compression or by direct involvement. A dilated ureter may occasionally be seen in a pelvic sonogram and the point of obstruction demonstrated. Examination of the kidneys is indicated to detect hydronephrosis in all instances of a large pelvic mass.

Ultrasound-guided biopsy

Ultrasound-guided transcutaneous aspiration of suspected adenopathy or intrahepatic metastases is readily performed using a Chiba type needle. Biopsy of cystic pelvic masses as a primary diagnostic procedure is probably not indicated, however, since there is a risk of dissemination of malignant cells. In ovarian carcinoma this would alter the stage of the disease. Where a patient presents with a recurrent mass after treatment and there is doubt as to whether this represents metastatic disease or some other process such as radiation fibrosis, ultrasound guidance biopsy may be very useful in differentiation.

Post-treatment follow-up

Particularly where radiation therapy has been used as an adjunct, the postoperative pelvis may be difficult to assess clinically. A detected pelvic mass may be secondary to recurrent disease, radiation fibrosis or adhesions with matting of bowel. Postoperative changes, especially in the size and shape of the vaginal cuff, may be quite variable, so a sonogram should be obtained approximately two months after surgery to serve as a baseline for each patient. Early detection of recurrent disease, perhaps with sonographically guided biopsy, greatly influences the prognosis of the disease.

Radiotherapy planning

At present ultrasound is not widely used in the planning of external or internal radiotherapy although it offers a superior and, in many ways, an ideal method.[1]

External radiotherapy. In the planning of field size and tumour dose, ultrasound allows visualisation of the margins of the tumour, tumour depth, body contour and the precise localisation of adjacent organs such as bladder, rectum and kidneys which need to be protected.

Intracavitary radiation. In the treatment of endometrial carcinoma and in the more advanced stages of cervical carcinoma, intracavitary radiation is commonly used. Although traditionally the location of an intrauterine applicator is checked with radiography, this only allows demonstration of the axis of the device within the pelvis. Examination with ultrasound not only allows demonstration of the correct intrauterine location of the device, the device producing strong intra-uterine echoes, but also allows determination of its relationship to the pelvic tumour and also to the bladder and rectum allowing more precise dose calculation. Where perforation of the uterus has occurred this may be demonstrated with sonography but would be undetected radiographically.

Ultrasound in infertility

In the patient with primary or secondary infertility, sonography may be of value in determination of the presence of uterus and ovaries, diagnosis of uterine anomalies, diagnosis of uterine masses (e.g., fibroids) which may interfere with implantation, diagnosis and management of endometriosis and diagnosis and induction of ovulation.

With the exception of uterine anomalies and ovulation induction, each of these entities are described earlier in this chapter.

Uterine anomalies

Whilst hysterosalpingography remains the primary imaging method of investigation of uterine anomalies, ultrasound does have a role in the following situations:

1. Determination of presence or absence of the uterus or the ovaries (or both). Where vaginal agenesis is present, sonographic examination of the genital tract aids in diagnosis.

2. Diagnosis of bicornuate, septate and uterus didelphys. Bicornuate uterus may be visualised as a widening of the uterine fundus. Two endometrial cavities may be seen. The cervix is single. With septate uterus, again two endometrial echoes may be obtained but the external configuration of the uterus should be normal. In uterus didelphys, two separate hemiuteri and two cervices are visualised.

3. Imperforate hymen. Where an adolescent presents with primary amenorrhoea and cyclical abdominal pain, imperforate hymen is a significant possibility. Examination of the hymen confirms the diagnosis. Sonography is of value in determining the degree to which the vagina and uterus are distended with blood, as a haematometrocolpos.

Induction of ovulation

The recognition of the developing follicle and the process of ovulation is described earlier. In the patient who is having ovulation induction, serial sonographic examination allows prediction to be made regarding timing of ovulation and the success of the induction.

Assessment of postoperative complications

When performed for treatment of malignancy, pelvic surgery is often quite radical and is associated with a significant rate of postoperative complications. Postoperative fever or haematocrit drop are readily assessed sonographically.

A persistent postoperative fever may be an indication of an abscess related either to the pelvic surgery or to the abdominal incision. Sonography reveals evidence of an irregular predominantly cystic mass, usually containing some low-level internal echoes in the pelvis. Occasionally, a fluid-fluid level is seen. Abscesses related to the incision may be subcutaneous, associated with the rectus muscle or they may be subfascial. It is often difficult or impossible to differentiate sonographically between an abscess and a haematoma. Occasionally, features may be seen which aid in the differentiation, e.g., the presence of septa within the mass are suggestive of loculations within an abscess. When studied over several days, haematomas are seen to follow tissue planes whereas an abscess tends to cross these planes and sonographically make them less well defined. An increase in size over several days also suggests an abscess.

An haematocrit drop in the immediate postoperative period, greater than expected from revealed blood loss, may indicate intraperitoneal or retroperitoneal bleeding. Sonographically a fresh intraperitoneal bleed may have an identical appearance to ascites, with echo-free collections of fluid within the pelvis or abdomen.

Where pelvic surgery has involved lymph node dissection, pelvic collections of lymph or lymphocysts may occur postoperatively. Lymphocysts appear sonographically as well defined cystic masses within the pelvis, occasionally containing low-level echoes. Fluid collections adjacent to the bladder postoperatively may also be secondary to a urinoma, from an incidental cystotomy.

Gynaecological urology

Various techniques, including cystoscopy, cysto-metry and cystography, are available to assess the structure and function of the bladder. Ultrasound offers a noninvasive method which allows the following aspects to be studied:

1. Size, shape and position of the bladder and its relation to other pelvic organs.

2. Assessment of the capacity of the bladder and postvoid residual volume, where catheterisation needs to be minimised.

3. Assessment of bladder wall and detection of invasion by adjacent malignancies.

4. Assessment of relationship of the ureterovesical junction at the bladder base and investigation of urinary stress incontinence.

5. Presence of urethral diverticula.

Bladder volume

Bladder volume may be estimated by using the formula:

$$\text{Volume} = \text{length} \times \text{width} \times \text{AP diameter}/2$$
$$(2.3)$$

Ultrasound measurement of the bladder volume after voiding is a noninvasive method of estimating postvoid residual and is of value where repeated catheterisation needs to be avoided, e.g., in patients with multiple sclerosis.

Bladder wall invasion

The bladder may be secondarily involved by spread of tumour from adjacent pelvic organs, particularly from cervical carcinoma. Sonographically, bladder wall invasion appears as thickening and irregularity of the wall, perhaps with protrusion of the tumour into the bladder lumen.

Urinary incontinence

It has been proposed by R. D. White and D. R. Ostergard that ultrasound may be used in the assessment of the patient with urinary stress incontinence. In serial longitudinal scans of the bladder, the urethrovesical junction may be visualised in approximately 30 per cent of cases. When catheterised, the urethrovesical junction may be seen in all cases. Patients with urinary stress incontinence may be assessed with sonography by measuring the degree of descent of the bladder. In the normal patient descent averages 7 mm and is always less than 15 mm, while in the patient with stress incontinence, the average descent is 31 mm and is always more than 25 mm. Since the test is simple, rapid and noninvasive, it may be performed in the early postoperative period to assess results obtained by bladder neck surgery to correct incontinence.

Urethral diverticula

Urethral diverticula may be diagnosed sonographically as a cystic structure beneath the bladder base.

REFERENCES

1. Brascho D J 1980 Ultrasound in gynecologic malignancy. In: Sanders R C, James A E Jr (eds) Ultrasonography in obstetrics and gynecology. 2nd edn. Appleton-Century-Crofts, New York
2. Campbell S 1974 The assessment of fetal development by diagnostic ultrasound. Clin Obstet Gynec 1:41
3. Campbell S 1977 Early prenatal diagnosis of neural tube defects by ultrasound. Clin Obstet Gynec 20: 351–359
4. Campbell S, Newman E B 1971 Growth of the foetal biparietal diameter during normal pregnancy. J Obstet Gynaec Br Commonw 78: 513–519
5. Campbell S, Thoms A 1977 Ultrasound measurement of the fetal head to abdomen circumference ratio in the assessment of growth retardation. Br J Obstet Gynaec 84: 165–174
6. Clewell W H, Dunne M G, Johnson M L, Bowes W A Jr 1980 Fetal transfusion with real time guidance. Obstet Gynec 57: 516–520
7. Crespigny L Ch de, O'Herlithy C, Robinson H P 1980 Ultrasound observations on the process of ovulation. Proc 25th ann mtg AIUM, p 85
8. Denkhaus H, Winsberg F 1979 Ultrasonic measurement of the fetal ventricular system. Radiology 131 : 781–787
9. De Vore G R, Mayden K, Tortora M, Berkowitz R, Hobbins J C 1980 Umbilical vein dilatation in erythroblastosis fetalis. Proc 25th ann mtg AIUM, p 88
10. Filly R A, Golbus M S, Carey J C, Hall J G 1981 Short-limbed dwarfism: ultrasonographic diagnosis by mensuration of fetal femoral length. Radiology 138: 653–656
11. Fleischer A C, Boehm F H, James A E 1980 Sonographic evaluation of ectopic pregnancy. In: Sanders R C, James A E Jr, (eds) Ultrasonography in obstetrics and gynecology, 2nd edn. Appleton-Century-Crofts, New York
12. Gohari P, Berkowitz R L, Hobbins J C 1977 Prediction of intrauterine growth retardation by determination of total intrauterine volume. Am J Obstet Gynec 127:255–260
13. Grannum P T, Berkowitz R L, Hobbins J C 1979 The ultrasonic changes in the maturing placenta and their relation to fetal pulmonic maturity. Am J Obstet Gynec 133:915–922
14. Grannum P T, Bracken M, Silverman R, Hobbins J C 1980 Assessment of fetal kidney size in normal gestation by comparison of the ratio of kidney circumference to abdominal circumference. Am J Obstet Gynec 138:249–254
15. Guttman P H 1977 In search of the elusive benign cystic ovarian teratoma: application of the ultrasound 'tip of the iceberg' sign. J Clin Ultrasound 5:403–406
16. Hadlock F P, Deter R L, Garcia-Pratt J, Athey P, Carpenter R, Hinkley C M, Park S K 1980 Fetal ascites not associated with Rh incompatibility: recognition and management with sonography. Am J Roentg 134:1225–1230
17. Hellman L M, Kobayashi M, Fillisti L, Lavenhar M 1969 Growth and development of the human fetus prior to the twentieth week of gestation. Am J Obstet Gynec 103:789–800
18. Hobbins J C, Grannum P T, Berkowitz R L, Silverman R, Mahoney M J 1979 Ultrasound in the diagnosis of congenital anomalies. Am J Obstet Gynec 134:331–45
19. Johnson T R, Graham D, Smith N, Saunders R C, Wynn K 1982 Ultrasound guided paracentesis of fetal ascites. J Clin Ultrasound 10:140–142
20. King D L 1973 Placental migration demonstrated by ultrasonography. Radiology 109:167–170
21. Kleinman C S, Hobbins J C, Jaffe C C, Lynch D C, Talner N S 1980 Echocardiographic studies of the human fetus: prenatal diagnosis of congenital heart disease and cardiac dysrhythmias. Pediatrics 65:1059–1066
22. Kurtz A B, Wapner R J, Kurtz R J, Dershaw D D, Rubins C S, Cole-Beuglet C, Goldberg B B 1980 Analysis of biparietal diameter as an accurate indicator of gestational age. J Clin Ultrasound 8:319–326
23. Levi S 1976 Ultrasonic assessment of the high rate of human multiple pregnancy in the first trimester. J Clin Ultrasound 4:3–5
24. Manning F A, Platt L D 1980 Antepartum fetal evaluation: development of a fetal biophysical profile. Am J Obstet Gynec 136:787–795
25. Milunsky A 1980 Prenatal detection of neural tube defects. Experience with 20,000 pregnancies. J Am Med Ass 244:2731–2735
26. National Institute of Child Health and Development 1979 Antenatal diagnosis. NIH publication No 79-1973
27. Queenan J T, O'Brien G O, Campbell S 1980 Ultrasound measurement of fetal limb bones. Am J Obstet Gynec 138:297–302
28. Robinson H P 1972 Detection of fetal heart movement in first trimester of pregnancy using pulsed ultrasound. Br Med J 4:466–468
29. Robinson H P, Fleming J E E 1975 A critical evaluation of sonar 'crown-rump length' measurements. Br J Obstet Gynaec 82:702–710
30. Sabbagha R E, Hughey M, Depp R 1978 Growth adjusted sonographic age (GASA): a simplified method. Obstet Gynec 51:383–386
31. Sample W F 1980 Pelvic inflammatory disease and endometriosis. In: Sanders R C, James A E Jr (eds) Ultrasonography in obstetrics and gynecology. 2nd edn. Appleton-Century-Crofts, New York

32. Sample W F, Lippe B M, Gyepes M T 1977 Gray-scale
    ultrasonography of the normal female pelvis. Radiology
    125:477–483
33. Seashore J H 1978 Congenital abdominal wall defects.
    Clin Perinat 5:61–77
34. Warsoff S L, Gohari P, Berkowitz R L, Hobbins J C 1977
    The estimation of fetal weight by computer-assisted
    analysis. Am J Obstet Gynec 128:881–892

35. Weiner S N, Flynn M J, Kennedy A W, Bank F 1977 A
    composite curve of ultrasonic biparietal diameters for
    estimating gestational age. Radiology 122:781–786
36. Zemlyn S 1978 The effect of the urinary bladder in
    obstetrical sonography. Radiology 128:168–175

3

The abdomen

LIVER AND BILIARY TREE

D. O. Cosgrove

Introduction

In the liver and biliary tree[104] ultrasound has emerged as one of the most useful imaging techniques. It is usually the first radiological procedure selected and is often sufficient alone to enable a clinical decision to be made. However there are also problems which are more effectively sorted out by employing a judicious combination of imaging techniques. Good results with ultrasound depend critically on expert scanning technique coupled with an understanding of tomographic anatomy and, of course, an appreciation of the clinical significance of any findings.

Techniques and instrumentation

Access to the liver by ultrasound is somewhat restricted because of the protective thoracic cage that overlies its greater part. The scanning techniques employed are designed to circumvent this problem by using sectoring movements of the probe in order to obtain maximum display of tissue through the restricted access ports.[42, 94] The most useful technique uses a subcostal sector movement in which, with the patient supine and in held inspiration, the probe is contacted on the skin just below the costal margin. With the operator's hand almost horizontal the probe is pointed towards the patient's head, tucked up under the costal margin and is then swept smoothly in an inferior-running

arc to the vertical. The scanning section is then changed to a linear, sliding motion along the anterior abdominal wall. Thus a sagittal (often termed longitudinal) slice through the liver and adjacent structures is produced. A sequence of such scans at about 5 mm intervals provides near complete coverage of the liver and biliary tree. Regions that are poorly imaged by this approach include the extreme lateral part of the right lobe and the anterior liver surface. For these parts alternative sector scans through the posterior intercostal spaces with the patient prone for the anterior surface and obliquely through the anterior abdominal wall approximately at right angles to the costal margin are helpful. Anterior intercostal views may also be required for those cases where the abdomen is very gassy or where scarring interferes with scanning. Although in this way only limited views can be obtained, they may be sufficient for diagnosis.

The sagittal subcostal views are often supplemented with transverse views where the probe is swept across the epigastrium in the intercostal angle. This view is especially helpful for the left and caudate lobes and leads on to pancreatic scans, though only part of the right lobe can be reached.

A technique that is particularly useful for the biliary tree is to scan in the longitudinal direction using subcostal sector scans but with the patient partly rotated to bring the right side up (i.e. left decubitus). This has the effect of moving the porta hepatis and related structures closer towards the midline and thus out from under the right costal margin, and also of rotating the vessels in the hepatic pedicle so that the common duct lies above the portal vein thus facilitating its recognition.

No patient preparation is required for liver

scanning but for the biliary tree fasting is an essential requirement to ensure that the gall bladder is filled. Ideally biliary examinations should be scheduled for the morning after an overnight fast; this saves the patient the discomfort of fasting during the day and may also reduce aerophagia which can exacerbate the problem caused by abdominal gas.

Most studies of the liver and biliary tract may be performed with static B scanners. For heavily built subjects a 3.5 MHz probe is required to reach the deeper lying posterior portion of the right lobe of the liver, but where possible a 5 MHz probe is preferred since better resolution and tissue differentiation is achieved. This is especially important for the gall bladder where stone shadowing may be inapparent with lower frequencies (since usually they have a wider beam). It may be possible to use 7.5 MHz here especially in children. The focal zone of the transducer should be chosen to optimise the resolution in the region of interest. A 5 MHz medium focus (5–7 cm) with a long focus 3.5 MHz (7–11 cm) and a medium focus 7.5 MHz (2–5 cm) is a useful set. The 5 MHz should have a small face diameter (13 mm) to facilitate intercostal access.

Real time scanners are invaluable in liver studies since they speed the examination and allow more certain identification of vascular structures.[86] Because of the inaccessibility of the liver, sector systems are much more useful than linear systems despite the limited anatomy that can be displayed on a single frame. Mechanical or electronic 'phased array' sector systems are equally suitable, the resolution available usually determining the choice.

Normal anatomy of the liver and biliary tree

The liver comprises a large right lobe and a smaller triangular left lobe, the line of demarcation being the attachment of the falciform ligament which carries the ligamentum teres (remnant of the umbilical vein) in its free margin. The anterior and superior surfaces are smooth and rounded to fit the right hemidiaphragm, but the postero-infero (or visceral) surface is complex since it bears the hilum of the liver (the porta hepatis) and is distorted by impressions from contiguous organs such as the stomach and duodenum, the right

kidney and the hepatic flexure of the colon (Fig. 3.1).[39,61]

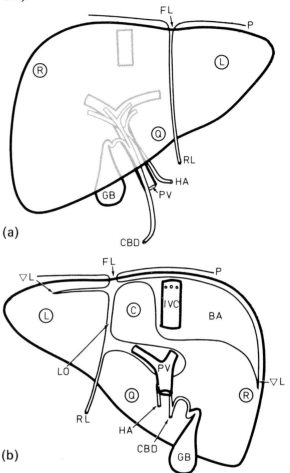

(a)

(b)

Fig. 3.1. Diagrammatic representation of the liver. (a) Anterior view. (b) Posterior view. The two major lobes of the liver are shown (L = left, R = right) together with the two named parts of the right lobe, the quadrate (Q) and the caudate (C) lobes. The attachment of the lesser omentum (LO) separates the caudate from the left lobe. The round ligament (RL, also called the ligamentum teres) running from the umbilicus, attaches to the junction between the left and right lobes and ends at the left portal vein. The falciform ligament (FL) which invests it, runs across the anterior surface of the liver and joins the peritoneum (P) at the diaphragm. On the visceral surface it continues as the triangular ligament (L) enclosing the bare area (BA) of the liver. The major structures at the porta are shown. The portal vein (PV) lies posteriorly with the hepatic artery (HA) antero-medially. The bile ducts, joining to form the common bile duct (CBD), lie antero-lateral with the cystic duct draining the gall bladder (GB) joining it. These structures are shown in grey on the anterior projection where they are covered by liver tissue. The posterior lying inferior vena cava (IVC) is also shown in grey on the anterior projection; in the posterior projection it is shown cut open to reveal the three main hepatic veins which drain into it.

The porta is arranged approximately transversely and contains three sets of vessels. The largest is the portal vein which divides into right and left branches at the porta just before entering the liver substance. Anterior to the portal vein lie the hepatic artery medially and the biliary system laterally. These also have left and right branches and tributaries respectively within the porta. The common hepatic duct is joined by the cystic duct draining the gall bladder at a variable point between the porta and the duodenum where the bile is discharged. Below the junction of the cystic duct it is called the common bile duct; since this point is usually not definable on ultrasound, the deliberately vague term 'common duct' is often used.

The gall bladder is a thin-walled pear-shaped sac attached to the porta by the narrow neck. Its fundus projects below the inferior margin of the liver and lies lateral to the porta. The neck is often tortuous, forming a spiral 'valve' though this term is misleading since it does not function as a valve.

Two other portions of the liver require mention because they may cause confusion. The quadrate lobe is a square shaped portion of the inferior part of the liver that lies between the gall bladder fossa and the falciform ligament and is bounded superiorly by the porta itself. The caudate lobe is a projection from the superior visceral surface of the right lobe towards the left. It passes anterior to the inferior vena cava and is in contact with the posterior surface of the left lobe, the peritoneal layers that intervene forming the interlobar fissure.

The veins draining the liver pass directly into the IVC just below the diaphragm. There are three major hepatic veins, the left and right draining those parts of the liver and an intermediate draining the mid portion.

Sagittal ultrasound sections near the midline show the left lobe as a triangular structure with the diaphragm and heart superiorly (Fig. 3.2).[56] The caudate lobe may be visualised posteriorly in contact with the IVC or the aorta. Moving to the right the portal vein is encountered as a C-shaped tubular structure curving into the liver (Fig. 3.3). This is actually the point at which the main portal vein gives rise to the left portal vein; further to the right the right portal vein is cut across and so is seen as a ring. The portal vein usually has a markedly reflective cuff of surrounding tissue

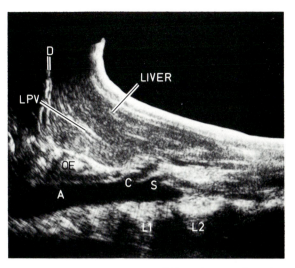

Fig. 3.2 Normal left lobe of liver. This section runs through the aorta and shows the left lobe of the liver overlying it. A portion of the left portal vein has been cut across. Additional structures shown are two major branches of the aorta, the coeliac axis and the superior mesenteric artery, the terminal oesophagus just as it penetrates the diaphragm to join the stomach and the diaphragm itself with the heart superior to it. The aorta overlies the spine. A = aorta, C = coeliac axis, D = diaphragm, L1, L2 = 1st and 2nd lumbar vertebrae. LPV = left portal vein. OE = oesophagus, S = superior mesenteric artery.

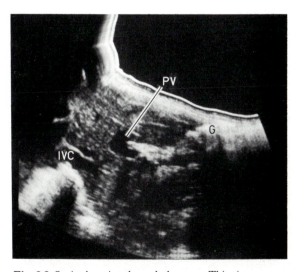

Fig. 3.3 Sagittal section through the porta. This view passes across a portion of the inferior vena cava and cuts through the liver at the point at which the portal vein enters. It is shown here as it passes anteriorly giving off the left portal vein which forms the typical C-shaped configuration. Inferior to the liver gas-containing bowel, probably here antrum of stomach, returns very strong echoes. IVC = inferior vena cava, G = gut, PV = portal vein.

representing the associated arterial and biliary vessels together with fat and fibrous tissue. Sections to the right show the major part of the right lobe bounded superiorly by the strongly reflective diaphragm and the right kidney posteriorly. The gall bladder may appear in this region (Fig. 3.4). Inferiorly the gas-containing bowel gives strong echoes with marked shadowing. Transverse sections show the right lobe of the liver in close contact with the diaphragm and the wedge-like left lobe extending anterior to the aorta and the stomach or pancreas (Fig. 3.5).

The liver parenchyma is seen as a uniform sponge-like mesh of low-grey spots. Passing through it are the blood vessels which are seen as branching tubular structures that can be traced towards the porta (portal veins) or IVC (hepatic veins). Identification of these two groups is facilitated by the fact that portal vein branches are usually surrounded by reflective tissue whereas the hepatic vein tributaries are not, though there are exceptions to this rule of thumb.

Normal biliary radicals and hepatic arterial branches are not visualised in the liver parenchyma but at the porta they can be located anterior to the portal vein as tubular structures measuring up to 4 mm in diameter. They may be distinguished from each other by the fact that the artery lies medially while the duct lies laterally, a feature that is exploited in the left decubitus technique already described.

The filled gall bladder is seen as a cystic space related to the porta (Fig. 3.4). Its walls are thin (less than 3 mm) and smooth, but at the fundus this may be difficult to appreciate since the gas-containing gut that is usually in immediate contact with the gall bladder gives strong echoes that spread on the sonogram and in-fill the adjacent bladder lumen. Similarly, reverberations may render definition of the anterior wall difficult. The neck ('spiral valve') region often contains enough fibrous tissue to cast quite marked shadowing over the distal liver. This is a normal finding.

The ligamentum teres is another troublesome portion of the liver since it often returns high level echoes which can appear as a focal lesion in the left lobe. It may also cause shadowing. The interlobar fissure may attenuate sufficiently to make the posterior-lying caudate lobe appear relatively echo poor and thus be mistaken as pathological.

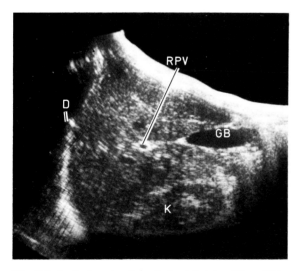

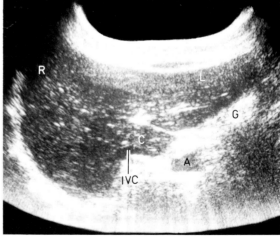

Fig. 3.4. Sagittal section through the right lobe of liver. This section passes through the liver just lateral to the porta and shows the gall bladder with a portion of the right kidney lying posteriorly. The right portal vein is cut across. The diaphragm here is bounded by lung superiorly, thus producing very strong reflections. D = diaphragm, GB = gall bladder, K = kidney, RPV = right portal vein.

Fig. 3.5 Transverse section through the liver. This section shows the right, left and caudate lobes, the latter two separated by the interlobar fissure where the lesser omentum attaches. The IVC and aorta are shown and bright reflections from immediately posterior to the left lobe probably originate from the stomach. A = aorta, C = caudate lobe, G = gut, IVC = inferior vena cava, L = left lobe, R = right lobe.

Focal liver disease

Simple liver cysts are common incidental findings. They appear as rounded, echo free spaces with marked distal enhancement and smooth walls. They are quite commonly loculated (Fig. 3.6).[92]

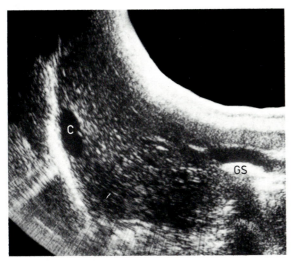

Fig. 3.6 Simple cyst of the liver. In this section through the right lobe of the liver an echofree smooth walled lesion is shown in contact with the diaphragm. Its position here makes it somewhat difficult to assess distal enhancement but the other typical features of a cyst are present. An incidental finding in this case is a collection of gall stones. C = cyst, GS = gallstones.

Polycystic liver disease is also asymptomatic, though the renal cysts that usually co-exist lead to progressive renal failure. The liver cysts have the same features as solitary cysts, except that they are often so packed together as to become polyhedral rather than rounded. Some may show internal echoes due to haemorrhage (Fig. 3.7). The summed distal enhancement of these multiple cysts makes the normal parenchyma appear patchy so that interpretation of other significant lesions can be impossible. (A similar problem applies in gross biliary tree distension.)

Hydatid disease can produce a cystic lesion that is quite indistinguishable from a simple cyst.[35] When daughter hydatids bud from the wall of the mother, however, irregularities of the wall are found. When mature, these form complete daughter cysts that ultrasonically look like cysts-within-a-cyst. Eventually the mother cyst may become

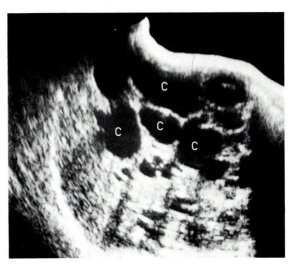

Fig. 3.7 Polycystic liver disease. In this patient with a large liver (note the convexity of the skin line) the anterior portion of the right lobe was replaced by a collection of cysts. These tend to be rounded except where they are in close contact with each other and the walls are flattened, They show marked distal enhancement which obscures the texture of the posterior lying parts of the liver. C = cyst.

completely filled with daughters, giving a 'cartwheel' pattern where the spokes represent the walls of the compressed daughter cysts.

The use of ultrasound in liver abscesses has changed the management of this dangerous condition.[19] Since the majority are due to infection ascending the biliary tree consequent on cholecystitis, or are spread up the portal vein from pelvic and appendiceal sepsis, the right lobe is more commonly affected than the left. The lesion is seen as a cortex of subtly altered liver parenchymal echoes (darker than normal) of perhaps 1 cm thickness, surrounding a more or less echo free central region representing the pus. The shaggy walled 'capsule' is inflamed but viable liver tissue (Fig. 3.8). The affected region of the liver is tender under the probe. Subphrenic and subhepatic abscesses are similar in appearance but lie in the spaces around the liver. Ultrasound is useful not only in detecting these abscesses but in guiding needle aspiration of their contents for bacteriological culture (anaerobic as well as aerobic) and when appropriate, for therapeutic drainage.

Liver trauma may also be detected on ultrasound.[96] When there is a parenchymal haematoma the appearance varies with the state of organisation.

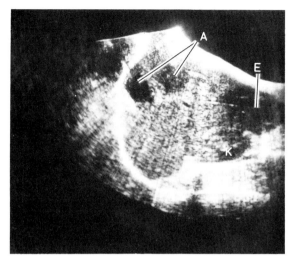

Fig. 3.8 Liver abscess. In this section through the right lobe of liver two abnormal foci are noted in the supero-anterior portion. The more superior has an echofree centre with distal enhancement. There is a poorly defined irregular 'capsule' surrounding the lesion. The inferior lesion has a low level of echoes. This represents the wall of the inflammatory capsule of the second abscess; adjacent tomograms show that this too had a fluid centre. The inferiorly placed echo poor mass (E) is an empyema of the gall bladder with surrounding infection. This was the cause of the liver abscess. A = abscess cavities, E = empyema of gall bladder, K = kidney.

Initially an echo poor region, as clotting and then fibrosis occur, internal echoes appear, eventually leading to a linear scar. Sometimes the blood products are resorbed to be replaced by serous fluid and produce a post-traumatic cyst. When the liver capsule is ruptured, the haemorrhage is into the peritoneal spaces surrounding the liver where they may be indistinguishable from ascites.

Haemangiomas are the commonest benign liver tumours, being a congenital malformation. Indeed when extensive, they may present with high-output heart failure in the neonate, but otherwise are asymptomatic. Two histological types occur, the capillary, consisting of numerous malformed small vessels, and the cavernous, consisting of lakes of blood. On ultrasound the capillary type returns strong echoes (Fig. 3.9) while the cavernous type is relatively echo free. They tend to be situated near the liver surface and do not show distal enhancement nor shadowing except if they calcify.[101]

Benign tumours of the liver substance have become increasingly important since at least one type, the true adenoma, is induced by long term exposure to oestrogens especially in contraceptives. The other type, focal nodular hyperplasia, is a somewhat puzzling lesion, perhaps representing a thwarted repair-response to ischaemia rather than a true tumour. On ultrasound these lesions are seen as ill defined foci of altered echo level, either lower or more usually higher level than the liver itself.[85] The echo texture is rather heterogeneous. On no account should such lesions be biopsied since this may cause catastrophic bleeding.

Metastatic tumours are much commoner than primary hepatocarcinoma (malignant hepatoma) but the ultrasonic features are indistinguishable. By far the commonest appearance is of one or several focal lesions of altered reflectivity.[14] The margins are poorly defined and there is neither shadowing nor enhancement. The echoes are more often of lower intensity than the liver (Fig. 3.10, 3.11) but up to one-third of cases show more strongly reflective or echogenic lesions (Fig. 3.12), this latter group usually being deposits from gastrointestinal or urogenital tract primary tumour. When the tumour is extensive the liver takes on a generalised patchy appearance often referred to as the 'moth-eaten liver' (Fig. 3.13). It may be difficult to decide which part of the tissue on the scan represents liver and which is tumour. Other patterns also occur, the bulls-eye or target lesion being the most dramatic. Here an echogenic central region is surrounded by a rim of echo poor tissue (Fig. 3.14). Fluid filled deposits arise following central necrosis and in functioning metastases from mucin secreting primaries such as carcinoma of ovary and pancreas. The necrotic type have a shaggy wall and contain debris, the appearance being indistinguishable from an abscess. The secretory type may simulate simple cysts.

Reliable detection of liver malignancy is one of the more demanding applications of ultrasound but attention to the effects of the mass of the lesion is helpful. The gross mass effects of hepatomegaly and rounding of the free margin of the liver are non specific and certainly do not always occur, but their detection raises the suspicion of some type of liver pathology. More subtle effects are the presence of surface nodules overlying superficial deposits, distortion of blood vessels by masses and obstruction to the biliary tree of lobar or lobular distribution (usually not accompanied by jaundice).

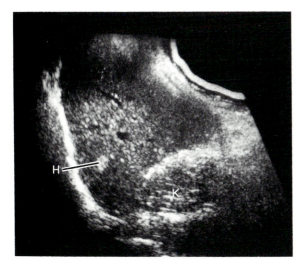

Fig. 3.9 Haemangioma. A small echogenic focus in the posterior portion of the right lobe of this liver represents a haemangioma. There is no distal enhancement nor shadowing and its margins are ill defined. These are the typical appearances of a capillary haemangioma; the cavernous type appear with a low level of echoes. H = haemangioma, K = kidney.

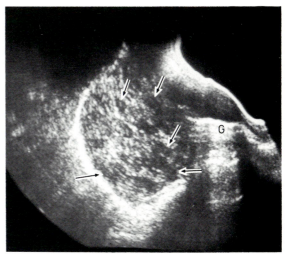

Fig. 3.10 Echo poor metastases. In this section through the right lobe of the liver the texture is disrupted by small regions of reduced reflectivity (arrows). They have ill defined margins and produce neither shadowing nor distal enhancement. This pattern of small multiple lesions is typical of carcinoma of breast from which these deposits were seeded. Note that the liver is not particularly enlarged. G = gas, probably in the hepatic flexure of colon.

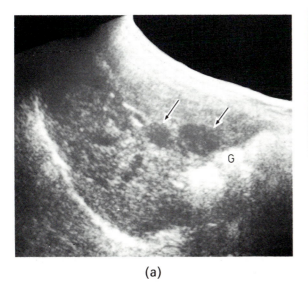

(a)

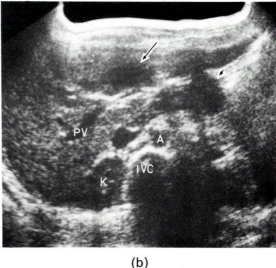

(b)

Fig. 3.11 Echo poor metastases. (a) Longitudinal section through the right lobe of liver. (b) Transverse section that shows several echopoor deposits (arrows). These are from a carcinoma of bronchus. In (b) note also the enlarged pre-aortic lymph node (arrowhead). A = aorta, G = gas in the hepatic flexure of colon, IVC = inferior vena cava, K = right kidney, PV = portal vein.

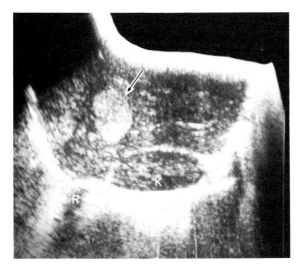

Fig. 3.12 Echogenic deposit. In this enlarged liver with a rounded inferior margin there is a clearcut echogenic lesion (arrow) which was a deposit from a carcinoma of pancreas. This type of metastasis is most commonly seen from tumours in the gastrointestinal and urogenital tracts. K = kidney, R = rib.

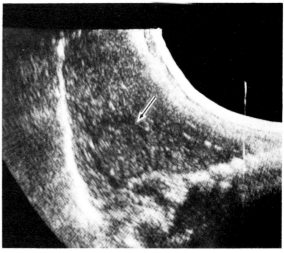

Fig. 3.14 Target lesion. This rather subtle lesion in the posterior portion of the right lobe of liver has a central region that is iso-reflective with the surrounding liver but is detectable because of the echopoor margin (arrow). Often in target lesions the central region is more reflective than the liver itself. This type of deposit is common in larger lesions from any primary site.

Using all of these criteria and careful scanning technique with a properly aligned scanner, some 85 per cent of cases with liver metastases can be detected with an almost negligible false positive rate. Since ultrasound is a labour intensive and time consuming method of imaging the liver, however, radionuclide scintigraphy is preferred as a screening technique. Ultrasonography should be used to evaluate equivocal and abnormal areas.

Diffuse diseases

Fatty change is a common and non specific response of the liver to a wide variety of injuries especially due to toxic compounds, alcohol being the commonest. It produces hepatomegaly that can be recognised on ultrasound and also an increase in reflectivity that seems sometimes to be accompanied by increased attenuation of the sound beam which makes for difficulty in imaging the more distant parts of the liver (Fig. 3.15).[26] The high reflectivity gives subjectively obvious brightening of the liver parenchyma on the scan, which may be verified by noting an exaggeration of the normally slight contrast between renal cortex and liver at the same depth. A further manifestation of this

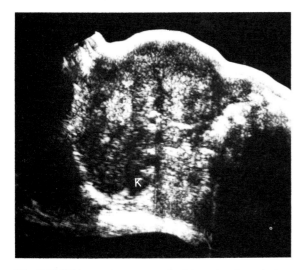

Fig. 3.13. Widespread metastases. In this section through the right lobe of the liver there is gross disruption of the echo texture with interspersed regions of high and low reflectivity which have become confluent. This 'motheaten' pattern is typical of extensive malignant involvement either primary or secondary. Note that it is difficult to decide in many parts of this liver which of the strongly or weakly reflective portions represents residual normal liver tissue. The liver is greatly enlarged, producing a mass in the right upper quadrant as evidenced by the convexity of the skin line. K = right kidney.

phenomenon is the disappearance of the bright cuff normally seen around the portal vein branches, due simply to masking by the strong liver reflections. These are features of the 'bright liver pattern'. It must be noted, however, that it is found in only about 50 per cent of patients with fatty change, the remainder appearing quite normal. Presumably this is related to the severity of the fatty infiltration but this has not as yet been established.

In addition, this is a non specific finding since it is also seen in some cases of cirrhosis, especially the micronodular type (80 per cent of cases).[18] In cirrhosis the liver size changes with the state of the disease, being enlarged early on and then contracting to the shrunken knobbly liver of end stage cirrhosis. The caudate lobe is often spared in this shrinking process and may be genuinely enlarged. The regenerating nodules are difficult to discern when they are deeply buried within the liver parenchyma, but when superficial can be made out as surface irregularities, particularly when there is

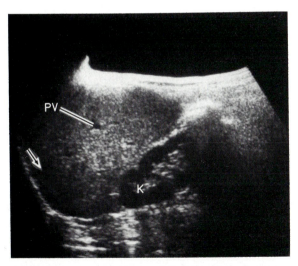

Fig. 3.15 The bright liver pattern. In this moderately enlarged liver with a rounded inferior margin, the echo level was greatly increased above normal. The brightness has been reduced in photographing the image but it can be deduced from the greatly enhanced contrast between the liver and the kidney which should normally be more similar in reflectivity and by the masking of the normally apparent echogenic tissue surrounding the portal vein. In addition the attenuation of this liver was abnormally high as evidenced by the inability to demonstrate the more posterior portion (arrow). In this particular case the appearance is due to gross fatty change but a similar pattern can be found in cirrhosis especially of the micronodular type. K = kidney, PV = portal vein.

associated ascites. Apart from ascites, the effects of portal hypertension may be evidenced sonographically as a portal vein larger than 15 mm in diameter and as varicosities around the spleen or pancreas. These are seen as convoluted tubular structures resembling a 'bag of worms'. Opening up of the umbilical vein can be detected as a tube lying within the ligamentum teres.

Thus in at least some cases of cirrhosis a specific diagnosis can be be suggested based on the ultrasonic bright liver together with characteristic ancillary findings. As with the fatty liver, however, many cases give quite normal findings, possibly due to the insensitivity of the eye at gauging changes in the image. This is an area where computer analysis of the echoes, so called tissue characterisation, may prove valuable.

The other main pattern seen in diffuse liver diseases is the 'dark liver' often found in acute hepatitis of all causes, the congestive liver and in the lymphomas and leukaemias. The changes are the reverse of those seen in the bright liver, with reduced echo intensity so that the liver echoes are weaker than those from the renal parenchyma. The periportal echoes appear exaggerated. As with the bright liver, these changes only occur in a proportion of cases and they are non specific. Sometimes additional clues may suggest the precise aetiology, for example dilatation of the hepatic veins points to liver congestion.

Many other disorders produce subtle texture changes that are rather subjective. When the process is patchy, irregularity of the liver echoes is produced giving a heterogeneous 'moth eaten' pattern. Examples include focal necrosis due to liver toxins (e.g. alcohol) and focal fatty change. These patterns may be difficult to distinguish from multiple metastases.

Jaundice

Ultrasound has a central role in the investigation of jaundice because of its great reliability in detecting dilatation of the biliary tree. This distinguishes between 'medical' and 'surgical' varieties, the mechanical blockage to the larger bile ducts in the second group being suitable for operative treatment.

The ultrasonic criteria for assessing biliary tree

dilatation can be divided into two groups, those concerning the ducts in the liver parenchyma and those of the extrahepatic ducts. Normally the intrahepatic biliary vessels are too small to be resolved; they run parallel to the portal vein branches (the flow, of course, being in opposite directions) and probably contribute to the reflective periportal cuff. When obstructed they dilate and then can be imaged as duplicate vessels running along the portal veins.[82] Cut longwise this produces the 'parallel channel sign', cut across the 'double barrel shotgun sign' (Fig. 3.16). In both projections there are 'too many tubes' and though it may not be possible to determine which is the vein and which the bile duct at any given site, the pattern is obvious. The only trap to beware of is that of mistaking one of the pair of portal veins just distal to a branch point as a duct, a difficulty easily overcome if the quick-search facility of a real time scanner is available. If, as is commonest, the obstructing lesion is extrahepatic, then the ducts throughout the liver are dilated, but a lobar or segmental pattern of dilatation points to a lesion in the porta or liver parenchyma.

For the extrahepatic biliary tree measurements are relied on in the detection of dilatation though in practice the degree of dilatation is so marked in almost every case that the measurements are rather academic.[12] Indeed in some cases the common duct becomes so large that one may have difficulty in recognising it for what it is. Within the porta the common hepatic duct or the right hepatic duct lies across the right portal vein. In a sagittal section the right portal vein appears as a ring of up to 1 cm in diameter with the duct as an anterior lying tube of up to 4 mm in diameter. Between these two structures the right hepatic artery can sometimes be discerned as a ring of a couple of millimetres diameter. When the right hepatic duct dilates its lumen exceeds the 4 mm and usually does so rather markedly, often becoming larger than the right portal vein itself (Fig. 3.17). This tomogram then has a rather characteristic appearance with the dilated and tense-looking duct draped across the right portal vein ring, almost as though it is too heavy to support itself and is sagging at either end.

The mid portion of the common duct is most easily identified with the patient in the partial left decubitus as a vessel parallel to the main portal

vein and anterior to it. At this level the common duct may measure up to 7 mm in diameter and again usually grossly exceeds this when under pressure. Inferior to this the common bile duct passes posterior to the duodenum and is usually lost to ultrasonic view due to the duodenal gas. Within the pancreas the common bile duct can

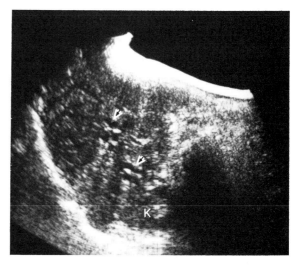

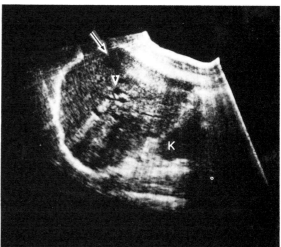

Fig. 3.16 Obstructive jaundice — intrahepatic features. In these two examples from different patients of the appearances of dilated bile ducts within the liver parenchyma, regions where doubling up of the number of apparent vessels are indicated with arrowheads. These either appear as paired rings in the 'double barrel shotgun sign' when they are cut across or as parallel tubular structures when they are cut longways in the 'parallel channel sign'. In the bottom scan, note also the echo poor metastasis in the supero-anterior part of the liver — this patient had a carcinoma of head of pancreas.

most easily be identified in transverse section where it is found in the posterior part of the head of the pancreas, usually overlying the IVC.

In an obstruction to the lower biliary tree the dilatation first affects the extra hepatic system and this is the most sensitive index of obstruction. Later and progressively the ducts at the porta and then the intra hepatic ducts dilate up. For obstruc-

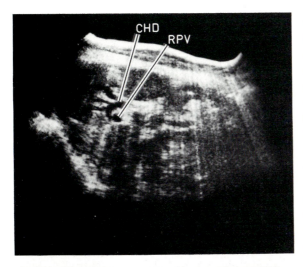

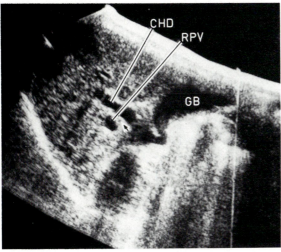

Fig. 3.17 Obstructive jaundice — features at the porta hepatis. Two examples of dilated common hepatic ducts seen in the typical orientation at the porta hepatis. The duct appears as a tense tubular structure, often dividing superiorly (as in the top scan) draped across the right portal vein which is cut in cross section. Sometimes the right hepatic artery can be made out between these two structures (arrowhead in bottom scan). CHD = common hepatic duct, GB = gall bladder, RPV = right portal vein.

tions at other sites the precise distribution of the dilatation depends on the ducts involved. The involvement or otherwise of the gall bladder is so unreliable as an indicator of the precise level or aetiology of obstruction that it is best ignored. It is all too easy to be lured to a diagnosis by the demonstration of stones in the gall bladder only to miss the tumour in the head of pancreas that has really caused the jaundice.

These features are reliable with a few exceptions that require particular attention. A false negative is uncommon but can occur if the scan is carried out too soon after the onset of jaundice. In any clinical problem-cases a repeat scan after a week is recommended. Rarely, fibrosis of the duct walls may so splint them that dilatation cannot occur. False positives again are uncommon but can arise because ducts may remain dilated permanently after relief of a previous obstruction so that duct dilatation (correctly observed) is falsely interpreted as due to ongoing obstruction. A somewhat similar problem applies following a cholecystectomy; for reasons that are not clear the common duct dilates somewhat after cholecystectomy and an upper limit of 10 mm must be allowed in these cases.

Careful use of these criteria and attention to possible exceptions allows the near-certain distinction between cases without large duct obstruction (medical) and those with a large duct lesion who in general require surgical intervention. The surgeon would like to know more about the location of the obstructing lesion and its nature to plan the operative procedure. Ultrasound can usually distinguish between a high lesion, i.e., in the porta, and one in the mid or lower portion of the duct but the duodenal gas that obscures the lower duct makes detailed anatomical display difficult. The nature of the lesion can only be detected by ultrasound when it is large. Since a stone impacted at the narrow lower end of the common bile duct may measure only 3 mm in diameter and an ampullary tumour can obstruct at the same sort of size, these are usually undetectable. Similarly, strictures also are often too localised to be detectable. Larger stones can be located, as can pancreatic tumours above about 2 cm in diameter but the detectable causes amount to only about 50 per cent of cases. Where the surgeon needs this information and it is not available from ultrasound then the next most useful test is usually a transhepatic cholangiogram.

The gall bladder

The gall bladder acts as a temporary store for bile which is produced continuously by the liver but only required episodically for digestion. Thus it fills during fasting and empties progressively with meals. When distended its wall is so stretched as to appear only as a line on the sonogram but when empty it thickens up to a maximum of about 3 mm. The dimensions of the full gall bladder itself are too variable to be of much use clinically. Maxima of 4 cm diameter and 10 cm lengthwise have been suggested but many abnormal gall bladders are in fact smaller than this. A tense, rounded shape is more helpful clinically but this is a subjective assessment.

Distension of the gall bladder may occur as part of general biliary tree obstruction, though if it is itself abnormal the gall bladder may be unable to dilate up though it be under pressure. (This is Courvoisier's sign which is not very reliable in practice.) The commonest cause for isolated gall bladder dilatation is a stone impacted in its neck. This usually leads to acute cholecystitis, but the gall bladder may dilate to enormous proportions and the bile be replaced by clear liquid. This is called a mucocele. It is usually an incidental finding with a large 'cyst' being demonstrated in the abdomen. A mucocele may become super-infected and produce an empyema when the gall bladder becomes very tender.

Gall stones are common in Western society, as many as 70 per cent of the over 70s being affected, often without symptoms. They occur in different chemical forms, only the calcific stones (20 per cent) being visible on a plain X-ray. Their composition however has no effect on the ultrasonic appearance for all stones return strong echoes and cast shadows beyond (Fig. 3.18). Provided there is bile in the gall bladder and the stones are not attached to the wall, postural mobility can be demonstrated, the stones settling with gravity as the patient's position is changed. This is most conveniently assessed using real time.[13] The size of the stone tends to be over-estimated on ultrasound because the strong reflections spread to either side of its true margin due to the width of the scanning beam. For the same reason the shadows cast tend to be minimised and if, because of angle or the small size of the stone only part of the beam is interrupted, then shadowing is inapparent. This is thought to be the cause of the rare 'non shadowing stones' that are rather confusing. If a high frequency well focused beam is used, shadowing should be demonstrable for stones down to about 1 mm in diameter.

Using these criteria of ultrasound for gall stone detection is high (> 90 per cent), and, since the test is easy and quick, some departments have substituted cholecystosonography for cholecystography as the first test for gall stone disease.[15] The relative skill-dependence of ultrasound, however, and its inability to detect calcification, important for drug treatment of stones, has deterred many who prefer to use ultrasound as the back-up technique.

There are occasional difficulties in the use of ultrasound to detect stones. One problem is the neck of the gall bladder; the strong echoes and shadowing from this part of the gall bladder can simulate a stone in this position, and there is no

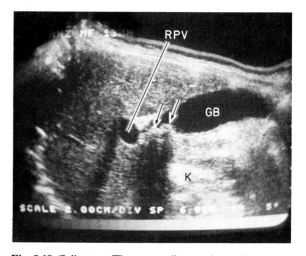

Fig. 3.18 Gall stones. These two gall stones (arrows) are seen in the superior portion of this gall bladder. They return strong echoes and cast complete shadowing over the upper pole of the right kidney. Note that the gall bladder wall is so thin as to appear only as a line on the sonogram. GB = gall bladder, K = kidney, RPV = right portal vein.

simple way to distinguish them. Another difficulty arises when the gall bladder is empty. In this case an irregular, echogenic and shadowing mass may be due to a loop of bowel in the gall bladder fossa or to a collection of stones onto which the gall bladder has contracted.

Careful scanning usually reveals at least a trace of bile but otherwise the mass should be studied with real time for evidence of peristaltic movement that allows its identification as gut. A repeat scan, again fasting, that still fails to demonstrate a normal gall bladder is very strong additional, though non specific, evidence of gall bladder disease. A minor difficulty is sometimes caused by misinterpreting the debris that quickly collects in any non-functioning gall bladder. Debris however is always free of shadowing. It usually layers out, sedimenting to the lower part of the gall bladder and may form after surprisingly short intervals such as a 48 hour fast.

The usual consequence of a stone impacted in the neck of the gall bladder is an attack of acute cholecystitis. Because of the difficulties of assessing the gall bladder neck referred to, the stone itself often escapes detection on ultrasound but the gall bladder wall becomes thickened and characteristically develops a thin echofree line around it, presumably representing oedema (Fig. 3.19).[46] Local tenderness, the ultrasonic Murphy's sign, is present. These findings are highly reliable and form the best method of making this diagnosis in the acute phase. Since many surgeons prefer urgent to interval cholecystectomy, this is a useful application of ultrasound. In chronic cholecystitis the wall may also be thickened and often rather strongly reflective, but the findings are less specific and cholecystography is usually required in addition. The wall of a normal gall bladder may seem thickened in ascites; this should not be confused with inflammatory thickening.

The uncommon carcinoma of gall bladder has a strong aetiological association with gall stones. On ultrasound this tumour is seen either as a gall bladder filled with irregular, poorly echogenic material, or as a mass extending from the gall bladder into the surrounding liver, in which case there is usually associated obstruction to the biliary tree. Gall stones can usually be demonstrated in either case.

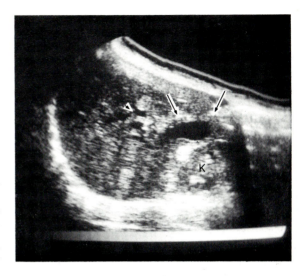

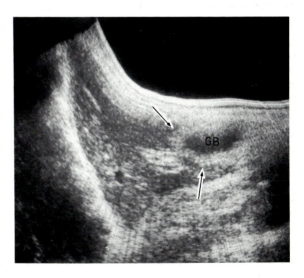

Fig. 3.19 Cholecystitis. In these two examples from different patients the gall bladder wall is thickened (the outer margins are marked by the arrows). These are the typical features of cholecystitis and in the right clinical context allow a confident diagnosis to be made. In the top scan, the 'double barrel shotgun sign' is present (arrowhead); this patient had coexisting cholecystitis and obstructed lower common bile duct both due to gall stone disease. GB = gall bladder, K = kidney.

PANCREAS

H. B. Meire

Introduction

The recent resurgence of interest in investigation of the pancreas has been precipitated by the progressive increase in the incidence of pancreatic disease throughout the civilised world. Although the causes for the increase in both inflammatory and neoplastic disease of the pancreas are obscure recent statistics suggest that cancer of the pancreas is now twice as common as it was in 1970 and that there has been a similar increase in the incidence of inflammatory disease of the pancreas. Although the latter may be associated with increased alcohol consumption this is certainly not the sole cause and it has to be admitted that the majority of cases are of uncertain aetiology. A recent report has suggested that cancer of the pancreas is twice as common in those who drink more than three cups of coffee a day and this claim has been supported by one further paper, though an additional study has found no association between coffee consumption and pancreatic cancer.[1]

Recent improvements in medical care have reduced the mortality and morbidity from acute inflammatory pancreatitis and its complications, but establishing the correct diagnosis in the first place remains a problem in many patients. For this reason the use of imaging techniques to assist in establishing the diagnosis has recently received added impetus. Unfortunately modern medical care has not as yet improved the outlook for patients with cancer of the pancreas: over 95 per cent of these patients die within one year of their disease being diagnosed. In other forms of malignant disease, however, the success of chemotherapy and the possibility of ultimate cure can be shown to be related to the size of the tumour at the time of presentation. If we assume that the same is true of cancer of the pancreas then any technique which enables the diagnosis to be established earlier should be expected, in the fullness of time, to improve the current appalling prognosis in this disease.

Early attempts at ultrasound visualisation of the pancreas with the old bistable display were unsuccessful and it was firmly believed that the normal pancreas could not be seen by ultrasound imaging.[24] When grey scale display became available, however, it soon became clear that the pancreas could indeed often be seen. Subsequent work by many authors to establish the normal anatomical relations of the upper retroperitoneal structures greatly facilitated localisation of the pancreas,[53, 80] and from the early beginnings most experienced ultrasonologists and ultrasonographers now expect to visualise the pancreas in nearly 90 per cent of patients.[50] Unfortunately the precise anatomy and location of the pancreas is somewhat variable both from patient to patient and with progressive increase in age, but its vascular relations remain remarkably constant and an accurate knowledge of these should permit the pancreas to be identified in the majority of patients. Similarly, a knowledge of the normal echographic appearance of the pancreas is also helpful in its identification and for the diagnosis of pancreatic disease.

The anatomy of the pancreas

The pancreas is a retroperitoneal structure which lies approximately transversely in the upper abdomen. The organ is approximately 'comma' shaped, the relatively rounded head lying anterior to the vena cava at the level of D12 or L1 vertebral body. The neck and body of the pancreas pass transversely, and usually superiorly, anterior to the superior mesenteric vessels and aorta. There is a small portion of the head of the pancreas, however, which usually passes behind the superior mesenteric vein and, since it has the shape of a hook, is called the 'uncinate' process. As might be expected, the neck of the pancreas is generally the thinnest portion of the gland, and the size of the pancreas increases from the neck towards the body. In many textbooks the tail is shown to be a progressively thinning structure as it passes to the left of the midline, probably because this is its shape in the cadaver. In live humans, however, the tail of the pancreas is generally rather more globular and it may be the widest portion of the gland, often having an AP diameter of up to 35 mm. The tail continues from the body of the pancreas in a superior and leftward direction and passes posteriorly towards the anterior surface of the upper third of the left kidney and the hilum of the spleen. The long axis of the pancreas thus lies oblique to the true transverse plane, the head usually lying several centimetres caudal to the body and tail. The precise angle between the axis of the pancreas and the patient's spine is very variable and this must be borne in mind when attempting to produce longitudinal scans of the gland.

The splenic vein has a remarkably constant relationship to the pancreas, passing posterior and superior to the tail and body and anastomosing with the superior mesenteric vein where this passes behind the neck of the gland. These two vessels fuse to form the portal vein and the appearance of this confluence of veins is characteristic on ultrasound scans and is probably the single most useful landmark for the pancreas on transverse scans (Fig. 3.20).

The precise shape of the pancreas varies from one individual to another and these variations have been discussed in some detail by other authors.[99] Similarly the level at which the pancreas lies varies from one patient to another and it is interesting to note that whilst in the young adult it is usually at the level of D12, in children it may lie higher than

this and in patients beyond about their sixth decade it may descend as low as L2. Being a retroperitoneal structure, however, it is relatively immobile and therefore it is not greatly affected by respiration of the patient. Its level changes little when the patient is in the erect position.

In certain instances it may be necessary to measure the dimensions of the pancreas in order to determine whether or not there is focal or generalised enlargement. In order to permit this it is important to have at least some idea of the normal dimensions of the gland, but at the same time to bear in mind the fact that its shape varies enormously from patient to patient. In general, the maximum diameter of the head should be no more than 30 mm when measured on transverse scans. The neck of the pancreas is the narrowest portion and should normally be less than 20 mm in diameter whilst the body and tail are frequently greater in diameter than the head, but rarely exceed 35 mm in diameter. Obviously the precise sizes vary according to patient size — in paediatric patients the pancreas is relatively larger compared with other organs and may appear pathologically large to the unwary. Probably more important than absolute dimensional measurement of the pancreas is an assessment of its contour. On scans along the long axis of the organ the outline should be smooth with a gentle curvature. Any localised expansions must be regarded as pathological.

Normal ultrasound appearances

Although the size, shape and position of the pancreas are remarkably variable, the echo pattern from the normal pancreas is fairly constant for patients of any given age. In paediatric patients the echo pattern is remarkably similar to that of normal liver both in texture and overall echo amplitudes. By adulthood the pancreas assumes a rather more coarse ultrasonic texture and the mean echo amplitude is normally somewhat higher than the adjacent liver (Fig. 3.21). Beyond middle age, the morphology and histology of the pancreas alter considerably, the duct system becoming progressively more dilated, possibly with small cyst formation, the gland becomes more heterogeneous and frequently the quantity of fibrous tissue is increased. These features are probably at least

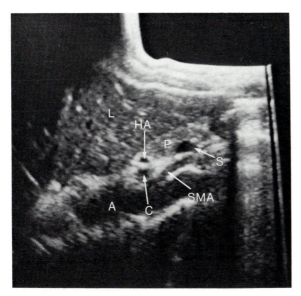

Fig. 3.20 Longitudinal midline scan showing the relationship between the left lobe of liver L, aorta A, coeliac axis C, superior mesenteric artery SMA, splenic vein S, hepatic artery HA and pancreas P.

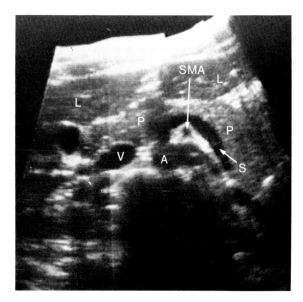

Fig. 3.21 Transverse scan along the long axis of the pancreas P, showing its relationship with the aorta A, vena cava V, superior mesenteric artery SMA, splenic vein S and liver L.

partly responsible for the progressive increase in echo amplitudes which occurs with advancing years, and it is vitally important to remember this when trying to exclude chronic pancreatic disease in patients beyond the age of 60 years or so. It is also important to remember that the echo pattern from the pancreas, and its relationship with that of the liver, is also determined by technical factors which are discussed in detail later. In general, however, incorrect increase in swept gain control settings produces apparent increase in pancreatic echogenicity. It has also been noticed that transducers with higher frequencies generally accentuate the difference between liver and pancreatic echo amplitudes.

The resolution obtained from most modern ultrasound scanners permits visualisation of portions of the pancreatic duct, though because the duct is rather tortuous only short sections are seen in any one image. Therefore the normal, slightly heterogeneous appearance of the gland may be expected to be interrupted in certain images by parallel bright structures encompassing a dark lumen.

The first section of this chapter deals with the

considerable value of ultrasound in investigation of the jaundiced patient, and the ultrasound equipment operator must remember the intimate relationship between the common bile duct and the pancreatic head. The normal common duct is well within the resolution of modern ultrasound scanners and can be expected to be seen in both longitudinal and transverse scans in the region of the pancreatic head.

There are two commonly seen artifactual appearances which may mislead the unwary sonographer. These relate to imaging of both the head of the pancreas and its duct system. The lateral relation of the pancreatic head is the second part of the duodenum. Although this can normally be correctly identified on static ultrasound images, duodenal contents may occasionally produce a heterogeneous appearance which may mask the correct outline of the pancreatic head and give the observer an erroneous impression of a pathological mass at this site. Great care must be taken to avoid this mistake. Therefore, it is essential to obtain multiple images of the pancreatic head when attempting to confirm the presence or absence of a mass at this site using a static scanner. The problem is obviated by use of real time scanners which readily show the active movement of duodenal contents. The anterior relation of the body of the pancreas in the majority of patients is the posterior wall of the stomach. When the stomach is relatively empty the muscular coat of its wall appears as an echo free linear structure with an area of irregular echogenicity anterior to this representing the convoluted mucosa and gastric contents. The thickness of the muscular coat of the stomach is similar to the diameter of the pancreatic duct and the echo pattern from within the stomach may be similar to that from the pancreatic body. Therefore it is frequently possible to produce an appearance similar to an enlarged pancreas with a parallel-walled duct running through its centre. This error can be avoided if the operator takes care to perform scans both above and below this level and in longitudinal as well as transverse planes. The apparent pancreatic duct can then be seen to extend over a wide area and to be a relatively flat plate and, therefore, clearly not a genuine duct. Once again real time scanning greatly eases the correct interpretation of this common finding.

Ultrasound scanning technique

The precise technique for imaging the pancreas varies according to the type of equipment employed. Greatest experience has been obtained to date with static B-scanners and these currently also produced the best definition of this organ. The recent developments in real time imaging, however, combined with the inherent advantages of the use of these systems, has produced a rapid increase in their use for pancreatic imaging.

One of the first factors that must be considered is preparation of the patient. Since the pancreas may well have to be visualised through the stomach and gas within the small bowel may interfere with the image it is preferable to avoid scanning the patient immediately after the patient has consumed a meal and carbonated fluids should be avoided by the patient during the two hours prior to the examination. If the pancreas is to be imaged together with the liver and biliary system it is necessary for the patient to be fasted for eight hours in order to optimise imaging of the gall bladder.

Transducer selection

Before starting the ultrasound examination it is essential to choose the transducer appropriate to the patient under examination. This varies considerably according to the make of equipment employed but with most modern apparatus a transducer frequency of 3.5 MHz is appropriate. It is also important to select a transducer in which the focal zone is placed at the depth of the organ to be imaged. Since this depth is not known until the examination has been initiated the operator should be prepared to change transducers during the examination in order to obtain optimal resolution in the region of interest. In the large or obese patient it may be necessary to employ a 2.25 MHz transducer, but this should be changed to a higher frequency, if at all possible, during the course of the examination.

The examination is normally initiated with the patient lying in the supine position. In view of the constant vascular relations of the pancreas it is probably advantageous to perform an initial longitudinal midline scan of the patient in an attempt to identify one or more of the relevant vessels.[62] On such a scan the aorta should be visualised and arising from its anterior surface either the coeliac axis or superior mesenteric artery may be seen (Fig. 3.20). The pancreas invariably lies inferior to the origin of the coeliac axis and anterior to the proximal few centimetres of the superior mesenteric artery. Provided either of these vessels can be identified the level at which the body of the pancreas is located can usually be discerned. The splenic vein may be seen in transverse section as it crosses posterior to the body of the pancreas and this vessel also gives precise localisation of the level of the body. Having identified this level on the longitudinal scan the transducer is moved along the midline of the patient whilst viewing the image and is held stationary over the level of the body of the pancreas. The position of the transducer on the surface of the patient is then noted and the gantry rotated to perform scans along the long axis of the pancreas. Although the angle between the long axis of the pancreas and the patient is rather variable it is seldom directly transverse and therefore it is preferable to commence this portion of the examination with the gantry rotated approximately 25° to the transverse plane, the left extremity of the plane being more cephalad.

Provided the plane of scan has been correctly selected a significant portion of the body of the pancreas should be identified on the initial transverse scans. If this is not the case then it is probably advisable to seek the relevant vascular anatomy once more in order to home in on the pancreas. All the vessels which can be seen on longitudinal scans can also be readily detected on transverse scans. The single most valuable landmark is the confluence of the splenic and superior mesenteric veins which produces a characteristic 'comma' appearance and is an invariable relation of the posterior aspect of the neck of the pancreas (Fig. 3.21). If these vessels are not identified with ease it is preferable to move the scanning plane cranially and to perform transverse scans of the liver. If the scanning plane is then incremented caudally at approximately 3 mm intervals the coeliac axis should be identified anterior to the aorta. Since this vessel is invariably superior to the pancreas further movement of the scanning plane by one or two centimetres in the caudal direction should bring the plane over the level of the pancreas. When performing scans inferior to the level of the coeliac axis the next major arterial structure to be imaged is the superior mesenteric artery. This also has a fairly characteristic appearance since it generally has a cuff of fat around it. This results in the vessel lumen being surrounded by a cuff of bright echoes (Fig. 3.21) even in patients with relatively small quantities of peritoneal fat. It should be remembered, however, that the superior mesenteric artery traverses a distance of several centimetres and is only related to the pancreas in the upper part of its course. Obviously it is not necessary to identify all of these vascular landmarks and so long as one can be correctly located a portion of the neck and body of the pancreas should be identified in the images.

In those patients where identification of the pancreas on transverse scans proves difficult it may be helpful to perform the scans during suspended inspiration in the hope that this will bring the liver down anterior to the pancreas. If this is not helpful it may prove necessary to examine the patient whilst sitting or standing as this will almost certainly cause the liver, stomach and small bowel to descend within the abdomen whilst the pancreas remains virtually unchanged in position. Since scanning the patient in the erect position with a static B-scanner is rather awkward it is probably advisable to use a high resolution real time system if one is available.

In patients in whom suspended inspiration is found unnecessary to permit imaging of the pancreas, scans can be performed during normal quiet respiration. It is important to remember, however, that structures in the region of the pancreas are subject to considerable movement during respiration and that all structures related to the diaphragm and aorta undergo pulsatile movements with the cardiac cycle. For this reason it is important to confine the scanning technique to 'single pass' scans in order to prevent loss of detail. The only exception to this rule is if a portion of the neck or head of the pancreas is obscured by acoustic shadowing from the ligamentum teres or duodenal contents; limited compounding may then be necessary to image the head and body on the same scan. Compounding should always be performed during suspended respiration.

The tail of the pancreas is the most difficult portion to identify adequately. This results from its position posterior to the stomach, the gastric contents almost invariably preventing adequate ultrasound penetration to this region. Several techniques have been devised for overcoming this problem. The first of these is the posterior scan through the left renal bed.[31] The left kidney can virtually always be identified on posterior longitudinal scans to the left of the spine and, since the tail of the pancreas lies anterior to the upper half or third of this kidney, an accurate knowledge of the anatomy of this region should permit the operator to identify the normal pancreatic tail in these views. Unfortunately, however, the ultrasound beam width characteristics are degraded by passage through the paravertebral muscles and the resolution obtained in this view is inferior to that pertaining on anterior scans. Nevertheless, abnormal enlargements of the pancreatic tail can be reliably identified. From a clinical point of view imaging of the pancreatic tail is seldom of direct importance since inflammatory disease generally involves the entire gland. If large pseudocysts are present at this site they are seldom difficult to identify. Carcinoma of the tail of the pancreas remains asymptomatic until a late stage in the disease process and therefore patients with this disease usually present when the tumour has already metastasised and become incurable.

For those who wish to improve their success rate for imaging the tail on anterior scans the technique of gastric fluid distension offers some help.[16] If the stomach is distended with fluid and the patient placed in a right anterior oblique or semi-erect position the fluid content may lie over the tail of the pancreas whilst the gastric air bubble floats free of this region. Scans can then be performed using the fluid-filled stomach as an ultrasound window to the tail and, so long as the swept gain is correctly adjusted to prevent inappropriate overamplification of the pancreatic echoes due to the non-attenuating nature of the gastric fluid, surprisingly good pictures can occasionally be obtained. The same technique may be used on erect scans using a real time scanner, though in this position the weight of the gastric fluid generally causes the body of the stomach to descend low in the abdomen and it is necessary to give the patient a much greater volume of fluid in order to distend that part of the stomach now related to the pancreas. Some authors have recommended the routine employment of 1.5 litres of fluid, but most ill patients will either not drink or will promptly return such quantities! The potability and ultrasound characteristics of the fluid can be improved by using orange juice rather than clear water; carbonated fluids must of course be avoided at all costs.

Examination of the pancreas using a real time scanner is both quicker and easier than using static imaging devices. With real time scanners the relevant vascular anatomy can be identified rapidly with the scanner placed initially in a plane assumed to be along the long axis of the pancreas (Fig. 3.22). As with static imaging, it may occasionally be necessary to examine a patient during suspended inspiration, with fluid in the stomach, or in the right anterior oblique, semi-erect or erect positions. The two major limitations of real time systems concern ease of access with linear array devices and the ultimate image quality. Many patients with pancreatic disease also suffer from weight loss and may have a scaphoid abdomen with consequent prominence of the lower costal margin. In many of these patients it may prove impossible to obtain adequate skin contact along the whole length of a linear array. This is less of a problem in the erect position when the liver tends to descend into the sub-costal region. Sector scanners do not suffer from the problem of patient contact and are probably preferable for pancreatic imaging. At the present time the image resolution and grey scale quality obtained with the majority of real time scanners is inferior to that of modern static B-scanners and the subtle changes in pancreatic echo pattern which may be seen in certain disease disorders may be overlooked by real time scanners. Since the pancreas is much easier to identify with these devices, however, and subtle changes are the exception rather than the rule, real time scanning of the pancreas should not be discouraged. The image quality of real time scanners is also improving at a rapid pace and will undoubtedly soon equal or surpass that of current static machines.

A further somewhat unconventional approach to pancreatic scanning is the use of the prone position for the patient. In this position the stomach can be distended by a small quantity of

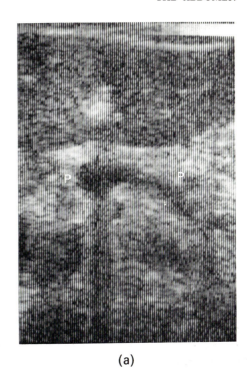

(a)

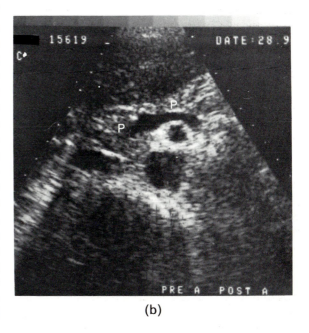

(b)

Fig. 3.22 (a) Real time scan of pancreas P, using a linear array scanner. (b) Real time scan of pancreas P, using a mechanical sector scanner.

fluid and the gas content rises free of the pancreatic bed. Access to the patient is difficult if not impossible with conventional static B-scanners, but becomes just possible with real time scanners. The technique was primarily devised for use with the 'Octoson' scanner in which the patient lies prone on a closed water bath and eight large ultrasound transducers automatically perform scans from within the water (see page 20). Using this apparatus, it is routinely possible to visualise the majority of the pancreas and in many patients it is possible to obtain cross-sectional images of almost the entire anatomy at the level of interest.

Ultrasound in pancreatic inflammatory disease

Inflammatory disease of the pancreas encompasses acute, sub-acute, recurrent and chronic pancreatitis. The pathology of pancreatitis is rather ill-understood and the nomenclature somewhat confused. Acute pancreatitis implies a severe painful illness in which the pancreas becomes inflamed and is usually enlarged and oedematous. In severe cases the pancreatic enzymes may begin to digest the gland itself and varying degrees of necrosis and haemorrhage may occur. Similarly leakage of the pancreatic juices into the peripancreatic tissues may set up an inflammatory reaction leading to the formation of pseudocysts. Pancreatic pseudocysts are a moderately common complication of acute pancreatitis and are partly responsible for the relatively high morbidity and significant mortality from severe acute pancreatitis. The management of this complication is greatly aided by ultrasound scanning and is discussed later. Patients with acute pancreatitis normally present with sudden onset of severe upper abdominal pain. In the very severe forms of the disease the patient may rapidly go into a state of shock mimicking the clinical findings in acute abdominal catastrophies such as ruptured viscus or leaking aortic aneurysm. In these patients, establishment of the correct diagnosis may be exceedingly difficult but it is essential for the management of the patient.

Relapsing and sub-acute pancreatitis are generally milder than the severe acute attacks but, as the name implies, run a rather more chronic course with exacerbations and remissions. During acute phases of the disease the pathology and ultrasonic appearances of the gland are similar to those in normal acute pancreatitis, and, in between attacks, the gland may apparently return to the normal. As the disease progresses, however, an increasing fraction of the gland is destroyed and replaced by either fibrous tissue or fat. Ultimately almost all the gland may be destroyed and the patient may present with symptoms of pancreatic insufficiency. At this stage the patient is said to have chronic pancreatitis and this term is often used as a misnomer for the end stage of pancreatitis in which the gland is destroyed and fibrotic, irrespective of the route by which it got to this condition. Patients with relapsing and chronic pancreatitis frequently have relatively mild symptoms and are commonly not diagnosed until the disease is very far advanced. A significant proportion of patients may have been in the care of the psychiatrists at some stage during the course of their illnesses. This fact emphasises the great difficulty in diagnosing disorders of the pancreas by normal clinical, biochemical and imaging techniques.

In acute pancreatic inflammation, the major ultrasound abnormalities are increase in size of the gland, accompanied by a substantial reduction in echo amplitudes to well below those of the adjacent normal liver (Fig. 3.23). When swelling of the gland is severe, the normal vascular landmarks already described may become difficult or impossible to identify, but this is seldom a cause of a failed examination since the sheer size of the gland now permits its identification and imaging. Only rarely is acute pancreatitis confined to a segment of the gland. The disorder may affect the tail alone if there is a stenosis or duct calculus whilst a portion of the head may be affected in isolation if this is drained by an accessory duct which becomes obstructed.

The development of a pseudocyst can usually be readily confirmed by ultrasound examination (Fig. 3.24). Although the majority of pseudocysts arise within the region of the pancreatic bed, it is important to remember that they may occur at sites considerably remote from the pancreas, including the lower abdomen and the chest or mediastinum. In addition to this, the patient may develop free peritoneal fluid or a pleural effusion during the acute phase of the disease. Aspiration of

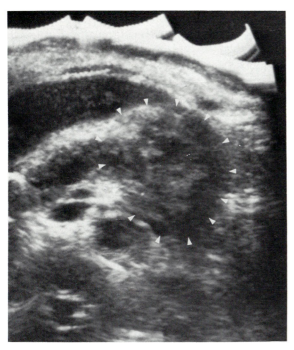

Fig. 3.23 Transverse scan showing enlargement of the tail of the pancreas (arrowed) due to acute pancreatitis.

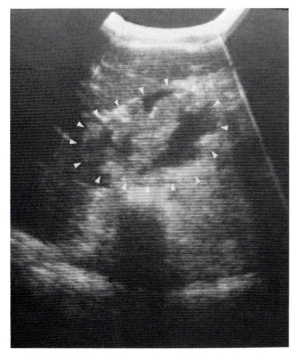

Fig. 3.24 Transverse scan showing early pseudocyst formation (arrowed) around the body and tail of the pancreas in a child with acute pancreatitis.

these fluid collections is immensely helpful in confirming the diagnosis since they are very rich in the enzyme amylase. Ultrasound is most helpful for both demonstrating the presence of these collections and guiding the aspirating needle to the correct site.

Ultrasound is a very useful technique for monitoring cysts once they have been demonstrated. Those which remain stable in size and appearance for two or three weeks and subsequently reduce in size need not be referred for surgical drainage. Conversely cysts which are increasing rapidly in size probably deserve prompt drainage either by open or closed surgical techniques. The experience of most workers currently suggests that approximately 50 per cent of all pseudocysts resolve spontaneously.

Recurrent and relapsing pancreatitis produce changes indistinguishable from those of acute pancreatitis if the patient is examined during an acute attack. The ultrasound appearance between attacks varies according to the degree of gland destruction and fibrosis. When this is fairly extensive the gland may become shrunken and the echo amplitudes abnormally high with a more heterogeneous appearance. The duct system may become the site of multiple stenoses with consequent dilatations and calculus formation. All the disorders can be identified with modern ultrasound scanners.[67]

Chronic pancreatitis predictably produces an ultrasound appearance of a small highly reflective gland (Fig. 3.25). The presence of both fibrous and fatty tissue causes an increase in the echo amplitudes whilst the presence of retention cysts, due to duct stenoses, may produce a heterogeneous appearance.

Ultrasound of pancreatic neoplasms

Benign tumours of the pancreas are extremely uncommon, but to the ultrasonologist the most important of these is the cystadenoma. These are slowly growing tumours which, as the name suggests, contain both solid and cystic components. Ultrasonically their appearance is remarkably similar to ovarian cystadenoma[55] and because of their slow rate of growth and noninvasive nature they may reach very considerable size before

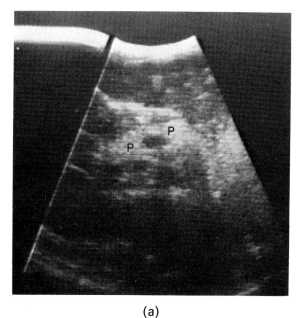

(a)

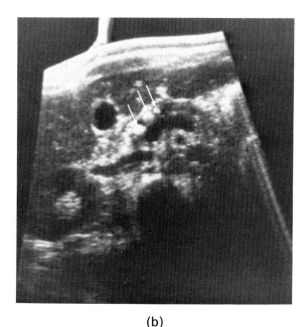

(b)

Fig. 3.25 (a) Transverse scan showing an abnormally echogenic pancreas P, in a child with pancreatic fibrosis secondary to cystic fibrosis. (b) Transverse scan showing calcium deposits (arrowed) in an adult with chronic pancreatitis.

diagnosis. The lesions usually present clinically as palpable masses and only exceptionally are they responsible for causing jaundice. On ultrasound examination the lesions are almost invariably spherical with clearly defined outlines and multiple cystic spaces with septa of varying thicknesses. There is usually very little sound attenuation in the lesions with consequent distal enhancement of ultrasound echoes. A single case has been reported in which the cystic components of the lesion were very small and the ultrasound appearance was then that of an intensely echogenic mass, an extremely unusual finding and quite unlike pancreatic carcinoma.[55]

Endocrinologically active benign tumours are also rare lesions, one of the commoner being the insulinoma. These lesions are almost invariably extremely small in size and may be microscopic. The chance of establishing the correct diagnosis by ultrasound therefore is somewhat limited.[17]

Malignant pancreatic tumours are almost invariably adenocarcinoma, though there is a malignant counterpart to the cystadenoma, the cystadenocarcinoma. Adenocarcinoma of the pancreas is now a very common disease and appears to be increasing in frequency. The typical ultrasound appearance of this disorder is that of a mass disturbing the outline of the pancreas in which the background echo amplitudes are very low compared with the normal, though multiple foci of apparently increased echo amplitudes may be present throughout the lesion, particularly centrally[103] (Fig. 3.26). The margin of the lesion is frequently lobulated and may be ill-defined. Approximately 70 per cent of carcinomas of the pancreas occur in the region of the head and may present relatively early due to obstruction of the common bile duct. When presented with a jaundiced patient, therefore, it is essential to attempt to exclude this particular cause for the jaundice.

The appearances of carcinoma of the pancreas are not dissimilar to those sometimes seen in chronic pancreatitis, and difficulty may arise in establishing the correct differential diagnosis. Several useful papers have been published recently reporting investigations into the relative success rates of different investigative techniques for establishing a correct differential diagnosis in such patients.[57,64] Despite some discouragement,[25] ul-

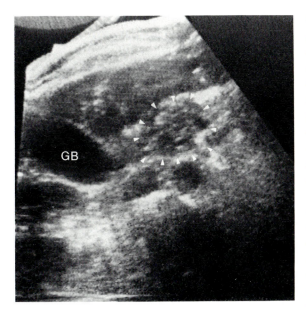

Fig. 3.26 Transverse scan showing a dilated gall bladder (GB) secondary to common bile duct obstruction by a carcinoma in the head of pancreas (arrowed).

trasonography seems to be the best noninvasive test for pancreatic carcinoma at the moment, provided that the operator has the proper training and expertise for the performance of adequate ultrasound examinations.

There are a number of additional signs which have been described to assist with the correct diagnosis of pancreatic cancer, but the majority of these have proved to be non-specific. Indentation of the anterior wall of the inferior vena cava by enlargement of the pancreatic head,[97] however, is a particularly useful sign to illicit in cases where there is doubt about the cause of an obstruction at the lower end of the common bile duct. This region can almost invariably be identified on longitudinal scans of the inferior vena cava with the patient in the right anterior oblique position.

Paediatric pancreatic ultrasound

Although the majority of pancreatic disease occurs in adults, the ultrasonographer should remain aware of the possibility of pancreatic disorders in children. Ultrasound has been reported as achieving correct diagnosis of carcinoma of the pancreas in a nine year old child.[60] At the Clinical Research Centre the diagnosis of acute pancreatitis with pseudocyst formation was diagnosed in a three year old child who had undergone many months of fruitless investigations for the cause of his illness. Children are also rather more prone to receive upper abdominal trauma and this may itself precipitate an attack of acute pancreatitis.[95] The disease may also occur during epidemics of mumps (epidemic parotitis) and certain other viruses may be responsible for precipitating acute pancreatitis in the paediatric population.

Certain of the modern chemotherapeutic agents used to treat leukaemia and lymphoma may themselves cause pancreatic inflammation and ultrasound has been useful for establishing the diagnosis of this disorder and helping with patient management.[83] When examining the normal paediatric pancreas it is important to remember that the size of the gland appears relatively larger than that of the adult and the echoes from within the gland may be similar to or slightly lower in amplitude than those of the adjacent normal liver.

Congenital cystic fibrosis (muscoviscidosis) eventually affects the pancreas in all patients with this disease, though to varying degrees. To some extent the degree of pancreatic involvement can be judged by ultrasound examination (Fig. 3.25) and disturbance in both echo amplitudes and gland size have been recorded.[88, 102] Similarly the formation of cysts within the gland can be demonstrated by ultrasound if the cysts are large enough to be resolved. Ultrasound is also valuable for demonstrating the associated changes within the patient's liver.

Ultrasound of the pancreatic duct

Since the introduction of grey scale display, some investigators have occasionally seen structures which appear to be segments of the pancreatic duct. Fortunately a number of authors have now been sufficiently confident to commit their thoughts to writing,[33] although one group were clearly still in doubt when they published their paper in 1979.[22] As experience has been gained and equipment has improved it is now reasonable to expect to identify at least a portion of the normal pancreatic duct in more than 50 per cent of patients. The diameter of the duct should not normally exceed 3 mm and, because of its extremely tortuous nature, it is normal only to image short irregular segments of it. When the duct is dilated it becomes easier to identify and the presence of this dilatation may be particularly helpful in establishing the presence of pancreatic disease (Fig. 3.27). Once again, real time scanning may be

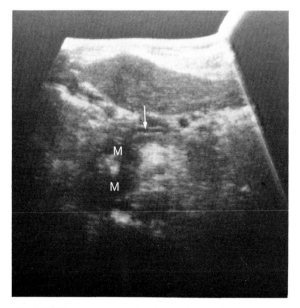

Fig. 3.27 Transverse scan showing moderate dilatation of the pancreatic duct (arrowed) due to an inflammatory mass in the head M.

helpful and instruments with fine resolution are capable of demonstrating pancreatic duct calculi with the associated acoustic attenuation distally.[67] The presence of pancreatic duct dilatation in patients with obstructive jaundice is a very useful sign if the precise cause and localisation of the jaundice cannot be established owing to duodenal gas obscuring the lower end of the common bile duct. Demonstration of a dilated pancreatic duct permits confident localisation of the obstruction to the region of the pancreatic ampulla and makes the possibility of tumour greater than that of calculus.

Ultrasound guided biopsy and aspiration

One of the major advantages of ultrasound is its entirely noninvasive nature. Since it is able accurately to localise specific anatomical structures and focal lesions, ultrasound scanning has now become well accepted as a means of guiding aspiration or biopsy needles to the target site. Many different systems for performing these manoeuvres are currently available, including passage of the needle through specially constructed transducers incorporating a needle guide. It is also possible to observe the passage of the needle towards its target using real time ultrasound. Ultrasound guided biopsy is particularly valuable in attempting to establish a definitive differential diagnosis in patients with irregular pancreatic masses which could be either malignant or a result of chronic inflammatory change. One of the largest series to be published recently has indicated that the technique of ultrasound guided fine needle aspiration biopsy of the pancreas can achieve a correct differential diagnosis rate in 81 per cent of patients with pancreatic masses.[40] The success of this technique does, however, depend upon the services of an expert cytologist who is prepared to attempt a diagnosis based on a few somewhat distorted cells scattered amongst tissue debris and bloody fluid. The technique can also be used for the aspiration of fluid collections including ascites, pleural effusions and pseudocysts and it is likely that many patients with progressive pseudocysts which might otherwise have required surgical drainage can be managed by repeated percutaneous aspiration under ultrasound control.

Conclusion

Ultrasound is rapidly assuming an almost indispensible role in the diagnosis and management of patients with pancreatic disorders. The diagnostic adequacy of the technique is highly dependent upon the skill of the operator and the diligence with which the normal anatomical structures are sought.

As the technical specifications of the equipment continue to be improved and as new and better ways of approaching the pancreas with ultrasound are developed, it seems likely that the role of ultrasound in this important field of medical diagnosis will continue to expand. It must be hoped that the steady improvement in the success rate for imaging the gland will soon be matched by an improvement in the ability to treat the disorders which afflict it.

RETICULOENDOTHELIAL SYSTEM AND SPLEEN

D. O. Cosgrove

Introduction

The reticuloendothelial system is an anatomically diffuse system defined by its function. It comprises cells lining blood and lymph vessels which are capable of ingesting particles by phagocytosis. The white blood cells are excluded since, although phagocytic, they may be mobile or fixed — they form essentially an intermediate group. The major sites of reticuloendothelial tissue are the bone marrow, the liver (here called Kupffer cells) and lymph nodes and spleen. It is these last two regions that are of particular interest anatomically since they may be imaged.

The main function of the reticuloendothelial system is to act as a filter, for blood in the case of the spleen, and lymph in the case of lymph nodes.[7,47] It gathers particles of foreign materials such as bacteria or injected materials, e.g., plastic from infusion bags, and also abnormal body cells, particularly damaged red blood cells. This last function is especially well developed in the spleen which has sinusoidal vessels through which blood percolates slowly in the red pulp. The phagocytic cells on the sinusoid walls remove effete red blood cells returning the breakdown products, including iron and bilirubin from haemoglobin, to the blood. The spleen also has a less well vascularised white pulp which consists of lymphoid tissue with germinal centres in which lymphocytes develop. These are immune competent cells which may produce antibodies or act in cellular immunity. The irregular interspersion of red and white pulp gives a mottled appearance to the cut surface of the spleen. Lymph nodes also have a dual function with centrally placed germinal centres where lymphocytes are formed and a marginal sinus into which afferent lymphatics empty to be exposed to the phagocytic reticuloendothelial cells which clear bacterial and cellular debris. Viable cells such as bacteria and tumour cells are often stopped in the first lymph node they encounter and may multiply there or form an abscess or malignant lymph-adenopathy.

Anatomy

The spleen is an ovoid structure which lies against the diaphragm in the left upper quadrant. Its lateral surface is smooth but medially it has a prominent hilum where a leash of blood vessels enter and leave. The sharper anterior margin often shows several notches from which fibrous septae extend into the otherwise very soft parenchyma. The spleen measures up to 10 cm in its long axis which usually lies in the tenth intercostal space but its size is very variable since it contracts on demand to return the blood sequestered in the sinusoids to the general circulation. Thus the normal spleen may measure as little as 4–5 cm in length. Medial to the spleen lies the stomach; inferiorly is the upper pole of the right kidney with the splenic flexure of the colon anterior to that. The spleen is an intraperitoneal organ, being attached to the posterior abdominal wall by the splenic vein and artery and the peritoneal folds that surround these vessels. Prominent lymphatics draining the spleen accompany the splenic vein as it courses along the superior margin of the pancreas.[39,56,61]

Lymph nodes are rounded or ovoid nodules ranging in size from about 1 cm in diameter to a few millimetres. One or several afferent lymphatics empty into the marginal sinus, the lymph then discharging through the solitary efferent lymphatic at the hilum where small arteries and veins also attach. In general lymphatics are disposed alongside the major blood vessels, with nodal groups at the branch points (Fig. 3.28). Thus in the abdomen nodes are prominent in the pelvis around the junction between the external and internal iliac veins, at the origin of the inferior cava and point of entry of the renal sinus. They also lie around the inferior vena cava along its length, often called para-aortic nodes here. The lymphatic chain empties into the cysterna chyli which lies just posterior to the crura of the diaphragm and leads directly into the thoracic duct to empty into the left innominate vein and thus joins the blood circulation.

A second important set of abdominal lymphatics drains the viscera and follows the portal vein radicles. Their nodes lie in the gut wall itself and in the mesentery. These drain to major node groups at the root of the mesentery around the

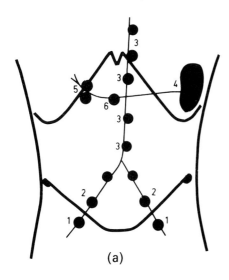

(a)

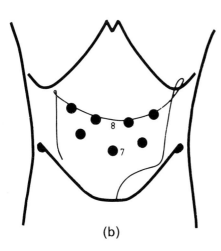

(b)

Fig. 3.28 Diagram of the abdominal lymph nodes. (a) Retroperitoneal group. The nodes associated with the aorta are disposed supero-inferiorly while those associated with the spleno-portal vein lie across the upper abdomen. The former group comprises inguinal (1), iliac (2) and para aortic (3) nodes. The latter drain centrally from the spleen (4) and the porta hepatis (5) nodes to the coeliac group (6) and thus join the paraortic chain. (b) Mesenteric groups. The nodes draining the gut lie in the mesentery, both peripherally (7) close to the gut and centrally (8) at its root. They drain to the para-aortic chain.

superior mesenteric artery and vein. There are also prominent nodes around the head of the pancreas (sometimes called the coeliac group) which receive lymph from the spleen and pancreas along the line of the splenic vein and also from the nodes draining the liver situated in the porta hepatis along the portal vein. For the most part the organs draining into particular node groups are self evident from the anatomy, but attention should be drawn to the unusual arrangement of the gonadal lymphatics. These accompany the testicular and ovarian veins which are long, lying in the retroperitoneum, and join the left renal vein and the IVC at renal level on the left and right respectively. Thus the first station node for gonadal lymphatics is in the upper abdomen, a rather curious arrangement that has its explanation in the embryological descent of the ovaries and testes.

Technique

The normal spleen, being protected laterally by the rib cage and obscured to ultrasound by the gas in gut medially and inferiorly, can be rather difficult to image. Oblique intercostal views in the tenth or eleventh intercostal spaces give the best exposures and allow the long axis to be measured. This view is most easily achieved with the patient in the right decubitus, the left arm being elevated to open up the rib spaces. Quiet breathing is best since a deep breath draws lung over spleen thus obscuring it. A sectoring action with a small-faced probe is required. When the spleen is enlarged it extends infero-medially and, because it is intra-peritoneal, it lies superficially so that it becomes easier to scan. Left subcostal sector sagittal scans analogous to those used for the liver, or transverse subcostal scans, may then be used. Although commonly the spleen is less echogenic than the liver, since this is the organ with which it is usually compared, it is important to note the spleen's appearance at settings correct for the liver before making gain adjustments to optimise the display of the spleen itself.[93]

For lymph nodes the technique depends on the site in question. Normal nodes are too small to be demonstrated on ultrasound, so attention is paid to the related blood vessels. Thus for the internal iliac nodes, the full bladder technique is used as for any pelvic scan and the vessels identified along the postero-lateral pelvic wall. The demonstration of pulsation on real time confirms the vessel's position. The caval nodes in the mid-abdomen are generally obscured by gas but in the epigastrium the inferior vena cava can be located in both sagittal and transverse views. When the cava is small, it can often be rendered more easily visible by scanning during a Valsalva manoeuvre using a real time scanner. The peri-pancreatic and coeliac regions are imaged in the conventional way for the pancreas and the same is true of the nodes in the porta. Mesenteric and intestinal node regions pose a particular problem due to the gas containing bowel around them. For this reason compounded transverse scans across the abdomen may be more useful than single pass scans. These nodes lie very anteriorly in the abdomen. They may also be distinguished from retroperitoneal nodes by observing on real time that they move with respiration and in concert with gut loops rather than with the more fixed retroperitoneal structures.[93]

The normal spleen

The spleen is seen as a poorly reflective semi-lunar region, its upper pole usually disappearing behind lung in the intercostal view (Fig. 3.29).[34] The

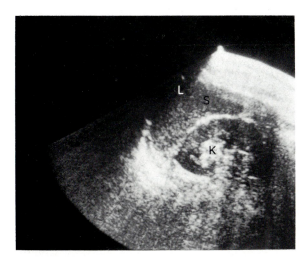

Fig. 3.29 The normal spleen. In this oblique view through the tenth intercostal space the spleen can be seen with the upper pole of the left kidney lying infero-medially. The upper pole of the spleen disappears behind lung. K = kidney, L = shadowing due to lung, S = spleen.

parenchyma has a uniform mottled texture. The hilar vessels may be identified, especially the larger veins. Care should be taken not to confuse the closely related upper pole of the left kidney which also has a low level of echoes, though in practice the reverse error is commoner, the spleen being misread as a renal mass. The nearby gut does not cause confusion when it is gas-filled, but when liquid-filled the stomach especially can simulate a cyst or a mass (inflammatory or malignant) associated with the spleen. If the patient is given a drink of water and the stomach viewed on real time the whirlpool of gas bubbles produced almost always resolves the difficulty.

Lymphadenopathy

Normal lymph nodes are too small to be visualised. When enlarged they are seen as masses with a low level of echoes and a characteristic lobulated margin (Figs. 3.30, 3.31).[3] Since attenuation is 'normal' for soft tissues there is neither shadowing nor echo-enhancement. These features are the same in inflammatory and malignant lymphadenopathy; ultrasound is currently unable to distinguish between these major pathologies. As they enlarge, nodes distort the nearby anatomy, especially the veins (Fig. 3.31), and may produce obstruction to them and, in the case of porta hepatis nodes, to the bile ducts, or to the ureter for a renal hilar and mid-para-aortic nodes. Pelvic nodes are seen as echo poor masses on the lateral pelvic side wall which distort the bladder when sufficiently large.

Para-aortic nodes surround the great vessels. When the pre-aortic group are large they may be so closely involved with the aortic wall as to efface the ultrasonic interface so that the aortic wall cannot be defined (Fig. 3.30). This is the silhouette phenomenon, and it produces a diagnostic problem since a similar appearance may be given by an aortic aneurysm with low level echoes from the thrombus simulating the nodal echoes. Careful scanning to define the aortic wall can be helpful (best done with real time) and a clue may be offered by the more lobulated margin characteristic of lymphadenopathy. Nodes posterior to the aorta may separate the artery from the spine to which it is usually closely applied. This effect is even more

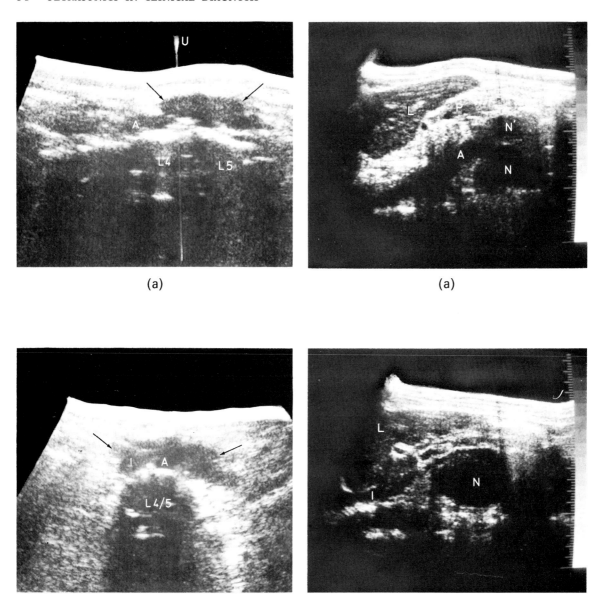

Fig. 3.30 Para-aortic lymphadenopathy. (a) Longitudinal section in the midline through the mid abdomen. The aorta is seen disappearing into a mass (arrows) which returns a low level of echoes. (b) Transverse section. The mass can be seen to have a lobulated outline and to extend laterally beyond the aorta and the cava (both seen here at the point of bifurcation and thus appearing prominent) which allows a clear distinction between nodes and thrombus within an aneurysmal aorta. A = aorta, I = inferior vena cava, L4,5 = 4th & 5th lumbar vertebrae, L4/5 = invertebral disc between L4 and 5, U = umbilicus.

Fig. 3.31 Lymphadenopathy around the great vessels. (a) Sagittal section through the aorta. (b) Sagittal section through the inferior vena cava. A prominent group of nodes of low reflectivity is situated posterior to the aorta and inferior vena cava lifting these vessels anteriorly and separating them from the spine. In the case of the cava there is marked narrowing also. A further group of nodes (N′) lies anterior to the aorta. The nodes show the typical low level of echoes. A = aorta, I = inferior vena cava, L = left lobe of liver, N = nodes, P = pancreas.

striking with post-caval nodes, perhaps because the cava is a less rigid vessel. Quite startling degrees of elevation of the cava can be produced, sometimes with sharp angulation (Fig. 3.31). Dilatation of the cava distal to the nodes implies obstruction and this may be confirmed by observing on real time the absence of pulsation in the dilated portion, in contrast with the normal pulsation of the cava upstream of the lesion. Lack of Valsalva response is an additional diagnostic feature. Surprisingly often quite marked obstruction to the cava goes unnoticed clinically.

Nodes around the pancreas may be quite difficult to distinguish from primary pancreatic tumour. The demonstration of widespread adenopathy in territories outside the drainage of the pancreas may be a helpful clue. Otherwise fine needle aspiration is required. Similarly porta nodes may so indent the liver that they seem to lie in the parenchyma and simulate liver metastases.

Mesenteric nodes are visualised as anteriorly placed lobulated echo poor masses. Due to the difficulty of imaging structures closely associated with gut, only large masses can be detected, 4 cm being a minimum compared to 2 cm for the epigastric nodes.

The echo poor characteristics of enlarged nodes have been mentioned as has the histological non-specificity of the ultrasound findings. Rarely, echo rich nodes are encountered when fibrosis or calcification occur. The former has been noted following successful treatment of malignant nodes. The latter may be encountered in healed tuberculous infections.

Splenic pathology

Splenomegaly is the overriding feature of most splenic disease and can readily be detected by ultrasound (Fig. 3.32). Assessment of the echo levels of the spleen is fraught with the same general problems of the variable effects of overlying tissue and the lack of an internal standard that causes difficulty with the liver. The liver is a poor yardstick since, though often used as a reference standard, it is also affected in splenic disease in many cases. These factors account for some of the confusion over the findings in the different causes of splenomegaly.[63,90] In general, the congestive

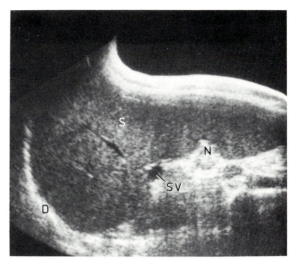

Fig. 3.32 In this subcostal parasagittal scan on the left the spleen can be seen extending well below the left costal margin. The texture looks remarkably like that of normal liver but, in this example, the hilum with the splenic vein and the notch make confusion less likely. This type of splenomegaly is seen in a wide variety of diffuse conditions and a differential diagnosis is in general not possible. D = left hemidiaphragm, N = notch, S = spleen, SV = splenic vein.

splenomegaly of heart failure and cirrhosis returns a medium level of echoes while the spleen in the leukaemias and lymphomas is echo poor but the echo levels may rise following treatment. In infections either high or low level echoes may be found, the former more often in chronic cases, the latter in acute infective splenomegaly. Likewise, in 'haematological' splenomegaly a great range of echo levels is found, in keeping with the range of pathologies. Where fibrosis is a feature, the spleen shows strong echoes as in myelofibrosis. Focal scarring in infarcts and sickle cell anaemia gives bands of strong echoes.

Focal lesions also occur in malignancy; those in the lymphomas are usually echo poor (Fig. 3.33), but echogenic lesions may be seen in carcinomatous deposits. A splenic abscess has the same features as an abscess elsewhere, as do the uncommon splenic cysts.[20]

Trauma to the spleen may produce an echo poor mass within the capsule or surrounding the spleen if it ruptures through.[4] As with infarcts, fibrotic healing would be expected to lead to an echogenic region (Fig. 3.34).

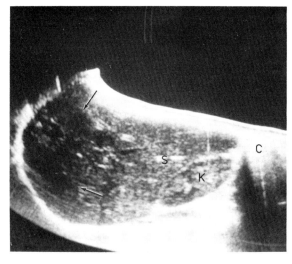

Fig. 3.33 Echo poor focal lesion. In this subcostal parasagittal scan (on the left) of an enlarged spleen, the upper portion can be seen to be replaced by an ill defined irregular region of predominantly low level echoes with interspersed echogenic regions. This was a large nodule of lymphoma. C = splenic flexure of colon, K = left kidney, S = spleen.

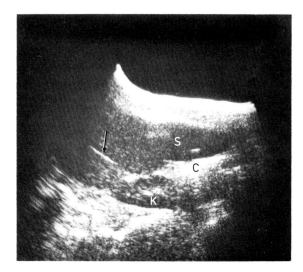

Fig. 3.34 Reflective focal lesion. In this coronal scan through the left flank the spleen can be seen to contain a linear reflective region (arrowed) which represents a scar. This sort of appearance is seen following splenic infarcts. C = splenic flexure of colon, K = left kidney, S = spleen.

The splenectomy bed can be very difficult to interpret on ultrasound.[52] It is often filled by haematoma in the post-operative period and then may look quite like a normal spleen! Organisation leads to higher level echoes of a heterogeneous pattern that is indistinguishable from an abscess. Since this is a common differential diagnostic problem, guided aspiration is required. The empty splenic bed becomes occupied by colon and stomach and again can simulate an abscess or a malignant mass. The demonstration of peristalsis on real time is very reassuring.

Clinical role

The value of splenic imaging is rather limited. While the spleen size can be estimated accurately, this can also be done by scintigraphy. Enlargement is non-specific and the echo levels seem to be poorly correlated with pathology. Focal lesions are more readily interpreted and this finding is helpful in an abscess and in the lymphomas.

Nodal imaging by ultrasound is especially useful for those regions that are not visualised on lymphography. These include the mesenteric nodes and those in the porta. Ultrasound can also indicate the maximal size of a nodal mass more reliably than lymphography, but this technique is more sensitive and follow-up studies are very simply obtained so that it retains pride of place as the first imaging technique for abdominal lymph nodes.

URINARY SYSTEM

B. B. Goldberg and A. H. Wolson

Kidney

Introduction

Ultrasound of the kidney is now well established as a safe, accurate method of diagnosis.[30, 72, 74] In general, an ultrasound examination is performed following intravenous pyelography, and the radiographs should be available to the sonographer whenever possible. Ultrasound most often is used as an adjunct to pyelography, but possesses certain inherent advantages. No ionising radiation is used, no iodinated contrast is required, no preparation is needed, and the status of the renal function is not significant.

Position

The position of the kidneys is variable but usually they are found between the lower ribs and the iliac crest. The normal kidney is easily recognised and absence of a kidney should prompt a search for an ectopic kidney, which is usually pelvic in location. The right kidney may be inferiorly and anteriorly displaced by the liver and present as an anterior abdominal mass. Ultrasound rapidly resolves this question.

Technique

Typically, the kidneys have been scanned in transverse and longitudinal planes in the prone position. While this position may be adequate for the lower poles, it often gives inferior visualisation of the upper poles due to overlying ribs, heavy musculature and air-filled lungs. In addition, the renal vessels cannot be identified, particularly where shadowed by the spine.

The right kidney is best examined with the patient supine, using the liver as an acoustic window. Transverse as well as longitudinal views should be obtained. Decubitus views are helpful in examining the right kidney but are essential for the left kidney for which the supine approach is often impossible due to gas-filled bowel. Both coronal (longitudinal) and transverse sections should be obtained in the decubitus position. This is particularly important for adequate demonstration of the renal pelvis and calyces and the vessels in the pedicle.[6,11,84,87]

Real-time scanning can be particularly helpful where access to the kidney is limited by body habitus. The narrow skin aperture needed with a real-time sector machine is particularly well suited to this application.

Anatomy

Grey scale ultrasound, high frequency transducers, and real-time equipment have all resulted in improved imaging of the kidneys. Subtle differences in renal parenchymal echoes are clearly demonstrated so that the distinction between cortex and medulla is easily made in over half of all examinations (Fig. 3.35). The cortex contains fine low level echoes of slightly lesser intensity than those of the spleen and liver. The medulla is relatively anechoic apparently due to the high fluid content of the collecting ducts and appears as lucent triangles with their apices pointing centrally and their bases closer to the surface. The arcuate arteries are sometimes seen as round bright reflectors at the cortico-medullary junction in longitudinal sections of the kidney (Fig. 3.35). The

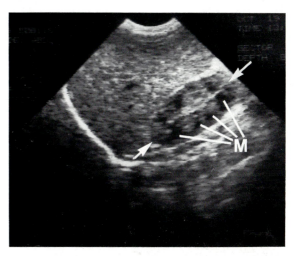

Fig. 3.35 Supine longitudinal ultrasonogram demonstrates a normal kidney (arrows) visualised through the anteriorly located liver. Note the difference in echogenicity between the cortex and medullary (M) areas.

cortico-medullary junction can be distinguished in transverse sections as well, though not as regularly.

It has been suggested that the textural pattern of the cortex and the sharpness of the cortico-medullary junction may be useful in distinguishing various renal parenchymal diseases.[79] The normal renal parenchyma is less echogenic than the adjacent spleen and liver and much less echogenic than the renal sinus. In Type I diseases, the echo amplitude of the cortex is equal to or greater than that of the spleen or liver when compared at the same depth.[77] The cortico-medullary definition is maintained or even enhanced. The more common renal diseases producing this pattern include acute and chronic glomerulonephritis, diabetic and hypertensive nephrosclerosis, and renal transplant rejection (Fig. 3.36).

Type II diseases produce diffuse disruption of renal parenchymal anatomy and are all processes not included in Type I. While renal masses can fall into this category, this pattern is most commonly

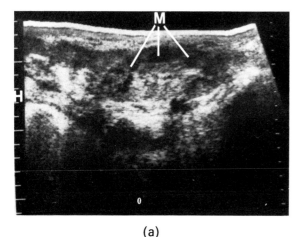

(a)

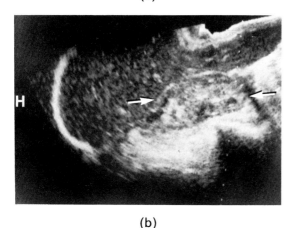

(b)

Fig. 3.36 (a) A longitudinal oblique image of a transplanted kidney shows an overall increase in echogenicity with prominence of the renal medullary areas (M) consistent with a diagnosis of renal transplant rejection. As the rejection increases, there is an increase in the overall echogenicity of the cortex equalling that of the renal pelvis. (H = direction of head.) (b) In another patient with advanced renal disease, there is a loss of normal architecture with the overall echogenecity of the kidney (arrows) being greater than that of the adjacent liver. The cortical areas are of similar echogenicity to the renal hilus. This represents more advanced renal disease.

seen in advanced renal disease. In fact, in severe cases it may not be possible to identify normal kidney structure.

The azotaemic patient

High dose urography should be avoided in patients with elevation of BUN and creatinine and concommitant congestive heart failure, diabetes, myeloma, or urate nephropathy. Even in the absence of these additional problems, azotaemia may result in such poor visualisation that no diagnostic radiographs can be obtained. Consequently, when renal function is insufficient (either unilaterally or bilaterally) for visualisation, or when laboratory data and clinical problems indicate that urography would be dangerous or unsuccessful, ultrasound is the examination of choice.

The possibility of bilateral hydronephrosis is the most common indication for an ultrasound study when renal failure is present. Several authors have suggested that ultrasound is a rapid, highly accurate screening procedure for obstruction.[23, 59] Fortunately, the diagnosis is generally not difficult, and often the aetiology may also be determined. This is discussed later in this section.

Other causes of azotaemia that can be detected include adult polycystic disease, infantile polycystic disease, medullary cystic disease, and diseases affecting the renal cortex.

The changes associated with the last group are, briefly, as follows. The most common of these cystic diseases is adult polycystic disease which presents a wide spectrum of changes from slight enlargement, through numerous small cysts, to many large cysts which distort the renal architecture (Fig. 3.37).[51] It is generally easily recognised and if cysts are present in the liver this helps to confirm the diagnosis. Since the disease can be detected prior to the onset of symptoms and earlier than with urography, screening of family members is suggested as an adjunct to genetic counseling.[76] Infantile polycystic disease produces large kidneys and distorted renal architecture often associated with increased echogenicity in the liver due to fibrosis.[8, 78] The individual renal cysts are usually too small to delineate. Medullary cystic disease presents instead with cysts of varying sizes and is usually unilateral (Fig. 3.38).

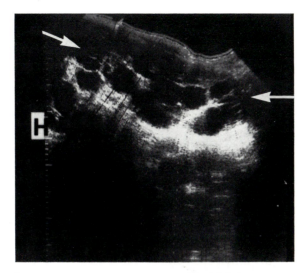

Fig. 3.37 Longitudinal prone image of the left kidney (arrows) shows it to be enlarged and distorted by numerous cysts of varying sizes consistent with a diagnosis of adult polycystic kidneys. A similar pattern was recorded from the opposite kidney.

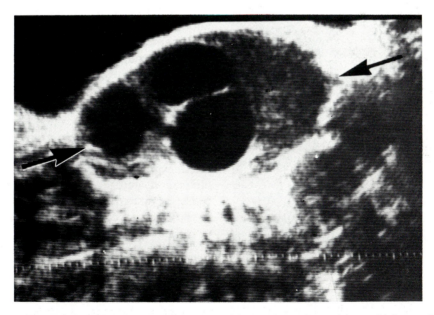

Fig. 3.38 Longitudinal left decubitus view demonstrates a large left kidney (arrows) containing multiple cysts. The opposite kidney was normal. These findings, along with non-visualisation on the urogram, in this newborn made a diagnosis of multicystic (medullary) disease possible. This was confirmed at surgery.

Hydronephrosis

Ultrasound is employed in the search for hydronephrosis either when azotaemia is present (as discussed already) or when there is nonvisualization on an intravenous pyelogram (Fig. 3.39). Although it is sensitive in detecting hydronephrosis

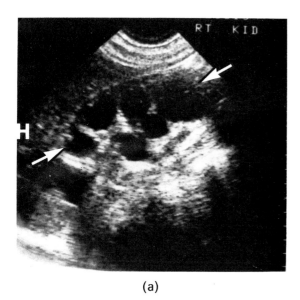

(a)

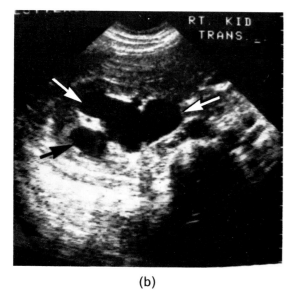

(b)

Fig. 3.39 (a) Longitudinal supine ultrasonogram of the right kidney (arrows) demonstrates multiple cystic areas. (b) A transverse view clearly shows the cystic areas to be dilated calices entering a dilated renal pelvis and consistent with a diagnosis of hydronephrosis (arrows).

— hence its usefulness as a screening procedure for obstruction — it is not very specific.[23]

To maximise the sensitivity of the procedure, meticulous technique is required. A coronal view is recommended for establishing the diagnosis since it displays the calyces, infundibula, pelvis and upper ureter on one scan.[6] The diagnosis is based upon visualisation of a dilated renal pelvis and as the degree of severity increases there is a corresponding increase in calyceal dilatation.[87]

A variety of causes for false positive examinations exists. These occur when there is dilatation on a non-obstructive basis or when the size of the calyces is at the upper limits of normal. Prior obstruction, reflux, infection, diuresis, congenital megacalyces, and extra-renal pelvis may all cause false positives. Distention of the bladder is also a well known cause of mild dilatation of collecting structures and ureters. Consequently, if the bladder is distended at the time hydronephrosis is found, rescanning after voiding is suggested.

False negatives are particularly undesirable in any screening procedure and can be minimised with a high index of suspicion. Several authors have reported accuracies from 95 to 100 per cent with the latter figure achieved in all but the minimal hydronephrosis.[59] Staghorn calculi, which are almost always seen on plain abdominal radiographs, are the major cause of false negative studies and can usually be recognised ultrasonically by their strong surface echogenicity and distal acoustic shadowing.[65] Ultrasound can be utilised for antegrade pyelography and percutaneous nephrostomy.[5] While many use fluoroscopy, ultrasound is the only contemporary method of visualising the collecting system in those kidneys which do not opacify. A combined approach may be useful, with a small gauge needle introduced under ultrasound guidance in order to opacify the collecting system with contrast. The remainder of the procedure is carried out under fluroscopy.[37]

Renal mass lesions

Generally, ultrasound evaluation of renal mass lesions follows intravenous pyelography in adults, or detection of a palpable mass in children, especially neonates. When a mass is detected, the major question is whether it is cystic or solid. This question can generally be resolved with a high degree of accuracy utilising ultrasound.[54]

Newer equipment has improved the resolution so that smaller structures can be visualised. Two centimetres is the lower limit of resolution for most solid masses, while cysts as small as five millimetres can be detected under ideal circumstances. As always, the highest frequency transducer focused at the area of interest produces the best results.

Cysts. The classic simple cyst is echo-free and displays enhanced transmission and a sharply defined backwall (Fig. 3.40). (The wall closer to

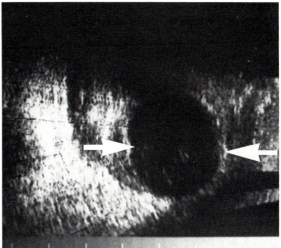

Fig. 3.41 Left lateral decubitus image of a renal cyst shows dependent echoes with a fluid-solid level (arrows). At aspiration, these echoes were found to be due to pus.

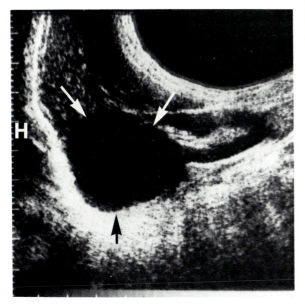

Fig. 3.40 Supine longitudinal image of the right kidney demonstrates an echo free mass (arrows) in the upper pole with increased sound transmission and smooth walls, consistent with a diagnosis of a renal cyst.

the transducer may be indistinct due to reverberations.) Such cysts are usually solitary but may be multiple and bilateral. Size is quite variable from under a centimetre to over 10 cm. Simple cysts generally project from the margins of the kidney but may also be central or parapelvic in location. On occasion, cysts may contain septations. Layering of echoes due to clot or pus can also be delineated (Fig. 3.41). Specific cystic diseases easily visualised by ultrasound include multicystic kidneys and adult polycystic disease. Irregular internal margins may suggest a necrotic tumour.[74]

Neoplasms. The most common neoplasm of the kidney is the hypernephroma. This tumour has a wide spectrum of ultrasonic patterns.[30] The factors which determine the appearance are not totally resolved. One report[58] suggested that it is related to the degree of vascularity and presence of haemorrhage and/or necrosis. Another report found that the degree of vascularity does not show a significant correlation with ultrasound appearance.[10] Hypernephromas may be more or less echogenic than the surrounding parenchyma. The smallest number of tumours in this series were more echogenic. The second largest group of tumours were minimally echogenic and could be mistaken for cysts. Meticulous scanning technique, however, can help to avoid these errors by noting any persistent internal echoes, irregular borders and signs of invasion, along with insufficient sound transmission for the size of the 'cystic' lesion (Fig. 3.42). The largest group of hypernephromas were moderately echogenic with internal echoes similar to surrounding normal parenchyma. The density was usually found to be slightly different from the normal parenchyma, however, with some coarse, bright echoes present among the finer, low-level echoes, making it possible to identify the extent of the tumour.[10] If the tumour becomes necrotic, the ultrasonic pattern becomes complex with an increase in sound transmission (Fig. 3.43). Evalua-

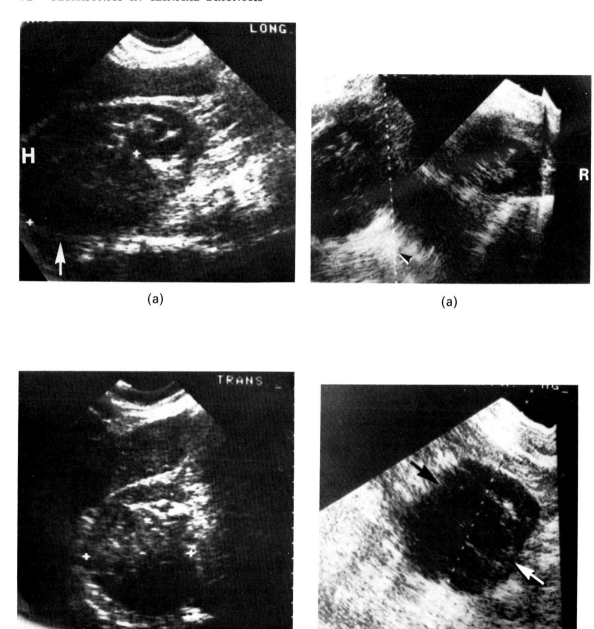

(a)

(a)

(b)

(b)

Fig. 3.42 (a) Supine longitudinal and (b) Transverse images demonstrate a weakly echogenic mass occupying the upper half of the right kidney. + on the images denotes the approximate extent of the mass in both places. Note the relatively weak distal wall echoes (arrow) which help to differentiate this weakly echogenic hypernephroma from a cystic mass. (R = right.)

Fig. 3.43 (a) Prone transverse and (b) Longitudinal images demonstrate a complex mass (arrows) arising from the lateral aspect of the left kidney. Note the increased through-transmission (arrowhead on (a)) and scattered internal echoes. This was a necrotic hypernephroma. A normal-sized right kidney is seen on the transverse image.

tion of the renal vein and vena cava for extension of tumour should be part of every examination when a renal mass is detected (Fig. 3.44).[70, 73]

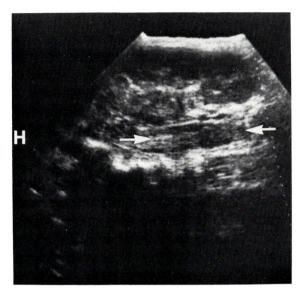

Fig. 3.44 Supine longitudinal image of the inferior vena cava demonstrates internal echogenicity (arrows) due to extension of renal tumour into the vena cava.

Angiomyolipoma. The ultrasonic features of this lesion are sufficiently typical to permit a presumptive diagnosis. These tumours are highly echogenic, apparently due to the mixed fatty-fibrous-vascular content. The finding of highly echogenic well-circumscribed mass with distal attenuation is typical of angiomyolipomas (Fig. 3.45). A radiolucent attenuation is typical on computed tomograms and confirms the ultrasonic diagnosis.[36, 69]

Wilms' tumour. Wilms' tumour is essentially a lesion of childhood although it is rarely seen in young adults. It is characterised by a complex pattern with the echoes of varying intensity as well as sonolucent areas which result from haemorrhage and necrosis.[29] The tumours usually are large and well circumscribed and can sometimes be bilateral. Serial ultrasonic examinations are useful in evaluating for recurrence as well as for development of tumours in the opposite kidney.

Lymphoma. Lymphomas and leukaemias may present as widely infiltrating or localised masses in one or both kidneys. They are very homogeneous with little vascularity and consequently tend to be relatively hypoechogenic.[82] There is modest sound transmission. These features can result in mistaken diagnosis of a cyst or even polycystic disease. Careful scrutiny, however, reveals low level and somewhat indistinct walls which should suggest the correct diagnosis. Metastatic tumours also present frequently as hypoechogenic masses.

Pseudotumours. The term pseudotumour usually refers to normal renal tissue which gives the appearance of an abnormal mass. (Inflammatory processes can also produce a pseudotumour when the process is localised.) These normal variants include cortical nodules, columns of Bertin, and fetal lobulations. The best method to determine if the mass is a neoplasm or normal tissue is radionuclide scanning. Normal activity is seen in the pseudotumour but not in neoplastic tissue. When an ultrasound study is performed, a solid mass is seen with an echo pattern identical to the normal cortex. When used alone, however, this can be an unreliable method of separating pseudotumours from pathological masses.[43]

Central masses. Filling defects within the collecting system seen on contrast urography can be due to numerous causes.[65] Radio-opaque stones present no diagnostic problem but non-opaque stones, tumours and clots may be impossible to separate using standard X-ray techniques. Stones, whether opaque or non-opaque, are echogenic and produce

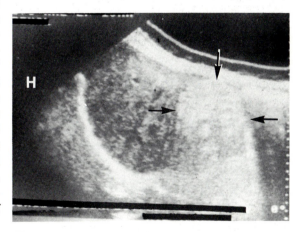

Fig. 3.45 Supine longitudinal ultrasonogram delineates a highly echogenic right renal mass (arrows) with distal attenuation. At surgery, this proved to be, as expected, an angiomyolipoma.

distal acoustic shadowing (Fig. 3.46). This distinguishes them from other filling defects within the pelvis. Appropriate gain settings and use of a transducer of as high a frequency as is possible, which is focused at the level of interest, enhance the acoustic shadowing.

Tumours arising from the urothelial lining are almost always transitional cell carcinomas. On

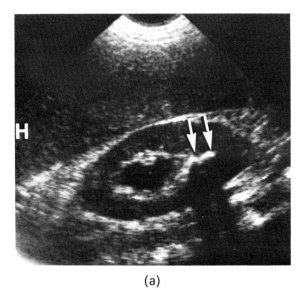

(a)

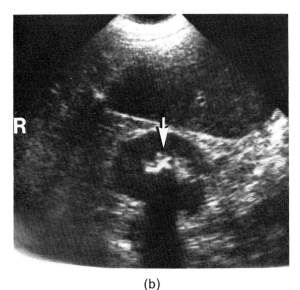

(b)

Fig. 3.46 (a) Supine longitudinal and (b) Transverse images demonstrate strong central echoes (arrows) and distal acoustic shadowing consistent with the diagnosis of renal stones. Note on the longitudinal image (a) the dilatation of the upper portion of the renal pelvis.

ultrasound, these tumours have low level echoes and are slightly less echogenic than the pelvocalyceal structures.[2] They cast no acoustic shadow. Clot tends to be present as moderately bright echoes with no acoustic shadowing.

Renal sinus lipomatosis is better diagnosed on good quality plain or contrast nephrotomograms, typically presenting as a large central lucent area with spidery stretched calyces. The ultrasound appearance is variable with a spectrum from being quite echogenic (most typical) to being almost echofree, depending on the degree of fatty-fibrous interfaces. When the fat is very lucent it may simulate hydronephrosis or small cysts, but there is usually no increased sound transmission; this is not the case when fluid is present.[106]

Renal transplants

The transplanted kidney is ideally suited to evaluation with ultrasound for numerous reasons.[41,49] The transplant is located very superficially with no interfering loops of bowel. Consequently, an anterior approach and high frequency transducers can be employed to produce high resolution images. Ultrasound is not dependent on renal function (as is urography). The region surrounding the transplant can also be readily assessed. Finally, serial studies can be performed as often as is needed.

A recent report suggests that ultrasound alone, or in combination with clinical data and radionuclide scanning, can diagnose acute rejection with sufficient accuracy to preclude biopsy.[27] The changes seen in acute rejection include increased size and echogenicity of the cortex with decreased echogenicity of renal pyramids, and focal zones of sonolucency in renal cortex or patchy sonolucent areas involving both medulla and cortex.[27,38] These changes occur in 92 per cent of patients with biopsy-proved acute transplant rejection. It is essential that a base line ultrasound examination be obtained immediately after transplantation, for comparison where later rejection is suspected, to avoid diagnostic errors.

Ultrasound is also valuable to distinguish other causes of decreased urine production from acute rejection, such as obstruction, lymphoceles, and perirenal collections. Perirenal fluid collections may compress the kidney or obstruct the collecting system. These include haematomas, urinomas, lymphoceles, and abscesses (Fig. 3.47). Urinomas

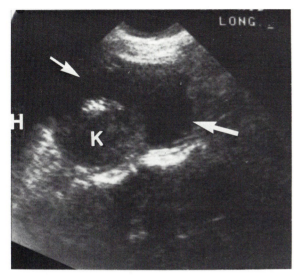

Fig. 3.47 Supine longitudinal image demonstrates a collection (arrows) anterior to a renal transplant (K). Note the diffuse uniform echogenecity of the kidney due to rejection. This collection proved to be a urinoma.

and lymphoceles are most often completely echo free, while haematomas and abscesses generally display low amplitude internal echoes. Ultrasonically guided aspiration has been successfully used to determine the type of collection.[9]

Ureter

The normal ureter is difficult to demonstrate.[72] When there is dilatation of the ureter, the upper portion may be visualised in continuity with the dilated pelvis (Fig. 3.48). The mid-portion is generally not seen except in megaureter, but the lower third may seen in the pelvis through the distended bladder when there is at least minimal dilatation.

Perinephric collections

Ultrasound is the method of choice for evaluation of suspected perinephric abscesses. After size and

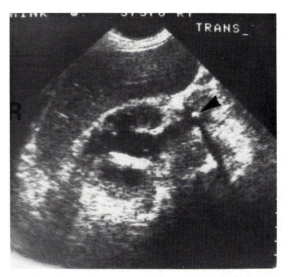

Fig. 3.48 Supine transverse ultrasonogram of the right kidney shows evidence of moderate hydronephrosis. The renal pelvis can be identified as well as a portion of the proximal ureter. Within the ureter there is evidence of a bright echo (arrowhead), producing distal acoustic shadowing, which represents a small calculus — the cause of the hydronephrosis.

location are determined, needle aspiration is easily performed under ultrasonic guidance for diagnosis and drainage. Perinephric haematomas are also easily delineated. The procedure is useful in cases of trauma, after renal biopsy, and in suspected retroperitoneal bleeding in patients on anti-coagulant therapy (Fig. 3.49).[45,100]

Renal infections

Acute pyelonephritis and ureteral obstruction may present with similar clinical and radiographic findings. These can include fever, flank pain, pyuria, and an enlarged kidney with delayed visualisation of the pelvocalyceal system. As noted elsewhere in this section, ultrasound can easily demonstrate hydronephrosis. In moderate to severe acute pyelonephritis, the sonographic findings include a large swollen kidney with an increased anechoic cortico-medullary area with multiple scattered low level echoes and an increase in sound transmission due to increased fluid content.[21] Infection may also be more localised in the form of acute focal bacterial nephritis which produces a mass without liquification. Clinically and radio-

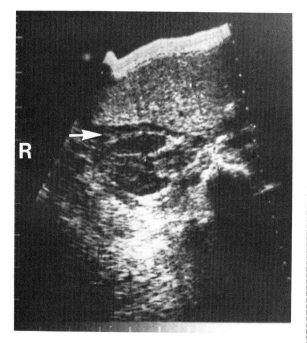

Fig. 3.49 Transverse supine ultrasonogram of the right kidney demonstrates a collection (arrow) located between the liver and right kidney which proved to be a small haematoma — the result of earlier trauma to this region.

graphically, this condition may simulate renal tumour or abscess. The characteristic findings include a relatively sonolucent ovoid mass which disrupts corticomedullary definition and produces some low level echoes, a solid-appearing mass on IVP, CT, or angiography and abnormal gallium uptake at the site of the mass.[75]

Adrenal and retroperitoneal masses

The adrenal glands can be visualised utilising special views.[81, 105] For the right adrenal the best position is with the patient supine, and for the left, the right lateral decubitus, resulting in a coronal section. Normal or minimally enlarged glands are seen only with considerable difficulty, however, and frequently not at all. Computed tomography is far superior for demonstration of normal or slightly enlarged adrenals, with bilateral visualisation in almost all patients. Ultrasound is most

useful to evaluate a mass seen on a prior radiographic examination, such as intravenous pyelography. In these cases it provides assessment of the solid, cystic or mixed nature of the mass, as well as determining the relationship of the mass to adjacent organs (Fig. 3.50).[32] Other retroperitoneal masses may also be visualised. Ultrasound can demonstrate

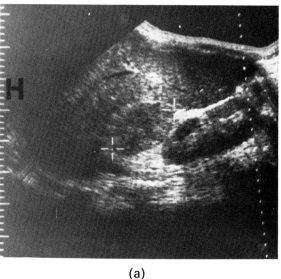

(a)

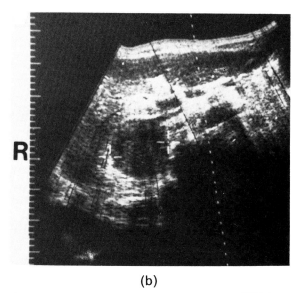

(b)

Fig. 3.50 (a) Supine longitudinal and (b) Transverse images demonstrate a solid mass (+ denotes extent of mass) located superior to the right kidney and beneath the right lobe of the liver. This is a typical location for an adrenal mass and at surgery this proved to be an adenoma.

lymphadenopathy as seen in lymphomas and in metastatic disease, particularly in the para-aortic region.[44]

Urinary bladder

Introduction

The distended bladder extends upwards from the pelvis into the abdominal cavity and can serve as an acoustic window for examinations of the pelvis. It displaces bowel from the pelvis and eliminates interference from bowel gas. The bladder and the adjacent pelvic structures are most often examined transabdominally; transrectal and transurethral techniques have also been used. Ultrasound provides a noninvasive means of measuring residual urine volume which is of particular importance when catheterisation is undesirable.[65]

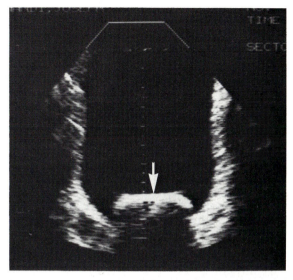

(a)

Bladder calculi

Stones not infrequently form in the presence of long-term indwelling bladder catheters or when the bladder chronically fails to empty. The stones are echogenic and produce distal acoustic shadowing. Clots are also echogenic though they do not cause shadowing. Both stones and clots are moveable (in contradistinction to bladder tumours) (Fig. 3.51).

Diverticula

Diverticula are most often seen when there is outlet obstruction or neurogenic bladder dysfunction. Diverticula larger than two centimetres can usually be visualised ultrasonically. The neck can be seen with very careful scanning especially using real-time equipment.

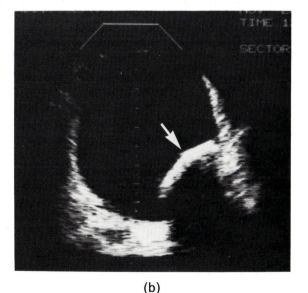

(b)

Bladder tumours

Bladder tumours are staged according to the extent of tumour invasion into or through the wall. Various staging procedures, including lymphangiography, arteriography and cystoscopy, have all proved to be unreliable. The latter two only display the mucosa and give no information about deeper layers of the bladder wall. Ultrasound of

Fig. 3.51 (a) Transverse and (b) Oblique images reveal a large echogenic structure (arrow), producing distal acoustic shadowing, which moved with change in position of the patient consistent with the diagnosis of a bladder calculus.

the bladder has been advocated for staging (Fig. 3.52).[91] Results with transurethral scanning are

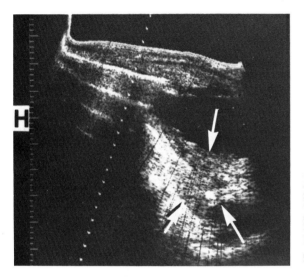

Fig. 3.52 Supine longitudinal ultrasonogram of the urinary bladder delineates a mass involving the posterior wall, which extends through the bladder into the region of the cul-de-sac indicating tumour spread (arrows).

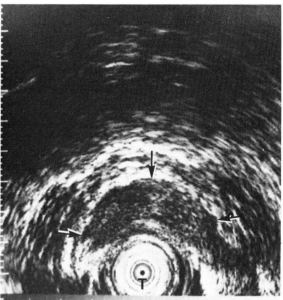

Fig. 3.53 Transverse transrectal image of a normal prostate. The arrows delineate the capsule.

now superior to previous transabdominal and transrectal approaches.[66,71] Using the transurethral approach, tumours are found to be slightly less echogenic than the wall. Noninvasive tumours display a normal thickness of the bladder wall. Superficial invasion shows slight indentation of the wall by the less echogenic mass, while deeply invasive tumours show a thickened wall with a less echogenic mass occupying the whole thickness.[66] Without using these special ultrasound approaches, however, computed tomography is the best approach for the staging of bladder tumours when they extend beyond the bladder walls.

Prostate

Technique and normal anatomy

Both transabdominal and transrectal scanning have been used to examine the prostate.[28,71,98] The normal gland has a triangular shape with a continuous, echogenic capsule (best seen with the transverse approach). The gland displays many fine, homogeneous echoes of lower intensity than the capsule (Fig. 3.53). The prostate is generally divided into five lobes (anterior, posterior, median and two lateral). Benign hypertrophy arises in the median and lateral lobes, whilst most malignancies originate in the posterior lobe.

Benign hypertrophy

Benign hyperptrophy produces a symmetrically enlarged gland with greatest involvement of the median and lateral lobes. Ultrasonic examination reveals a diffusely enlarged, more spherical gland with greatest increase in the anterior-posterior diameter (Fig. 3.54). With the transrectal approach, the capsule is often seen to be thickened and the parenchyma is filled with fine, homogeneous echoes which fade as attenuation is increased.

Carcinoma

In moderately advanced carcinoma, ultrasound demonstrates an enlarged irregular gland. With the transrectal approach, asymmetric enlargement and a deformed, interrupted, and irregular capsule are seen. There is increased echogenicity in the parenchyma, and these echoes do not fade with

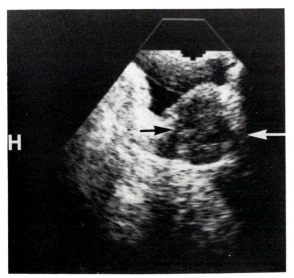

Fig. 3.54 Transabdominal supine longitudinal image of a symmetrically enlarged prostate (arrows) consistent with a diagnosis of benign hypertrophy.

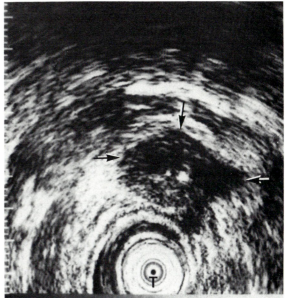

Fig. 3.55 Transrectal image of prostatic carcinoma showing irregularity in the internal echoes (arrows).

increased attenuation (Fig. 3.55). Ultrasonic studies are useful to assess local tumour extension and response to hormonal therapy.[71]

Prostatitis

Acute prostatitis produces diffuse enlargement with decreased echogenicity and increased sound transmission. In chronic prostatitis, symmetrical densities can sometimes be seen extending laterally

from the urethra and distorting or obscuring the capsule. These changes may not be distinguishable from carcinoma.[71]

ACKNOWLEDGEMENT

Figures 3.53 and 3.55 were kindly provided by C. J. Griffiths and K. T. Evans

REFERENCES

1. Anon 1981 Editorial: coffee drinking and cancer of the pancreas. Br Med J 283:628
2. Arger P H, Mulhern C B, Pollack H M, Banner M P, Wein A J 1979 Ultrasonic assessment of renal transitional cell carcinoma: preliminary report. Am J Roentg 132:407–411
3. Asher W M, Freimanis A K 1969 Echographic diagnosis of retroperitoneal lymph node enlargement. Am J Roentg 105:438–445
4. Asher W M, Parvin S, Virgilio R W, Harbor K 1976 Echographic evaluation of splenic injury after blunt trauma. Radiology 118:411–415
5. Babcock J R, Shkolnik A, Cook W A 1979 Ultrasound-guided percutaneous nephrostomy in the pediatric patient. J Urol 121:327–329
6. Behan M, Wixon D, Kazam E 1979 Sonographic evaluation of the non-functioning kidney. J Clin Ultrasound 7:449–458
7. Billinghurst J R 1978 The big spleen. Br J Hosp Med 20:413–424
8. Boal D K, Teele R L 1980 Sonography of infantile polycystic kidney disease. Am J Roentg 135:575–580
9. Brockis J G, Hulbert J C, Patel A S, Golinger D, Hurst P, Saker B, Haywood E F, House A K, Merwyk A van 1978 The diagnosis and treatment of lymphoceles associated with renal transplantation. Br J Urol 50:307–312
10. Coleman B G, Arger P H, Mulham C B, Pollack H M, Banner M P, Arenson R L 1980 Grey scale sonographic spectrum of hypernephromas. Radiology 137:757–765
11. Cook J H, Rosenfield A T, Taylor K J W 1977 Ultrasonic demonstration of intrarenal anatomy. Am J Roentg 129:831–835
12. Cooperberg P L 1978 High-resolution real-time ultrasound in the evaluation of the normal and obstructed biliary tree. Radiology 129:477–480
13. Cooperburg P L, Burhenne H J 1979 Real time ultrasound, diagnostic technique of choice in calculous gall bladder disease. N. Eng J Med 302:1277–1279
14. Cosgrove D O 1981 Liver. In: Goldberg B B (ed)

Ultrasound in cancer. Churchill Livingstone, New York, p 1–20

15. Crade M 1979 Comparison of ultrasound and oral cholecystogram in the diagnosis of gallstones. In: Taylor K J W (ed) Diagnostic ultrasound in gastrointestinal disease. Churchill Livingstone, New York, p 132–135

16. Crade M, Taylor K J W, Rosenfield A T 1978 Water distension of the gut in the evaluation of the pancreas by ultrasound. Am J Roentg 131:348–349

17. Daggett P R, Kurtz A B, Morris D V, Goodburn E, Le Quesne L P, Nabarro J D N, Raphael M J 1981 Is pre-operative localisation of insulinomas necessary? Lancet 1:483–486

18. Dewbury K C, Clark B 1979 The accuracy of ultrasound in the detection of cirrhosis of the liver. Br J Radiol 52:945–948

19. Dewbury K C, Joseph A E A, Sadler M, Birch S J 1980 Ultrasound in the diagnosis of early liver abscess. Br J Radiol 53:1160–1165

20. Dubbins P A 1980 Ultrasound in the diagnosis of splenic abscess. Br J Radiol 53:488–489

21. Edell S L, Bonavita J A 1979 The sonographic appearance of acute pyelonephritis. Radiology 132:683–685

22. Eisenscher A, Weill F 1979 Ultrasonic visualization of Wirsung's duct: dream or reality? J Clin Ultrasound 7:41–44

23. Ellenbogen P H, Scheible W, Talner L B, Leopold G R 1978 Sensitivity of gray scale ultrasound in detecting urinary tract obstruction. Am J Roentg 130:731–733

24. Filly R A, Freimanis A K 1970 Echographic diagnosis of pancreatic lesions. Radiology 96:575–582

25. Foley W D, Stewart E T, Lawson T L, Greenan J, Loquidice J, Mahler L, Unger G F 1980 Computed tomography, ultrasonography and endoscopic retrograde cholangiopancreatography in the diagnosis of pancreatic disease: a comparative study. Gastrointest Radiol 5:29–35

26. Foster K J, Dewbury K C, Griffith A H, Wright R 1980 The accuracy of ultrasound in the detection of fatty infiltration of the liver. Br J Radiol 53:440–442

27. Frick M P, Feinberg S B, Sibley R, Idstrom M E 1981 Ultrasound in acute renal transplant rejection. Radiology 138:657–660

28. Gammelgaard J, Holm H H 1980 Transurethral and transrectal ultrasonic scanning in urology. J Urol 124:863–868

29. Gates G F, Miller J H, Stanley P 1980 Necrosis of Wilms' tumors. J Urol 123:916–920

30. Goldberg B B (ed) 1977 Abdominal gray scale ultrasonography. Wiley, New York

31. Goldstein H M, Katragadda C S 1978 Prone view ultrasonography for pancreatic tail neoplasms. Am J Roentg 131:231–234

32. Gooding G A W 1979 Ultrasonic spectrum of adrenal masses. Urology 13:211–214

33. Gosink B B, Leopold G R 1978 The dilated pancreatic duct: ultrasonic evaluation. Radiology 126:475–478

34. Graaf C S de, Taylor K J W, Jacobsen P 1979 Grey scale echography of the spleen. Ultrasound Med Biol 5:13–21

35. Hadidi A 1979 Ultrasonic findings in liver hydatid cysts. J Clin Ultrasound 7:365–368

36. Hartman D S, Goldman S M, Friedman A C, Davis C J, Madewell J E, Sherman J L 1981 Angiomyolipoma: ultrasonic-pathology correlation. Radiology 139:451–458

37. Hay M S, Elyaderani M K, Belis J A 1981 Percutaneous approach to the renal pelvis: combined use of ultrasonography and fluoroscopy. S Med J 74:31–33

38. Hellman B J, Birnholz J C, Busch G J 1979 Correlation of echographic and histological findings in suspected renal allograft rejection. Radiology 132:673–676

39. Hollinshead W H 1971 Anatomy for surgeons. Vol. 2. Thorax and abdomen. Harper and Row, New York

40. Holm H H, Gammelgaard J 1978 Ultrasonically guided biopsy in malignant disease. In: Hill C R, McCready V R, Cosgrove D O (eds) Ultrasound in tumour diagnosis. Pitman Medical, Tunbridge Wells, p 249–257

41. Hricak H, Toledo-Pereyra L H, Eyler W R, Madrazo B L, Zammit M 1979 The role of ultrasound in the diagnosis of kidney allograft rejection. Radiology 132:667–672

42. Jacobsen P, Taylor K J W 1980 Biliary system. In: Taylor K J W, Jacobsen P, Talmont C A, Winters R, eds. Manual of ultrasonography. Churchill Livingstone, New York, p 111–125

43. Kahn P C 1979 Renal imaging with radionuclides, ultrasound and computed tomography. Semin Nucl Med 9:43–57

44. Karp W, Halfstrom L O, Jonsson P E 1980 Retroperitoneal sarcoma: ultrasonographic and angiographic evaluation. Br J Radiol 53:525–531

45. Kay C J, Rosenfield A T, Armm M 1980 Grey-scale ultrasonography evaluation of renal trauma. Radiology 134:461–466

46. Kazam E, Schneider M, Rubenstein W A 1980 The role of ultrasound and CT in imaging the gall bladder and biliary tract. In: Alavi A, Arger P H (eds) Multiple imaging procedures, Vol 3: abdomen. Grune and Stratton, New York, p 233–310

47. Kessell R G, Kardon R H 1979 Tissues and organs: a text atlas of scanning electron microscopy. Freeman, San Francisco

48. Kreel L, Sandin B, Slavia G 1973 Pancreatic morphology: a combined radiological and pathological study. Clin Radiol 24:154–161

49. Kurtz A B, Rubin C S, Cole-Beuglet C, Brennan R E, Curtis J A, Goldberg B B 1980 Ultrasound evaluation of the renal transplant. J Am Med Ass 243:2429–2431

50. Lawson T L 1978 Sensitivity of pancreatic ultrasonography in the detection of pancreatic disease. Radiology 128:733–736

51. Lawson T L, McClennan B L, Shirkhoda A 1978 Adult polycystic kidney disease: ultrasonographic and computed tomographic appearance. J Clin Ultrasound 6:295–302

52. Lee T G, Forsberg F G, Koehler P R 1980 Post-splenectomy: true mass and pseudomass ultrasound diagnosis. Radiology 134:707–711

53. Leopold G R 1975 Grey scale ultrasonic angiography of the upper abdomen. Radiology 117:665–671

54. Lingard D A, Lawson T L 1979 Accuracy of ultrasound in predicting the nature of renal masses. J Urol 122:724–727

55. Lloyd T V, Antonmattei S, Freimanis K 1979 Gray scale sonography of cystadenoma of the pancreas: report of two cases. J Clin Ultrasound 7:149–151

56. Lyons E A 1978 A color atlas of sectional anatomy. Mosby, Philadelphia.

57. Mackie C R, Cooper M J, Lewis M H, Moossa A R 1979 Non-operative differentiation between pancreatic cancer and chronic pancreatitis. Ann Surg 189:480–487

58. Maklad N F, Chuang V P, Doust B D, Cho K J, Curran J E 1977 Ultrasonic characterization of solid renal lesions: echographic, angiographic and pathologic correlation. Radiology 123:733–739
59. Malave S R, Neiman H L, Spies S M, Cisternino S J, Adamo G 1980 Diagnosis of hydronephrosis: comparison of radionuclide scanning and sonography. Am J Roentg 135:1179–1185
60. Masterson J B, Bowie J D, Port R B, Elahi C F, Burrington J D, Kranzler J 1978 Carcinoma of the pancreas occurring in a child: a case report with description of gray scale ultrasound findings. J Clin Ultrasound 6:189–190
61. McMinn R M H, Hutchins R T 1977 A colour atlas of human anatomy. Wolfe Medical, London
62. Meire H B, Farrant P 1979 Pancreatic ultrasound — a systematic approach to scanning technique. Br J Radiol 52:562–567
63. Mittelstaedt C A, Partain C L 1980 Ultrasonic-pathologic classification of splenic abnormalities: grey-scale patterns. Radiology 134:697–705
64. Moossa A R, Levin B 1979 Collaborative studies in the diagnosis of pancreatic cancer. Semin Oncol 6:298–308
65. Mulholland S G, Arger P H, Goldberg B B, Pollack H M 1979 Ultrasonic differentiation of renal pelvic filling defects. J Urol 122:14–16
66. Nakamura S, Nijima T 1980 Staging of bladder cancer by ultrasonography by transurethral intravesical scanning. J Urol 124:341–344
67. Ohto M, Saotome N, Saisho H, Tsuchiya Y, Ono T, Okuda K, Karasawa E 1980 Real time sonography of the pancreatic duct. Am J Roentg 134:647–652
68. Pedersen F J, Bartrum R J, Grytter C 1975 Residual urine determination by ultrasonic scanning. Am J Roentg 125:474–478
69. Pitts W R, Kazam E, Gray G, Vaughan E D 1980 Ultrasonography, computerized transaxial tomography and pathology of angiomyolipoma of the kidney: solution to a diagnostic dilemma. J Urol 124:907–909
70. Pussell S J, Cosgrove D O 1981 Ultrasound features of tumour thrombus in the IVC in retroperitoneal tumours. Br J Radiol 54:866–869
71. Resnick M I 1980 Ultrasound evaluation of the prostate and bladder. Semin Ultrasound 1:69–79
72. Resnick M I, Saunders R C 1979 Ultrasound in urology. Williams & Wilkins, Baltimore
73. Rosenberg E R, Trought W S, Kirks D R, Sumner T E, Grossman H 1980 Ultrasonic diagnosis of renal vein thrombosis in neonates. Am J Roentg 134:35–38
74. Rosenfield A T (ed) 1979 Genitourinary ultrasonography. Churchill Livingstone, New York
75. Rosenfield A T, Glickman M G, Taylor K J W, Crade M, Hodson J 1979 Acute focal bacterial nephritis. Radiology 132:553–561
76. Rosenfield A T, Lipson M H, Wolf B, Taylor K J W, Rosenfield N S, Hendler E 1980 Ultrasonography and nephrotomography in presymptomatic diagnosis of dominantly inherited (adult-onset) polycystic kidney disease. Radiology 135:423–427
77. Rosenfield A T, Siegel N J 1981 Renal parenchymal disease: histopathologic sonographic correlation. Am J Roentg 137:793–798
78. Rosenfield A T, Siegel N J, Kappleman N B, Taylor K J W 1977 Gray scale ultrasonography in medullary cystic disease of the kidney and congenital hepatic fibrous disease with tubular ectasia: new observations. Am J Roentg 129:297–303
79. Rosenfield A T, Taylor K J W, Crade M, Graaf C S de 1978 Anatomy and pathology of the kidney by grey scale ultrasound. Radiology 128:737–744
80. Sample W F 1977 Technique for improved delineation of the normal anatomy of the upper abdomen and high retroperitoneum with gray scale ultrasound. Radiology 124:197–202
81. Sample W F 1977 A new technique for the evaluation of the adrenal gland with gray scale ultrasonography. Radiology 124:463–469
82. Sample W F, Sarti D A, Goldstein L I, Weiner M, Kadell B M 1978 Gray-scale ultrasonography of the jaundiced patient. Radiology 128:719–725
83. Samuels B I, Culbert S J, Okamura J, Sullivan M P 1979 Early detection of chemotherapy related pancreatic enlargement in children using abdominal sonography. Cancer 38:1515–1523
84. Sanders R C, Conrad M R 1977 The ultrasonic characteristics of the renal pelvicalyceal echo complex. J Clin Ultrasound 5:373–377
85. Sandler M A, Petrocelli R D, Marks D S, Lopez R 1980 Ultrasonic features and radionuclide correlation in liver cell adenoma and focal nodular hyperplasia. Radiology 135:393–397
86. Scheible W 1980 Real time ultrasonography — clinical applications in abdominal disease. In: Alavi A, Arger P H (eds) Multiple imaging procedures, Vol 3: abdomen. Grune and Stratton, New York, p 413–436
87. Scheible W, Talner L B 1979 Gray scale ultrasound and the genito-urinary tract. Radiol Clin N Am 17:281–300
88. Shawker T A, Parks S I, Linzer M, Jones B, Lester L A, Hubbard V S 1980 Amplitude analysis of pancreatic B-scans: a clinical evaluation of cystic fibrosis. Ultrasonic Imaging 2:55–66
89. Shirkhoda A, Staab E V, Mittelstaedt C A 1980 Renal lymphoma imaged by ultrasound and gallium-67. Radiology 137:175–180
90. Siler J, Hunter T B, Weiss J, Haber K 1980 Increased echogenicity of the spleen in benign and malignant disease. Am J Roentg 134:1011–1014
91. Singer D, Itzchak Y, Fischelovitch Y 1981 Ultrasonographic assessment of bladder tumors: II; clinical staging. J Urol 126:34–36
92. Speigel R M, King D L, Green W M 1978 Ultrasonography of primary cysts of the liver. Am J Roentg 131:235–238
93. Talmont C A 1980 Spleen. In: Taylor K J W, Jacobson P, Talmont C A, Winters R (eds) Manual of ultrasonography. Churchill Livingstone, New York, p 157–164
94. Taylor K J W, Jacobson P 1980 Liver. In: Taylor K J W, Jacobson P, Talmont C A, Winters R (eds) Manual of ultrasonography. Churchill Livingstone, New York, p 97–110
95. Vallon A G, Lees W R, Cotton P B 1979 Grey scale ultrasonography and endoscopic pancreatography after pancreatic trauma. Br J Surg 66:169–172
96. Viscomi G N, Gonzalez R, Taylor K J W, Crade M 1980 Ultrasonic evaluation of hepatic and spenic trauma. Arch Surg 115:320–321
97. Walls W J, Templeton A W 1977 The ultrasonic demonstration of inferior vena caval compression: a

guide to pancreatic head enlargement with emphasis on neoplasm. Radiology 123:165–167

98. Watanabe H, Holmes J H, Holm H H, Goldberg B B (eds) 1981 Diagnostic ultrasound in urology and nephrology. Igaku-Shoin, Tokyo

99. Weill F, Schraub A, Eisencher A, Bourgoin A 1977 Ultrasonography of the normal pancreas. Radiology 123:417–423

100. Wicks J D, Silver T M, Bree R L 1978 Gray scale features of hematomas: an ultrasonic spectrum. Am J Roentg 131:977–980

101. Wiener S N, Parulekar S G 1979 Scintigraphy and ultrasonography of hepatic hemangioma. Radiology 132:149–153

102. Willi U V, Reddish J M, Teele R L 1980 Cystic fibrosis: its characteristic appearance on abdominal sonography. Am J Roentg 134:1005–1010

103. Wright C H, Maklad F, Rosenthal S J 1979 Grey scale ultrasonic characteristics of carcinoma of the pancreas. Br J Radiol 52:281–288

104. Wright R, Alberti K G M M, Karran S, Milward-Sadler G H 1979 Liver and biliary disease. Saunders, London

105. Yeh H-C 1980 Sonography of the adrenal glands: the normal and small masses. Am J Roentg 135:1167–1177

106. Yeh H-C, Mitty H A, Wolf B S 1977 Ultrasonography of renal sinus lipomatosis. Radiology 124:799–802

4

The heart

J. R. T. C. Roelandt

INTRODUCTION

Echocardiography is an essential tool for the practice of cardiology as it provides unique information about structure and function. No other diagnostic method based on modern electronic engineering provides the 'hands-on' characteristics that permit the clinician to examine anatomy in motion, confirm clinical impressions, direct further clinical evaluation, and in some cases to decide upon immediate treatment. The field has grown so rapidly that quite a number of the interpretations of investigations have been modified and require continuous adjustment with increasing experience and critical investigation.

This chapter provides an overview of the underlying principles of ultrasonic examination and a critical evaluation of the current utility of the method in the assessment of cardiac disease.

BASIS PRINCIPLES, INSTRUMENTATION AND DISPLAY TECHNIQUES

Echocardiographic examination and interpretation require a thorough knowledge of underlying physical principles, instrumentation and recording techniques. Most of these topics are discussed individually in chapter 1. Those aspects pertinent to ultrasonic analysis of the heart are reconsidered here to provide the reader with an integrated overview of the basic and clinical concepts involved.

BASIC PRINCIPLES OF ECHOCARDIOGRAPHY

Ultrasound refers to sound waves of a frequency above the range audible to man or 20 000 Hz. In

cardiology, frequencies ranging from 2 to 7 MHz are employed. To generate these ultrahigh frequency sound waves, piezoelectric crystalline material is used which converts electrical energy to mechanical (ultrasound wave) energy and vice versa. The diagnostic value of ultrasound waves is the result of partial reflections of these waves when they strike boundaries between media of different characteristic impedances.

Assuming a constant speed of $1580\,\mathrm{m\,s^{-1}}$ for sound in biological soft tissue, the time for a round trip from transducer to reflecting surface allows calculation of the distance. In fact, part of the sound energy is reflected from each subsequent tissue interface along the sound beam pathway and as a consequence distance measurements to all echoproducing interfaces encountered are available from individual pulses. The greater the difference in characteristic impedance between the media, the greater the amount of energy reflected. The intensity of the echoes is further dependent upon the angle at which the sound beam strikes the boundary, and the effect of increasing attenuation with increasing distance. Since ultrasound does detect minor biological tissue differences, soft tissue structures can be outlined in a manner which is not possible with conventional radiological and nuclear imaging methods.

Most M-mode instruments transmit ultrasound during 1 microsecond and receive echoes for the next 999 microseconds yielding a repetition rate of $1000\,\mathrm{s^{-1}}$. Because of this high repetition rate, resolution in time is excellent and motion analysis of cardiac events is more precise than with cineangiographic techniques.

Ultrasonic waves of a relative low frequency (2.25 MHz) penetrate to deep structures and are used in adults. In contrast, higher frequencies (5–7 MHz) have less penetration but a better axial resolution and are used in paediatric applications. The width of the ultrasonic beam determines the lateral resolution or ability to distinguish two adjacent structures in the directions perpendicular to the sound beam axis. Beam width is a function of the transducer size and the frequency at which it operates. As the ultrasound beam is formed it remains parallel over a distance termed the 'near field' and then begins to diverge in the 'far field' The narrow beam and concentrated energy make the near field desirable for recording. Beam divergence degrades the accuracy for examining structures in the far field by limiting the lateral resolution. A wide beam causes superimposition of laterally spaced targets on the display and may pose serious problems in correct interpretation of echocardiograms.[79] A partial solution to divergence of the beam is to use acoustic lenses. Unfortunately, the segment of the beam in focus where the resolution is good, is not very long. Therefore, transducers focused at 5, 7.5 or 10 cm are available

for more accurate data acquisition at the focal length indicated.

The power density used in diagnostic ultrasound systems is well within known limits of safety.[91,93] At present there is no clinical evidence of toxicity associated with ultrasound examination (see chapter 12).

INSTRUMENTATION AND RECORDING TECHNIQUES

M-mode echocardiography

Echoes returning to the transducer are electronically processed to produce three types of oscilloscope display (Fig. 4.1). In the 'A-mode', the height or amplitude represents the intensity of the echoes. The distance of the boundary to the transducer is in practice represented on the horizontal axis of the oscilloscope display but on the vertical axis on the diagram in Figure 4.1. The 'B-mode' is an intermediate step where 'A-mode' echoes are converted to dots of which the brightness

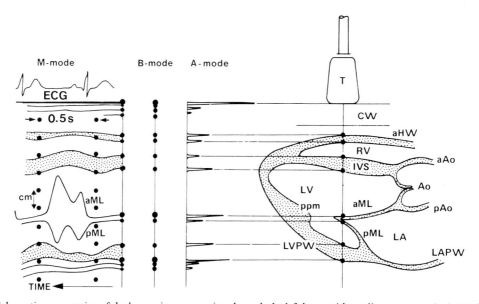

Fig. 4.1 Schematic presentation of the long axis cross-section through the left heart with cardiac structures included. The single element transducer (T) in front is aimed so that the sound beam traverses from anterior to posterior: the chest wall (CW), the anterior heart wall (aHW); right ventricle (RV), interventricular septum (IVS), left ventricular cavity (LV), the tips of the mitral valve leaflets and the left ventricular posterior wall (LVPW). The echoes which originate from these boundaries are classically represented in three types of oscilloscope display and are referred to as the 'A', 'B' and 'M-mode'. For further explanation see text. Abbreviations: Ao: aorta; aML and pML: anterior and posterior mitral valve leaflets; LA: left atrium; LAPW: left atrial posterior wall; aAo and pAo: anterior and posterior aortic walls; ppm: posteromedial papillary muscle; ECG: electrocardiogram.

indicates echo relative intensity. The 'B-mode' is well suited to produce a record of echo motion or 'M-mode'. This is obtained by moving a recording surface such as photographic paper at a constant speed past the B-mode display. The M-mode display permits recording of both the depth and motion pattern of intracardiac reflectors. Measurements are aided by centimetre depth markers which are displayed on the oscilloscope at calibrated time intervals of 0.5 second. Modern echocardiography utilises fibre optic strip chart recorders with high sensitivity and definition which allows echoes of cardiac structures to be recorded continuously on a handy paper strip and M-mode scanning to be performed (Fig. 4.2). An electrocardiographic tracing is recorded simultaneously for timing purposes. Using the M-scan technique it is possible by sweeping the sound beam from one area of the heart to another while recording, to derive information concerning the spatial relationship of cardiac structures[26,76] (Fig. 4.3). The need for more accurate spatial information concerning cardiac structures and function, however, led to the development of real time two-dimensional imaging systems.

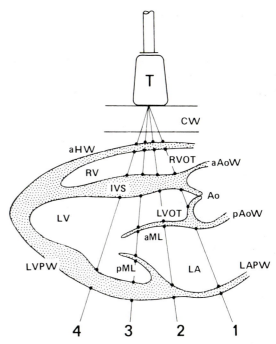

Fig. 4.2 Schematic cross section of the heart. A 'sector scan' or 'M-scan' is performed when the transducer is swept from the aorta (direction 1) towards the apex (direction 4). For abbreviations see Figure 4.1. RVOT and LVOT: right and left ventricular outflow tracts.

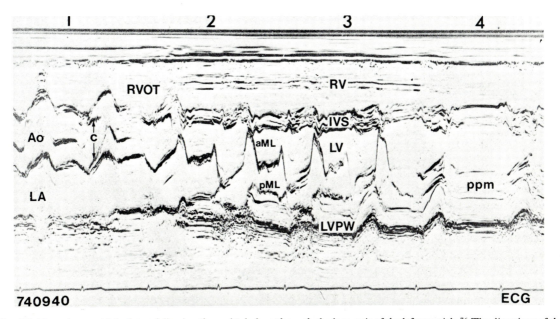

Fig. 4.3 M-mode scan of the heart following the sagittal plane through the long axis of the left ventricle.[76] The directions of the sound beam labelled 1 to 4 on the diagram of Figure 4.2 correspond to the areas labelled 1 to 4 on the record. For abbreviations, see Figure 4.1.

Two-dimensional echocardiography

The basic unit of information, the echo signal indicating the presence and the location in depth of a cardiac structure in the direction of the sound beam, is the same both for M-mode and two-dimensional echocardiography. The fundamental difference lies in the spatially-oriented display which permits information to be appreciated and utilised which is meaningless in the absence of a spatial reference. In all two-dimensional systems an acoustic beam scans a cardiac cross-section at a high rate in order to obtain instantaneous two-dimensional structure information and thus moving cardiac anatomy (Figs. 4.4 and 4.5). The

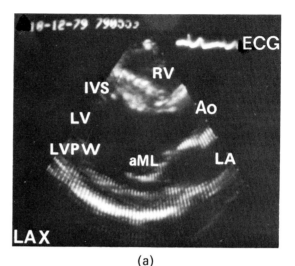

(a)

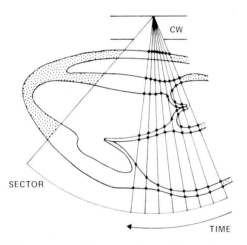

Fig. 4.4 A two-dimensional sector image of a cardiac cross section is obtained by sequentially changing the direction of the ultrasound beam at a constant speed. The image is simultaneously being created on the display screen. The echo information is represented in the B-mode in the same relative positions as each beam arising from the transducer. Rapid angling results in real-time visualisation of the cardiac cross section under study and thus moving anatomy.

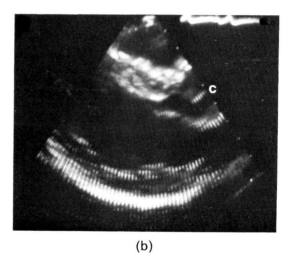

(b)

Fig. 4.5 (a) Systolic and (b) Diastolic long axis views (LAX) obtained from the parasternal transducer position with a phased-array electronic sector scanner in a normal subject. Note the posterior position of closed anterior mitral valve leaflet (aML) in systole and its anterior position when open in diastole. The aortic valve cusps (c) are seen in their closed position during diastole. Abbreviations: see Figure 4.1.

method is extremely useful in determining the shape of cardiac structures and in measuring distances and motion which are perpendicular to the sound beam axis. The information provided is different from that obtained with cineangiocardiography since it images a tomographic plane rather than a silhouette and has the advantage of clearly visualising valvular structure and function.

Three main types of two-dimensional ultrasonic imaging systems are available:

Mechanical scanners

The simplest and least expensive of the real-time systems are the mechanical sector scanners. One principle of image formation (Fig. 4.6a) relies on the rapid pivoting motion about a fixed axis of a single transducer by means of an electrical motor contained within the transducer assembly.[36] The resulting image is of a sector format with a maximum field of view of 45°. The frame rate is typically 30 frames per second. Larger fields of view are difficult to obtain because of skin contact problems at larger scan angles. As an alternative to mechanical crankdriven pivoting systems, magnetic deflection and rotation mechanism (spinning wheel scanners) are also used to angle the transducer. In the latter systems, three or four transducers are mounted on a rotating wheel in a fluid-filled scanning head to create a sector beam deflection at a high repetition rate (Fig. 4.6b). Each transducer is active when it passes an acoustic window in the housing. The transducer is usually large compared to that of the other real-time scanners but a wider field of view of up to 90° is obtained at a frame rate of 30 frames per second.

Advantages of the mechanical scanners are that they are easily adapted to existing M-mode units, the retention of the flexibility and resolution of the single crystal technique, and that only a small acoustic window on the heart is required.

Linear array scanners

These systems were among the first two-dimensional systems but have not gained wide acceptance for cardiac examination. They consist of a number of individual crystals arranged side by side in a single transducer assembly. Rectangular cross-sectional images are produced by sequentially transmitting on each of the array elements (or overlapping sub-groups of elements) and by receiving the echo information with the same

elements for each B-mode line in the final display[6] (Fig. 4.6c). Rapid electronic switching allows cardiac scanning at a high frame rate (50 per second). When using subgroups of elements, dynamic focusing can be realised resulting in a high resolution.[53] The large transducer in linear array systems makes recording from some positions on the chest wall difficult. Shadowing produced by the sound attenuation of ribs, sternum and lungs commonly reduce the fields of view in adults. In children, however, such difficulties are less frequent.

Phased array scanners

These types of scanners make use of a small stationary transducer with multiple small elements in a row (Fig. 4.6d). In contrast to the linear array scanners, all of the array elements are utilised in producing each of the individual B-mode lines which comprise the final sector shaped image. The acoustic beam is steered through the cardiac cross-section by electronic means which makes them the most complicated and, hence, the most costly.[75] These systems typically produce images at a rate of 30 frames per second with a field of view of up to 90°. Dynamic focusing can also be applied. Advantages of these systems are their good resolution and the small transducer size allowing excellent manoeuvrability within the confines of a small ultrasonic window. A major advantage of these electronic systems over mechanical scanners is that

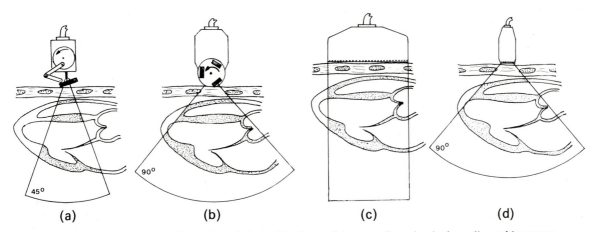

(a)   (b)   (c)   (d)

Fig. 4.6 Schematic representation of transducer designs of the four real time two-dimensional echocardiographic systems which are used for cardiac examination. The long axis cross section of the heart is shown and the areas indicate the relative fields of view. For details see text.

they can simultaneously display the two-dimensional image while sampling areas of the image for M-mode recording.

Resolution of motion in all real time two-dimensional systems is limited by the frame rate which is typically 30 frames per second and thus too low to record very rapid events such as fluttering and opening and closing events of the valves. Thus M-mode recordings are more likely to demonstrate functional abnormalities while two-dimensional techniques are better able to visualise anatomic abnormalities, making both methods complementary for a comprehensive analysis of cardiac disorders (Fig. 4.7). The relative advantages of M-mode and two-dimensional echocardiography are summarised in Table 4.1.

Two-dimensional echocardiographic images are recorded on videotape for playback and analysis. Stop-action video fields can be documented using Polaroid photography, but results in a significant degradation in image quality since only single frame fields are recorded. These contain only half of the information present during real-time study and playback. This problem can be overcome by the use of digital scan converters which allow direct transfer of stored images on light sensitive paper via a video-hardcopy recorder.

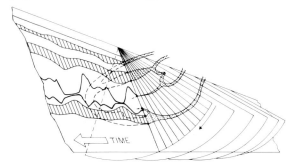

Fig. 4.7 Diagrammatic illustration of the relationship between two-dimensional images and the M-mode display. The M-mode echocardiogram is sampled at a high repetition rate (1000 transmit-receive cycles per second) and yields a clear definition of a small part of the mitral valve with an excellent resolution of motion; but little or no information on its anatomical relationship is available. This is present on the two-dimensional images. Building up these images by many individual sound beams, however, is time consuming and hence results in a frame rate of only 30 per second. This limits resolution of motion of cardiac structures. Most electronic phased-array scanners are presently capable of displaying two-dimensional images and M-mode recordings simultaneously following the principle as shown on this diagram, and this is a major advantage of these systems.

Table 4.1 Relative advantages of M-mode and two-dimensional echocardiography

M-mode echocardiography
1. Time resolution
2. Accurate measurements
3. Timing of events against other physiological parameters
4. Easy storage and retrieval

Two-dimensional echocardiography
1. Spatial information
     anatomical relationships
     shape information
2. Lateral vectors of motion
3. Easier to understand

Doppler echocardiography

A Doppler shift is a change in the frequency of a reflected sound signal due to motion of the target relative to the transducer. The Doppler shift caused by physiological blood flow velocities is in the audible range when the transmitted signal is in the 2–5 MHz range. Simple instruments use continuous wave Doppler. The Doppler shifts of the reflected sound in these instruments are converted to audio signals and make it possible to obtain information about velocity of blood flow towards or away from the transducer as well as an impression of its phasic nature. No information is available about the distance of the reflectors from the transducer. To remedy this defect, pulsed Doppler instruments have been developed. In these instruments a short pulse of high frequency sound is emitted and the Doppler shift is detected as a function of time. Since the speed of sound is roughly constant in biological materials, this approach allows calculation of the Doppler shift at various depths (various delays for round-trip of the sound pulse). Instruments which combine pulsed Doppler units with M-mode or sector scanning techniques are now available and allow interrogation of a small sample volume in an M-mode or two-dimensional image and to derive information about local blood flow.[37] Other promising new techniques include Doppler two-dimensional imaging, which has been applied to vascular flow profiles with some success,[72] and multigate display of the Doppler shift throughout all depths of an M-mode tracing.[9] The latter method can use colour-encoding to display maximum Doppler shifts towards or away from the transducer at each depth. Since the necessity to interrogate each depth separately is thereby eliminated, this may be a more efficient way of searching for localised flow disturbances.[89] The clinical utility of Doppler echocardiology is mainly in detecting abnormal flow patterns, such as those due to valvular stenosis and regurgitation, and intracardiac shunts. Normal intracardiac laminar blood flow produces nearly uniform velocity profiles within a given vessel or cardiac chamber, and little spread of the Doppler spectrum in each sample volume. Abnormal turbulence due to valvular disease can be detected as abnormally spread Doppler signals from the area of that valve and its outflow or inflow tract. For example, aortic stenosis causes a wide band of Doppler shifts from the valve orifice and aortic root in systole, while aortic regurgitation is associated with an abnormal diastolic spectrum of Doppler shifts in a direction opposite to the systolic signals when the sample volume is placed in the aortic root, valve area or left ventricular outflow tract. Similarly, atrial and ventricular septal defects can be detected and located by Doppler techniques. The interested reader is referred to a review of Doppler echocardiography.[3]

At present, Doppler studies cannot accurately quantify intracardiac flow velocity or the magnitude of valvular lesions. A major limitation remains the rather complicated display format of the data. Doppler echocardiography is in a rapid growth phase, however, and quantification of cardiac output and regurgitant flows may be available before too long.

CONTRAST ECHOCARDIOGRAPHY

Contrast echocardiography is the technique of injecting an echo-producing biologically compatible solution into the bloodstream and, using M-mode and two-dimensional techniques, observing the bloodflow patterns as revealed by the resulting cloud of echoes. The method makes it possible to derive information that is otherwise available only from cardiac catheterisation and angiocardiography. The source of ultrasound contrast is microbubbles of air in the injectate which are strong reflectors of ultrasound.[64] Many different agents have been reported to cause ultrasound contrast when injected in the bloodstream. They include indocyanine green dye, dextrose 5 per cent in water and saline solutions, carbon dioxide and the patient's own blood. Initial use of contrast echocardiography was for structure recognition and identification.[35] Right-sided structures are delineated by peripheral venous injections (Fig. 4.8), and left-sided structures are opacified by injections performed directly into the left heart chambers during cardiac catheterisation.

Within several years after the initial description of the ultrasound contrast technique, the method was being used during cardiac catheterisation for the identification of valvular regurgitation and intracardiac shunts.[47] Further work has shown that contrast echocardiography is a sensitive means of detecting atrial septal defects.[28] An important advantage of this technique is that it can be done using peripheral venous injections, since virtually all uncomplicated atrial septal defects (ASDs) have some right-to-left shunting. In fact, the method may be over-sensitive to atrial septal defects, since a recent report suggests that patients with patent foramen ovale can have right-to-left shunting, especially during Valsalva manoeuvre.[50] A most important methodological point is that a study cannot be considered to be negative unless adequate right heart contrast opacification has been achieved during multiple injections using multiple different echocardiographic views, and during several Valsalva manoeuvres.

In the absence of Eisenmenger physiology, peripheral contrast echocardiography is not useful in detecting ventricular septal defects.

The technique is important in complex congenital heart disease, helping to diagnose the various intracardiac connections and communications. A particularly useful approach has proved to be the suprasternal notch transducer position, largely because of the consistently superior position of the aorta compared to the pulmonary artery from this incidence, regardless of the presence or absence of transposition of the great vessels.[68] The contrast method is an important advance in the ability to diagnose tricuspid regurgitation.[60] Normally, contrast injected in an upper extremity flows from the superior vena cava into the right atrium and right ventricle without retrograde flow into the inferior vena cava. In patients with tricuspid regurgitation, however, it can be detected in the inferior vena

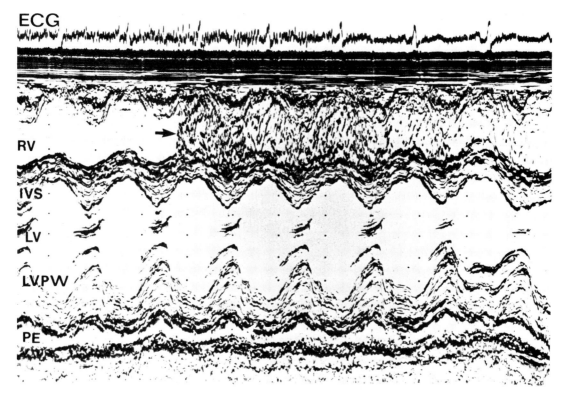

Fig. 4.8 A peripheral venous injection of dextrose 5 per cent in water demonstrating the production of echocardiographic contrast in a patient with a small pericardial effusion (PE). The contrast echoes appear in the right ventricle (RV) in early diastole at the moment of opening of the tricuspid valve (see arrow) and not on the left side of the heart. Abbreviations: IVS: interventricular septum; LV: left ventricle; LVPW: left ventricular posterior wall.

cava and hepatic veins during the 'v' wave on the right atrial pressure tracing. Timing of its appearance is much easier from M-mode than from two-dimensional studies. The clinical applications of contrast echocardiography are indicated in Table 4.2. Recent studies have shown that injections in the pulmonary wedge position can cause contrast to appear on the left side of the heart.[63] These preliminary reports are interesting, for until now

Table 4.2 Clinical usefulness of contrast echocardiography. M-mode and 2-D: indication for which M-mode or two-dimensional echocardiography offer relative advantages

1. Cardiac structure identification (2-D and M-mode)
2. Presence (or exclusion) of shunts (2-D and M-mode)
   Localisation (2-D)
   Direction (2-D)
   Timing (M-mode)
3. Complex congenital heart disease (2-D)
4. Valvular regurgitation (2-D and M-mode)
5. Intracavitary and transvalvular flow patterns (2-D)
6. (Improved quantitation of LV volume?) (2-D)

there was no method for causing transmission of ultrasonic contrast through the lungs.

There are several interesting additional future prospects for contrast echocardiography. It has been suggested that quantitative videodensitometric techniques can be used to measure cardiac output using peripheral contrast injections,[7] and similar techniques have been tried to quantify intracardiac left-to-right shunts.[41] Right-sided pressure measurements may eventually become available using complex ultrasonic analysis of microbubble resonant frequencies.[92]

EXERCISE ECHOCARDIOGRAPHY

Exercise echocardiography is an experimental technique. Its important limitations at present are that successful exercise studies are technically difficult to perform and that there is a large proportion of patients in whom adequate studies for analysis are unobtainable. Nevertheless there are some important advantages to this method. It makes it possible to study cardiac physiology during exercise[96] and to detect left ventricular dysfunction in some patients with coronary artery disease.[95] Because of the methodological difficulties, the method will not become a routine clinical screening method but it may gain wider application in the investigation of normal and abnormal physiology in selected patients.

ECHOCARDIOGRAPHIC EXAMINATION TECHNIQUES

In order to obtain and interpret echocardiographic data, it is imperative to have a fundamental knowledge of the cardiac cross-sections and motion patterns of structures.

Most of the diagnostic M-mode information obtained in adults is contained in a plane through the long axis of the left ventricle. This cross-section is diagrammatically shown in Figures 4.1 and 4.2. Echocardiographic examination is performed with the patient in the left lateral decubitus position.

The transducer is placed in the third or fourth intercostal space along the left sternal border and perpendicular to the chest wall while recording mitral valve motion. By angling the transducer towards the base of the heart (Figs. 4.2, 4.3 and 4.9; direction 1), the sound beam traverses successively the chest wall (CW), the anterior heart wall (aHW), the right ventricular outflow tract (RVOT), the root of the aorta (Ao) with the aortic valve cusps, the left atrial (LA) cavity and the left atrial posterior wall (LAPW). By directing the transducer inferiorly and laterally, the right ventricular (RV) cavity, the interventricular septum (IVS), anterior mitral valve leaflet (aML) and the LA cavity posterior to it are visualised (Figs. 4.2, 4.3 and 4.9; direction 2). The normal anatomical relationships between anterior Ao wall (aAo) and the IVS (septal-aortic continuity) and the posterior Ao wall (pAo) with the aML (mitral-aortic continuity) can thus be demonstrated and these structures are easily identified from their typical motion pattern (Fig. 4.3). Further tilting of the transducer to direction 3 (Figs. 4.2, 4.3 and 4.10) causes the ultrasonic beam to traverse the left ventricular (LV) cavity at the edges of both the aML and posterior mitral valve leaflet (pML).

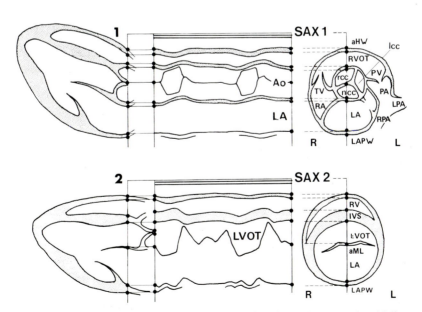

Fig. 4.9 Schematic long axis cross section of the heart with sound beam directions 1 (upper part) and 2 (lower part) and resulting M-mode patterns. The sound beam pathways relative to the structures in the short axis views (SAX) are represented to the right. lcc, ncc and rcc: left coronary, noncoronary and right coronary aortic valve cusps. PV: pulmonary valve; PA: pulmonary artery; RPA and LPA: right and left pulmonary artery; TV: tricuspid valve.

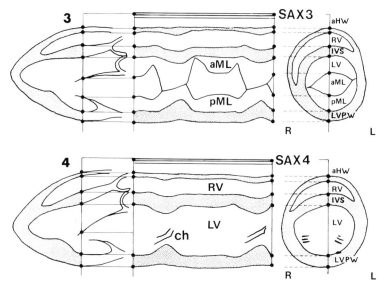

Fig. 4.10 Schematic long axis cross section of the heart with sound beam directions 3 (upper part) and 4 (lower part) and resulting M-mode patterns. The sound beam pathways relative to the structures in the short axis cross sections (SAX) are represented to the right. For abbreviations see Figure 4.1. R and L: right and left orientation.

The walls of the LV, anteriorly the IVS and posteriorly the LVPW, are now recorded. By further scanning in an inferior and lateral direction towards the apex, the chordae tendineae which are in continuity with the mitral valve and their merging into the posterior papillary muscle (ppm) are demonstrated (Figs. 4.2, 4.3 and 4.10; direction 4).

Localisation and recording of the tricuspid valve leaflets is somewhat more difficult, at least in the normal sized heart. Angulation of the ultrasonic beam medially and inferiorly from the aortic valve cusps generally displays the anterior tricuspid valve at about the same depth as the aAo wall. Angling the transducer in a superior and slightly lateral direction from the aortic valve allows recording of the posterior cusp of the pulmonary valve with the RVOT anterior and the atriopulmonic sulcus posterior to it. Other approaches for M-mode examination of the heart are with the transducer in the suprasternal notch[34] and subcostal (also called subxiphoid)[17] positions.

There are four main transducer positions for two-dimensional echocardiography study of the heart: These are the *parasternal*, the *apical*, the *subcostal* and the *suprasternal notch* (Fig. 4.11). Tomographic views of the heart and great vessels relative to their long and short axes are obtained from each of these transducer positions. Recently, recommendations were proposed by the American Society of Echocardiography to standardise these views with respect to right-left and superior-inferior orientation of the sector apex on the output screen and these are followed in this chapter.[44] Many of these images are not readily comprehended since they have little similarity to the views used in other diagnostic methods.

The parasternal long axis view (Figs. 4.1 and 4.5) is the easiest to understand, because it is obtained in the same plane as the standard M-mode scan (Fig. 4.2). A 90° rotation of the transducer yields the parasternal short axis views (Figs. 4.9 and 4.10, on the right; Fig. 4.12). The parasternal views are most useful for imaging structures in the base of the heart, including all four valves, the aorta, left atrium, and basilar portion of the left ventricle. For examination of the apical portion of the left ventricle, an area which is largely inaccessible to standard M-mode recording techniques and parasternal views, the apical views are best. These are obtained by placing the transducer directly upon the palpable cardiac apex, usually with the patient in a steeper left lateral decubitus position than used for the parasternal

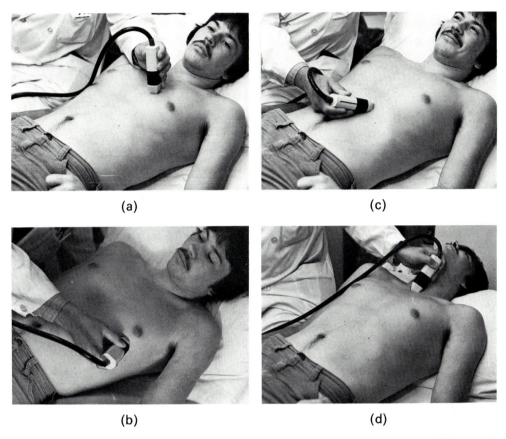

Fig. 4.11 The four standard transducer positions for two-dimensional echocardiographic examination. (a) Parasternal. (b) Apical. (c) Subcostal. (d) Suprasternal.

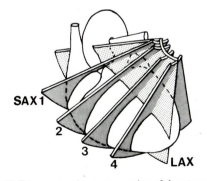

Fig. 4.12 Diagrammatic representation of the parasternal two-dimensional imaging planes and the relation of the long axis (LAX) to the short axis (SAX) views. The structures included in each of these views are diagrammed in Figures 4.9 and 4.10. The numbers 1 to 4 correspond to the short axis cross-sections represented on the right of these Figures.

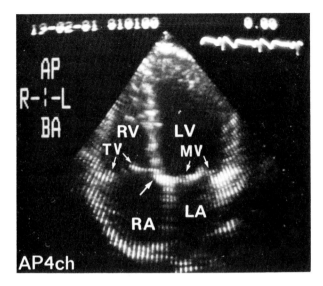 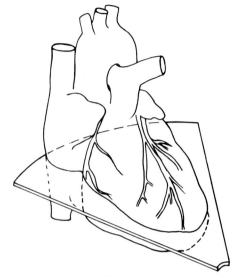

Fig. 4.13 Diagrammatic representation of the examining plane for the apical four chamber view (AP4ch) and stop-frame photograph obtained from a normal subject. Note lower insertion of septal leaflet of tricuspid valve on the interventricular septum as compared to anterior mitral valve leaflet. Abbreviations: LA and RA: left and right atrium; LV and RV: left and right ventricle; MV and TV: mitral and tricuspid valve. Orientation: AP: apical; BA: basal; R and L: right and left.

views, and directing the transducer towards the base (Fig. 4.13). The apical four chamber view is obtained by rotating the plane of imaging about 30° clockwise from the coronal plane. Aiming more posteriorly, the mitral and tricuspid valves are both seen, and aiming more anteriorly, the aortic root is imaged. The most important use of this view, as well as of the perpendicular apical long axis view (Fig. 4.14), is in outlining the left ventricular cavity in two perpendicular views, and in the detection of wall motion abnormalities due to coronary artery disease. Subcostal images are frequently useful in patients with lung disease and low diaphragms, in whom parasternal and apical

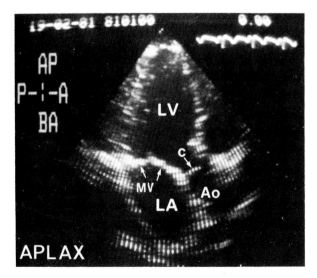 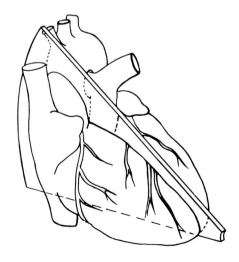

Fig. 4.14 Diagram demonstrating the examining plane for obtaining the apical long axis view (APLAX) which is shown on the stop-frame photograph obtained from a normal subject on the left. Abbreviations: Ao: aorta; C: aortic valve cusps; LA and LV: left atrium and ventricle; MV: mitral valve. Orientation: AP: apical; A: anterior; BA: basal; P: posterior.

views are difficult to obtain. The interatrial septum and inferior vena cava are well imaged from this transducer position. Suprasternal notch views are especially useful in complex congenital heart disease for working out the number and relationship of the great vessels. For a more complete discussion of these views, the interested reader is referred to recent review papers on ultrasonic cross-sectional cardiac examination.[62, 73, 90] All tomographic views as discussed cannot be obtained in every patient because of limitations of available ultrasonic windows, presence of calcified ribs and costal cartilages, variable body habitus, chest deformity, intervening lung tissue, and, exceptionally, lack of patient cooperation.

M-MODE VERSUS TWO-DIMENSIONAL ECHOCARDIOGRAPHIC EXAMINATION

The M-mode examination procedure is mainly confined to the parasternal 'ultrasonic window' since the information obtained via other 'windows' is limited or meaningless because of the lack of a spatial reference. As a result, the M-mode technique is restricted to scanning of the long axis of the heart and recording the tricuspid and pulmonary valves. Thus, M-mode echocardiography is readily standardised and can be performed by a well-trained technician. Another advantage of M-mode echocardiography is that a stripchart record is obtained. This can be rapidly analysed and allows simple but diagnostically important measurements. The analysis can be done off-line in ideal circumstances and the data can be readily stored and retrieved. Two-dimensional studies of the heart are much more demanding than M-mode, despite the more familiar-looking display which anatomically resembles the heart. Four 'ultrasonic windows', instead of one, can be used, and the number of cross-sections which can be studied, is, in principle, unlimited.[62, 73, 90] Some cardiac structures, such as the interatrial septum, right atrium and right ventricle, are imaged to much greater advantage by two-dimensional than by M-mode techniques. Thus, the diagnostic yield of two-dimensional echocardiography is potentially much higher. The complexity of the examination, during which the patient's history and other clinical findings should be taken into account, however, requires the interaction of a trained clinician. Only he can assess the medical problem and redirect the examination in order to obtain the diagnosis. This situation is similar to that in the catheterisation laboratory. The examination when done by a technician takes a longer time, since many unnecessary views are recorded in order not to miss potentially relevant information. Furthermore, off-line analysis of two-dimensional images is not very practical. Retrieving the data from video recordings is cumbersome and the analysis is longer than the real study itself. Therefore, it appears that the multitude of imaging possibilities requires an integrated approach effectively to solve a patient's clinical problem in a short period of time. Thus, ideally a clinician should perform the two-dimensional examination. This may be hard to realise in practice. Indeed, it means that full-time clinicians would have to be available in laboratories with high patient loads.

CLINICAL APPLICATIONS OF ECHOCARDIOGRAPHY

The aorta and aortic valve

The normal aorta and aortic valve

The aortic root is a relatively easy structure to demonstrate echocardiographically and is characterised by two nearly parallel moving echoes with an end-dialostic dimension normally less than 40 mm (Fig. 4.15). In systole, both Ao walls move anteriorly. Diastole can be divided into three phases: an initial posterior motion during the rapid LV filling period, a plateau that disappears with faster heart rates, and a presystolic dip in patients in sinus rhythm. Ao valve cusps yield thin, linear echoes which move briskly toward the periphery of the aortic root in systole and occupy a position in the middle of the aortic lumen in diastole. The right coronary cusp is the closest to the transducer and the most distant is the noncoronary cusp. Fine systolic fluttering of the cusps is recorded in most normals. Two-dimensional echocardiography permits a more precise characterisation of the morphology of the Ao, Ao valve and subaortic area. On parasternal short axis views, the Ao root appears as

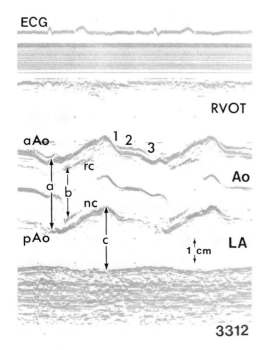

Fig. 4.15 Echogram of a normal aorta (Ao) and aortic valve. aAo and pAo: anterior and posterior aortic walls; LA: left atrium; nc: noncoronary cusp; rc: right coronary cusp; RVOT: right ventricular outflow tract. The aortic root dimension is measured as the distance between the anterior echoes of the anterior and posterior aortic walls at the Q wave of the electrocardiogram (measurement a). Separation of the aortic valve cusps is the distance between the anterior (right coronary) and posterior (noncoronary) cusps in early systole (measurement b). Measurement c indicates LA dimension and 1, 2 and 3 the three distinctive parts of aortic wall motion that occur in diastole. For further explanation, see text.

a circle and the trileaflet Ao valve has the appearance of the letter Y during diastole (Fig. 4.9, top right and Fig. 4.16). Occasionally, the origins of the right and left main coronary arteries can also be seen in this view.

Abnormalities of the aortic valve

The diagnosis of calcific aortic valve stenosis can be considered whenever the cusps appear thickened, multilayered and their movements are restricted. Completely obscured cusp motion and Ao valve echoes wider than one-third of the Ao root diameter indicate heavy calcification and severe stenosis (Fig. 4.17a). The severity of the stenosis cannot at present be reliably assessed from M-mode and two-dimensional echocardiography.[98]

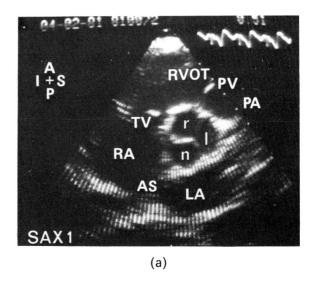

(a)

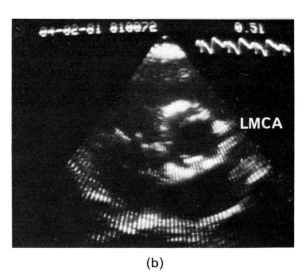

(b)

Fig. 4.16 Stop-frame images in (a) Diastole and (b) Systole of the short axis view at the level of the great arteries (see diagram of Fig. 4.9 — SAX 1). The aorta appears as a circular structure and the cusps are in a closed position (l: left, n: noncoronary, r: right cusp). Left atrial (LA) cavity is posterior to the aorta. Right atrium (RA) and tricuspid valve (TV) are to the left and the right ventricular outflow tract (RVOT) curves around the aorta anteriorly. The pulmonary valve (PV) and pulmonary artery (PA) are to the right. The interatrial septum (AS) is visualised. During systole, the aortic valve cusps are in an open position and the left main coronary artery (LMCA) is well recorded. Orientation: A: anterior; L: left; P: posterior; R: right.

Mild aortic valve stenosis is characterised by slightly thickened cusps showing decreased systolic separation and diminished opening and closing velocities. On two-dimensional echocardiograms, differentiation between a valve with normal thickness and one which is mildly thickened is extremely difficult. Two-dimensional echo is the method of choice, however, for visualisation of the doming valve in congenital aortic stenosis which may appear normal on M-mode recordings.

Eccentricity of the early diastolic position of the cusps of more than 30 per cent within the aortic lumen is seen with congenitally bicuspid aortic valves (Fig. 4.17b). This is a specific but not very sensitive sign in young patients for diagnosing congenital aortic deformity. Multilayered diastolic cusp echoes may also be seen in this condition. M-mode records in patients with bicuspid aortic valve may be ambiguous and analysis is more satisfactory from two-dimensional studies. Shaggy, non-uniform thickening of normal moving valve cusps indicates infective vegetations. These vegetations may reflect prominent echoes so as to fill the aortic root in systole (Figs 4.17c and 4.18). They may be seen prolapsing into the left ventricular outflow tract in diastole on both M-mode and two-dimensional echocardiograms (Fig. 4.19). Cusp perforation may result in diastolic cusp flutter.

Echocardiography is helpful at defining the level of nonvalvular LV outflow obstruction. In supravalvular stenosis (hypoplastic aorta), cusp motion is normal. In fixed subaortic stenosis (discrete membrane or muscular tunnel type), one or both valve cusps may immediately reclose after systolic opening and the subvalvular area is usually narrowed (Fig. 4.17d). The complex anatomy, however, does not permit accurate prediction of the degree of obstruction.

Closure of the valve cusps in mid-systole followed by late systolic reopening is a frequent occurrence in idiopathic hypertrophic subaortic stenosis and possibly reflects flow dynamics over the valve (Fig. 4.17e). Coarse fluttering and gradually closing cusps during systole are seen in patients with low cardiac output (Fig. 4.17f). Diastolic cusp vibration and dilated aortic root suggest aortic regurgitation. Echocardiographic studies of prosthetic valves in the aortic position are difficult to perform. The method is useful,

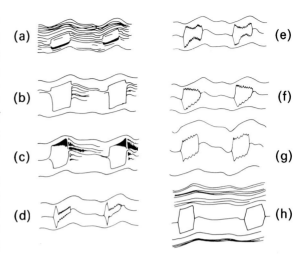

Fig. 4.17 Diagrams of aortic root and aortic cusp motion as visualised in different cardiac conditions. They are representative examples and significant variability may occur. (a) Calcific aortic valve stenosis. (b) Congenital bicuspid valve. (c) Vegetative endocarditis. (d) Fixed subaortic stenosis. (e) Idiopathic hypertrophic subaortic stenosis. (f) Low cardiac output. (g) Aneurysmatic dilatation of aortic root. (h) Dissecting aneurysm. For further details see text.

however, in demonstrating valvular or paravalvular regurgitation seen either as fluttering or as premature closure of the mitral valve leaflets, or both together. Echocardiography allows an accurate measurement of the aortic annulus, permitting the selection of a proper size valve prosthesis preoperatively.

Abnormalities of the aorta

Significant enlargement of the aortic root image is seen with aneurysms, the most dramatic widening being observed in patients with Marfan's syndrome (Figs, 4.17g and 4.20). In aortic root dissection, marked widening of the aortic walls, maintenance of parallelism between the elements of the dissected wall and normal appearing cusp echoes are commonly seen, but false positive patterns may occur (Fig. 4.17h). Hypoplasia of the aortic root may be diagnosed in neonates with the width ranging from 2–7 mm in contrast to the usual aortic dimensions of approximately 10 mm. The narrowed supravalvular area in the hourglass variety of supravalvular aortic stenosis is more reliably demonstrated with two-dimensional analysis than by M-mode.

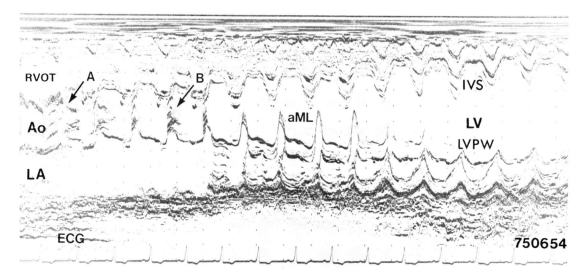

Fig. 4.18 M-mode scan from a patient with infective endocarditis and vegetations of the aortic valve. There are additional peculiar shaggy echoes at the level of the aortic valve seen in both systole and diastole (arrow A). They prolapse into the left ventricular outflow tract in diastole (arrow B). The left ventricle is dilated (7 cm) and the amplitude of motion of the interventricular septum (IVS) is larger than that of the left ventricular posterior wall (LVPW), which is increased as well. This pattern indicates LV volume overload. There is presystolic mitral valve closure as a result of the acute increase in severity of the regurgitation. Abbreviations, see Fig. 4.1.

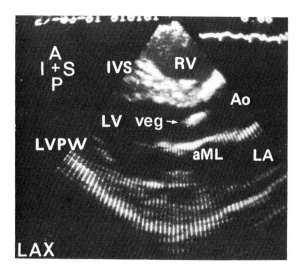

Fig. 4.19 Parasternal long axis view (LAX) of a patient with vegetative endocarditis of the aortic valve. The vegetation (veg) is recorded in the left ventricular outflow tract during diastole. Orientation: A: anterior; L: left, P: posterior; R: right. For abbreviations see Fig. 4.1.

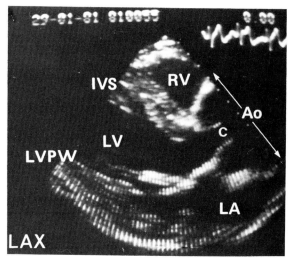

Fig. 4.20 Systolic stop-frame photograph of the parasternal long axis view (LAX) of a patient with an extremely dilated aorta (Ao) apparently encroaching on the left atrium (LA). The dimension of the aortic root is 70 mm. Note the open aortic valve cusps (c). For abbreviations and orientation see Fig. 4.1.

The left atrium

Echocardiography is one of the best methods for evaluation of LA size and is superior to fluoroscopy, chest X-ray and barium swallow. Correct identification of the LAPW from M-mode recordings requires an M-scan in order to demonstrate its continuity with the atrioventricular junction and LVPW. The Ao valve cusps must be seen on the recording in order to standardise the single sound beam pathway as intra-atrial landmarks are lacking. The largest LA dimension is measured and occurs at the moment that the Ao is in its most anterior position (Fig. 4.15). The normal LA dimension is less than 40 mm. Determination of ratio of the LA to Ao root dimensions is commonly utilised for the assessment of left atrial size and should be less than 1.3.[14]

Two-dimensional apical views are extremely helpful for the study of both the right and left atrial cavities (Figs. 4.13 and 4.14). The interatrial septum is best seen from the subcostal transducer position[5] or from right parasternal positions (Figs. 4.9 and 4.16). Examination of the left atrium is a very efficient and accurate method for the early detection of mobile left atrial tumours. There is always a short interval between the opening of mitral valve and the appearance of the tumour echoes in the orifice (Fig. 4.21). With the combined use of two-dimensional imaging, the diagnosis of a LA tumour is sufficiently obvious in most cases to permit surgery without other investigations. The ability to visualise left atrial thrombi has been somewhat disappointing, partly because the atrial appendage, the site of most atrial thrombi, is relatively inaccessible to echocardiographic examination.

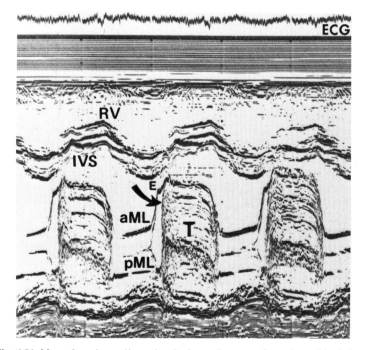
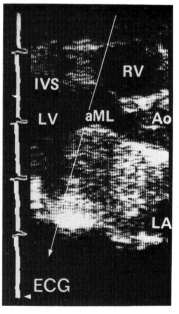

Fig. 4.21 M-mode and two-dimensional echocardiograms of a patient with a left atrial myxoma. The two-dimensional study was performed with a linear array electronic scanner (see Fig. 4.6). The arrow at the bottom of the electrocardiogram (ECG) indicates the moment in diastole at which the stop-frame photograph was taken. The tumour mass fills the mitral valve orifice and largely the left atrial cavity. The arrow represents the approximate sound beam direction from which the M-mode echocardiogram was recorded. Note the interval between mitral valve opening and appearance of tumour echoes (arrow), which results in a slight additional opening of the valve after the E point. These events are difficult to appreciate from two-dimensional studies. Abbreviations: aML and pML: anterior and posterior mitral valve leaflets; Ao: aorta; IVS: interventricular septum; LA: left atrium; LV and RV: left and right ventricles.

The mitral valve

The normal mitral valve

The recorded mitral valve motion pattern on M-mode echocardiograms is a combination of the movement of the heart as a whole, and the motion of the mitral valve itself. This principle applies to all other cardiac structures. It has a systolic phase during which the echoes of the closed valve gradually move towards the transducer as the mitral valve annulus moves towards the cardiac apex and chest wall during ventricular ejection (C–D interval), as shown in Figure 4.22. At valve

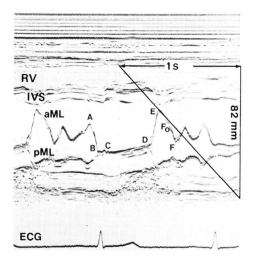

Fig. 4.22 M-mode recording of the mitral valve. The anterior (aML) and posterior (pML) mitral valve leaflets have a reciprocal motion pattern. There is early closure (at point B) of the valves in end-diastole which is seen with a prolonged PR interval and results from atrial relaxation. The echoes of the closed valve gradually move anteriorly towards the transducer (C–D interval). Principle of measurement of rate of early diastolic posterior motion of aML is indicated. The E–F$_0$ slope (or E–F slope with faster heart rates), is extended to equal one second of continuous motion, and the total displacement is here equal to 82 mm s^{-1}. For further explanation see text. IVS: interventricular septum; RV: right ventricle.

opening in diastole, rapid anterior leaflet motion and posterior leaflet motion, reaching their maximum excursions nearly simultaneously, are recorded. During the filling of the LV, the initial phase (E–Fo interval) represents annular motion without independent leaflet motion and reflects the rate of LV filling. The second phase (Fo–F interval) represents leaflet motion towards a mid-

closed position which is maintained during slow ventricular filling. With atrial systole, reopening of the valves (A wave) is recorded, followed by rapid closure terminating at point C. Occasionally, an intermediate point B is observed.

Rheumatic mitral stenosis

Characteristic findings include reduced amplitude of opening, reduced E–F slope, obliteration of the response to atrial systole, and the posterior leaflet moving in the same direction as the anterior leaflet in diastole (Figs. 4.23a and 4.24). There is little correlation between the E–F slope and severity of mitral stenosis. A slope greater than 50 mm s^{-1} virtually excludes mitral stenosis. Thick heavy multilayered echoes indicate heavy calcification, suggesting the need for valve replacement rather than commissurotomy. Echo appears to be more sensitive than fluoroscopy in the detection of mitral valve fibrosis and calcification. The M-mode pattern of rheumatic mitral valve disease may be simulated by other conditions (e.g. primary pulmonary hypertension) but these are readily recog-

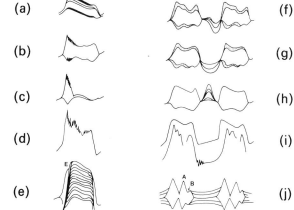

Fig. 4.23 Diagram of mitral valve motion as recorded in different cardiac conditions. They are representative examples and significant variability may occur. (a) Rheumatic mitral valve stenosis. (b) Chronic aortic regurgitation. (c) Acute (severe) aortic regurgitation. (d) Flail anterior mitral valve leaflet. (e) Prolapsing left atrial tumour. (f) Midsystolic mitral valve prolapse. (g) Holosystolic mitral valve prolapse. (h) Systolic anterior motion (SAM) in obstructive cardiomyopathy. (i) Flail posterior mitral valve leaflet. (j) Dilated left ventricle and impaired LV function. For further details see text.

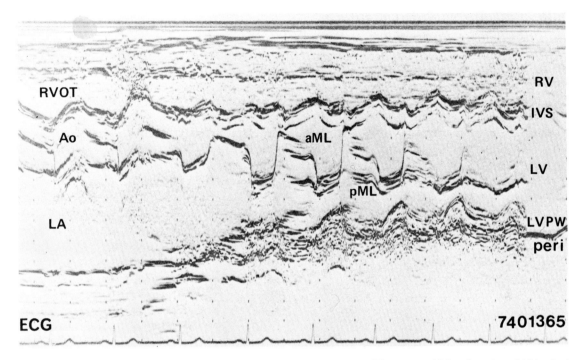

Fig. 4.24 M-mode scan obtained from a patient with severe mitral valve stenosis. There are multiple valve echoes (thickening) and the posterior mitral valve leaflet (pML) moves with the anterior leaflet (aML). The A wave is almost absent. Note changing amplitude of opening of the aML along the scan. The left atrium (LA) is dilated and the right ventricular (RV) dimension is within the normal range. For abbreviations, see Figures 4.1 and 4.2. Peri : pericardium.

nised by two-dimensional studies since more direct qualitative information about thickness and mobility of the valve is obtained (Fig. 4.25). In addition, measurement of the mitral valve orifice size is possible from the parasternal short axis views and good surgical and pathologic correlation have been reported. The method may eliminate some difficulties in the estimation of this value by cardiac catheterisation that are encountered in the presence of mitral regurgitation.[57] In some patients, cardiac catheterisation can thus be avoided.

Mitral regurgitation

Although echocardiography has not been uniformly helpful in the analysis of mitral regurgitation as it has been in mitral stenosis, the method has been invaluable in the assessment of a number of specific entities that result in mitral regurgitation. Typical echocardiographic abnormalities have been observed in mitral prolapse syndrome, ruptured chordae tendineae, mitral annulus calcification, bacterial endocarditis of the mitral valve,

dilated ventricle, idiopathic hypertrophic subaortic stenosis and left atrial myxoma. As a result, echocardiography has proved to be particularly valuable in the differential diagnosis of patients known to have mitral regurgitation.[67] The clinical suspicion of mitral prolapse is a very common reason for making an endocardiogram. The criteria for the diagnosis from M-mode and two-dimensional studies of mitral valve prolapse, however, are one of the principal areas of controversy, since their precise sensitivity and specificity remain unresolved.[33, 54] Characteristically, the gradual anterior systolic motion of the leaflets (C–D) is replaced by an abrupt mid-systolic posterior buckling, often coinciding with either a click or with the onset of a late systolic murmur, or with both (Figs. 4.23f and 4.26). A 'hammock-like' pansystolic posterior bowing is considered to be equally diagnostic for this disease (Figs. 4.23g and 4.27).

The redundant leaflets due to myxomatous changes demonstrate a larger amplitude of opening than normally seen and may appear thickened. M-mode echocardiograms are negative in approxi-

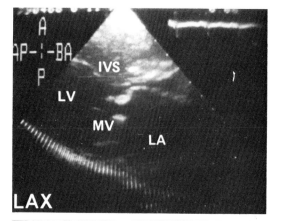

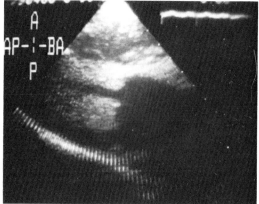

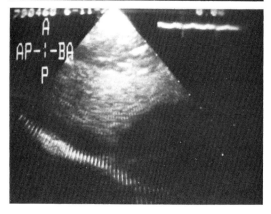

Fig. 4.25 Stop-frame photographs of parasternal long-axis views (LAX) of a patient with mitral valve stenosis before (top photograph) and after injection of echo contrast via a catheter in the left ventricle (LV). Note the orifice size and thickened mitral valve (MV) leaflets. The middle photograph is a frame recorded during diastole. The negative shadow caused by the noncontrast blood flowing from the dilated left atrium (LA) into the LV visualises the transmitral blood flow pattern. During systole (bottom photograph), the echo contrast does not pass into the LA thus excluding incompetence. A: anterior; AP: apical or inferior; BA: basal or superior; P: posterior; IVS: interventricular septum.

mately 10–15 per cent of the patients with the clinical syndrome, perhaps due to undetected localised prolapse. In patients with only auscultatory evidence of mitral valve prolapse with no or minimal symptoms, the sensitivity of M-mode echocardiography is even less. Several studies have recently enhanced the understanding of the echocardiographic manifestations of mitral valve prolapse and both the auscultatory and ultrasound manifestations have been demonstrated to originate at a relatively constant LV dimension.[58]

A coarse and erratic type of aML motion in diastole is seen with flail anterior leaflet due to ruptured chordae tendineae (Fig. 4.23d). Chordal rupture of the pML is identified by a 'flicking' anterior motion of the leaflet shortly after the anterior leaflet, from a point that is very posterior near the LAPW in early diastole (Figs. 4.23i and 4.28).

Mitral annulus calcification, a degenerative disorder not uncommonly found in elderly, is seen as a dense band of echoes posterior to otherwise normal mitral valve leaflets.[61] Patients with calcification limited to the lateral or medial portions of the annulus are better identified from parasternal short-axis views. Echoes of the calcific annulus may appear as an intracavitary LV structure, an artifact related to the beam width.

'Shaggy' non-uniform thickening of the leaflets without restriction of leaflet motion is seen with infective vegetations (Fig. 4.29). Unfortunately, the findings are not sensitive and a normal mitral valve is recorded in many patients with infective endocarditis. Two-dimensional echocardiography may improve sensitivity to vegetations, but this is still controversial.[32, 66]

In dilated ventricles, the mitral valve apparatus is stretched and valve mechanics is altered, which may lead to valvular incompetence. The valve leaflets have a limited amplitude of excursion and may assume a 'double-diamond' pattern of motion, often with a 'B-notch' during closure (Fig. 4.23j).

Alterations in motion pattern of otherwise normal mitral valves

Mitral valve motion may be altered in conditions that do not involve the mitral valve apparatus and is seen with aortic regurgitation, idiopathic hyper-

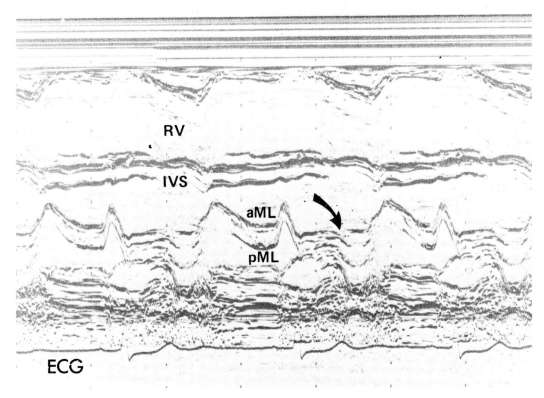

Fig. 4.26 M-mode echocardiogram from a patient with mitral valve prolapse syndrome. The anterior mitral valve leaflet (aML) appears thickened and demonstrates a mid-systolic posterior buckling (see arrow). ECG: electrocardiogram; IVS: interventricular septum; RV: right ventricle; pML: posterior mitral valve leaflet.

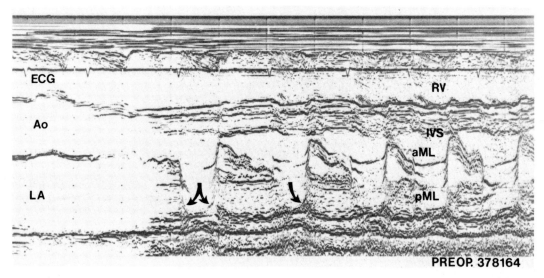

Fig. 4.27 Pan-systolic deep prolapse of the mitral valve in a patient with severe mitral insufficiency (see arrows). Multiple non-uniform leaflet echoes result from valve thickening by myxomatous infiltration of the valve. Ao: aorta; aML and pML: anterior and posterior mitral valve leaflets; IVS: interventricular septum; LA: left atrium; RV: right ventricle.

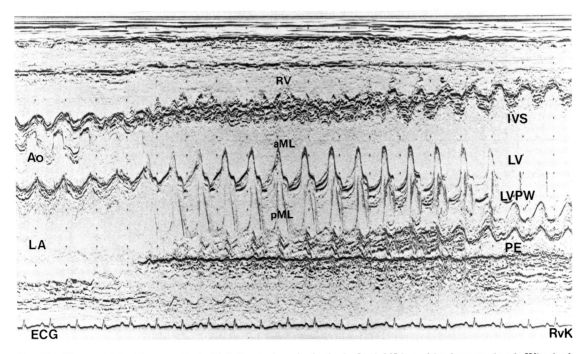

Fig. 4.28 M-scan obtained from a patient with flail posterior mitral valve leaflet (pML) resulting in severe (grade IV) mitral incompetence. The left ventricle (LV) and left atrium (LA) are dilated. There is pericardial effusion (PE). Ao: aorta; IVS: interventricular septum; LVPW: posterior wall; RV: right ventricle.

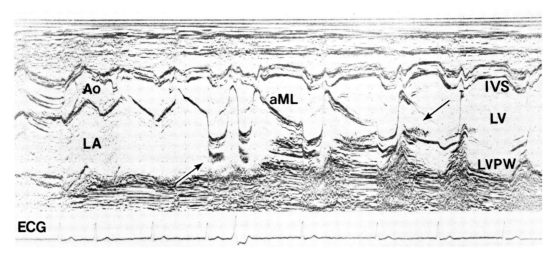

Fig. 4.29 M-scan obtained from a patient with vegetative endocarditis of the mitral valve. The vegetation appears in the left atrium (LA) in systole and in the left ventricle (LV) during diastole (see arrows). At surgery, torn chordae with vegetations were found. Severe mitral regurgitation was present which is seen on the echocardiogram as a dilated LV with increased amplitude of motion of its walls.

trophic subaortic stenosis (IHSS), elevated diastolic LV pressure and cardiac arrhythmias. Aortic regurgitation is detected by noting high frequency fluttering on the mitral valve leaflets which is related to the direction of the regurgitation jet rather than to the degree of regurgitation (Fig. 4.23b). The septum and chordae tendineae have also been shown to flutter occasionally, and this finding assumes importance when the mitral valve is calcified. In severe acute aortic regurgitation there may be mid-diastolic mitral valve closure (Figs. 4.18 and 4.23c). In hypertrophic obstructive cardiomyopathy there is abnormal systolic anterior motion (SAM) of the aML at a rate greater than that of the endocardium of the LVPW (Fig. 4.23h). SAM of the aML causes a dynamic outflow gradient. A Venturi effect due to rapid early ejection and excessive traction by a thickened papillary muscle have been invoked as possible factors responsible for the SAM. A prolonged A–C interval and a 'B-notch' indicate left ventricular dysfunction and loss of compliance with elevated end-diastolic pressure in patients with dilated LV.[20] In conditions where the LV filling rate is impaired, the E–Fo slope is reduced and the pattern may mimic mitral stenosis. The aML echo is thin, however, the A wave is well preserved, and the pML moves in the opposite direction.

Prosthetic mitral valves

Direct assessment of bioprosthetic valves is an important application of two-dimensional echocardiography since masses within these valves and their detachments are often quite easily recognised.[84] In addition to visualisation of the implanted prosthesis, a complete echocardiographic analysis should be made since deterioration of myocardial performance rather than malfunction is the cause of heart failure in most patients following valve replacement.

The simultaneous recording of M-mode echocardiograms with external pulse tracings and phonocardiogram is often helpful for optimal evaluation of prosthetic valves.[10] A baseline postoperative study should be routine practice. This approach has proved to be of value in the determination of a variety of types of complications arising in artificial prostheses including mechanical malfunction as a result of thrombus or vegetation, ball or disc variance, valve dehiscence and paravalvular leak.[46]

The tricuspid valve

Although the tricuspid valve morphology and orifice shape are different from those of the mitral valve, the echocardiographic motion patterns of the two valves are quite similar. In general, this valve is difficult to study with M-mode echocardiography. Thickened valve echoes and a reduced diastolic E–F slope are seen in tricuspid stenosis. The latter parameter, however, is highly influenced by extrinsic forces. Normal diastolic slopes are useful in excluding the presence of tricuspid stenosis. The systolic appearance of contrast in the IVC after the rapid antecubital vein injection of dextrose 5 per cent in water is a sensitive and specific sign of tricuspid regurgitation[52, 60] (Fig. 4.30).

Tricuspid valve prolapse has occasionally been diagnosed and presents the same echocardiographic picture as mitral valve prolapse. False patterns are common, however, and at present no reliable criteria are available. The valve may be extensively involved in patients with bacterial endocarditis (drug addicts). Differentiation of infective vegetations from right atrial masses extending into tricuspid orifice may be difficult. Vegetations generally move in concert with the leaflets to which they are attached, while a prolapsing mass trails the opening leaflets by a short interval. Fluttering of the leaflets during diastole may be seen in patients with pulmonary insufficiency especially in the presence of pulmonary hypertension or large volume flow. Examination of the tricuspid valve is useful in patients with congenital heart disease. The increased amplitude of motion as compared to the mitral valve results in a larger volume flow across the valve in atrial septal defects. The abnormal spatial orientation of the tricuspid valve echoes, increased amplitude of motion, decreased E–F slope and, most specific, a marked delay in valve closure relative to the mitral valve are seen in Ebstein's anomaly. Two-dimensionally, the apical four-chamber view allows direct visualisation of the apical displacement on the IVS of the right sided

Fig. 4.30 M-mode echocardiogram showing the inferior vena cava (IVC) and a 'v-wave synchronous' appearance of echo-contrast after antecubital vein injection. This pattern indicates tricuspid regurgitation. Timing of appearance of echo-contrast is easier from M-mode than from two-dimensional echocardiograms.

atrioventricular valve in this condition.[74] Cardiac arrhythmias and an abnormal pressure in the right ventricle alter the pattern of tricuspid valve motion similar to those observed in mitral valve motion in response to abnormal LV pressures.

The pulmonary valve

Changes in pulmonary valve motion result from alterations in pressure and flow rather than from structural abnormalities and are further affected by respiration and fluctuations in cardiac cycle length. The left cusp is recorded normally and shows an oblique position in diastole, a circumscribed posterior deflection with atrial systole ('a' dip) and a relatively slow valve opening velocity which relates to the small diastolic gradient present across the valve. The increasing return of blood to the right heart during inspiration increases the 'a' wave depth (Fig. 4.31a). In valvular pulmonary stenosis, a decreased RV compliance and forceful atrial contraction frequently cause the RV end-diastolic pressure to exceed the simultaneous pulmonary artery pressure. This positive gradient causes opening or doming of the pulmonary valve before RV systole and is seen as a marked increase in 'a' wave depth[97] (Fig.4.31b). The two dimensional approach simplifies the diagnosis of pulmonary valve stenosis because the enlarged field of vision facilitates location and recording of both the pulmonary artery and valve. This makes it possible to view the entire doming valve during systole. Patients with pulmonary hypertension, on the other hand, present with a general straightening of the diastolic valve image, a reduced 'a' dip, and rapid valve-opening slopes.[69]

Mean pulmonary artery pressures exceeding 50 mmHg regularly result in complete 'a' dip obliteration which is a useful indicator of moderate to severe pulmonary hypertension. Delayed valve opening (greater than 100 ms after onset QRS) in the absence of bundle branch block generally indicates pulmonary hypertension (Fig. 4.31c). Midsystolic closure of the valve is frequently

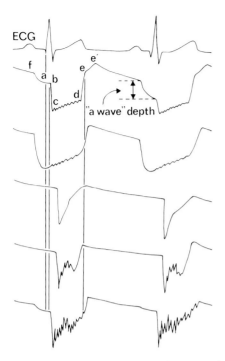

Fig. 4.31 Diagram of pulmonary valve motion as seen in a normal subject and in different cardiac conditions. They are representative examples and significant variability may occur. (a) Normal echo patterns of the posterior pulmonic cusp. The depth of the 'a-wave' may change significantly during respiration. Lettering of the pulmonary valve echo with lower case letters similar to those used to describe mitral valve motion facilitates discussion. (b) Pulmonary valve stenosis. (c) and (d) Patterns of the midsystolic closure or systolic notch as seen in pulmonary hypertension. (e) Coarse systolic fluttering in infundibular pulmonic stenosis. For further details see text.

recorded with higher pulmonary artery pressure and reflects an alteration in the systolic flow pattern (Fig. 4.31d). In patients with infundibular pulmonary stenosis, the turbulent bloodstream strikes the pulmonary valve cusps, producing a very high frequency chaotic fluttering (Fig. 4.31e).

The ventricles

The right ventricle

Reliable RV intracavitary landmarks are not available. Therefore dimensional measurements are made along the sound beam axis used to obtain a standard LV study. Despite a large variability in RV dimensional measurement, it is a most useful parameter in identifying patients with dilated RV (normally less than 30 mm at end-diastole). RV

volume overload causes RV dilatation and paradoxical septal motion. This echocardiographic complex is seen with atrial septal defect, partial and total anomalous pulmonary venous drainage, tricuspid incompetence, Ebstein's anomaly and pulmonary insufficiency.

Pulmonary stenosis, primary pulmonary hypertension and cardiomyopathy may occasionally be associated with dilated RV especially when RV failure develops. Absent or diminutive RV cavity dimensions occur in patients with tricuspid atresia and type I pulmonary atresia (hypoplastic right heart syndrome). Compression of the RV cavity may occur as a result of pericardial effusion, pectus excavatum and, exceptionally, as a result of tumours within the anterior mediastinum.

The left ventricular outflow tract

The LV outflow tract (LVOT) is the subaortic area between the IVS and anterior mitral (aML) bounded by the aortic valve cusps superiorly and the area where both mitral leaflets are recorded inferiorly. Its dimension in early systole should be greater than 20 mm. Conditions associated with a narrow LVOT include: fixed subaortic stenosis (Fig. 4.32), hypertrophic obstructive and nonobstructive cardiomyopathy, endocardial cushion defect and, in some patients, mitral stenosis. The membrane in the membranous type of subaortic stenosis is only occasionally seen as extra linear echoes within the LVOT on M-mode recordings. Because of the tubular nature of the LVOT, apparent decrease in diameter may result from eccentric beam angulation. Therefore, a more precise characterisation of the morphology of the LVOT and its size is better obtained by the two-dimensional approach.[13] Dilated LVOT (greater than 35 mm) is seen in conditions with severely dilated LV such as primary dilated and ischaemic cardiomyopathy.

The interventricular septum

On M-mode echocardiograms, the IVS may be divided into two general areas. These are the basal portion or outflow septum, which moves similarly to the anterior aortic root, and the muscular septum, which contracts and moves to the LV

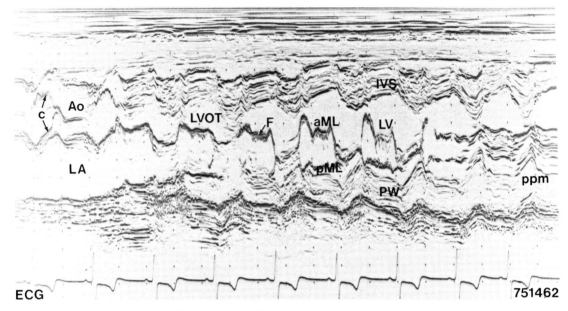

Fig. 4.32 M-mode scan from the aorta (Ao) to the apex of the left ventricle (LV) performed with a uniform transducer speed in a patient with membraneous subaortic stenosis. Note the fluttering of the aortic valve (small arrows), the extra echoes in the left ventricular outflow tract (LVOT) due to the membrane, the high frequency flutter on the mitral valve (F) and the concentric LV hypertrophy.

posterior wall. Septal thickness and motion are interpreted on the apical side of the mitral valve free edge and their analysis mainly deals with the left-sided rather than right-sided echoes. The analysis of the IVS is unique to echocardiography and not possible with any other method.[2] The normal IVS moves slightly anteriorly as a result of atrial contraction (some additional anterior motion may be seen after the onset of QRS during isovolumic contraction for 40–60 ms) followed by a smooth posterior motion, concave to the chest wall and ending at the time of the T wave of the ECG (Fig. 4.33). With the onset of relaxation, the IVS moves anteriorly for 40–50 ms. This anterior motion is interrupted by a notch of posterior motion which reaches its maximum slightly after the E point of the mitral valve motion. Thereafter, anterior movement occurs during rapid LV filling and is followed by slow filling. The explanation of the different phases of IVS motion is complex and still unclear. During systole, thickening occurs with increase in distance between the right and left sided echoes of the IVS. The normal thickness increase should be more than 30 per cent of diastolic thickness. Abnormalities of motion and

thickening may occur separately or in combination. Hyperkinesis with normal contraction and dilated LV is seen with LV volume overload conditions due to mitral and aortic regurgitation (Fig. 4.18), VSD, and so on . Secondary hyperkinesis is seen in patients with inferoposterior or apical ischaemic aneurysms (Fig. 4.34). Hypokinesis and diminished thickening occur in dilated cardiomyopathy and antero-septal myocardial ischaemia or infarction, while akinesis and paradoxical motion indicate extensive fibrosis. A large number of conditions may produce abnormal or paradoxical IVS motion. They can be brought together in four categories: RV volume overload, intraventricular conduction disturbances, coronary artery disease and postoperative conditions. The differential diagnosis is presented in Table 4.3. Hypertrophy of the IVS may be divided into symmetric and asymmetric forms. The classic example of the latter is cardiomyopathic asymmetric septal hypertrophy (ASH), the thickness being greater than 15 mm and the IVS to LVPW thickness ratio exceeding 1.3. ASH may also be seen in RV pressure overload and occasionally in aortic stenosis and systemic hypertension. Symmetric hypertrophy occurs in

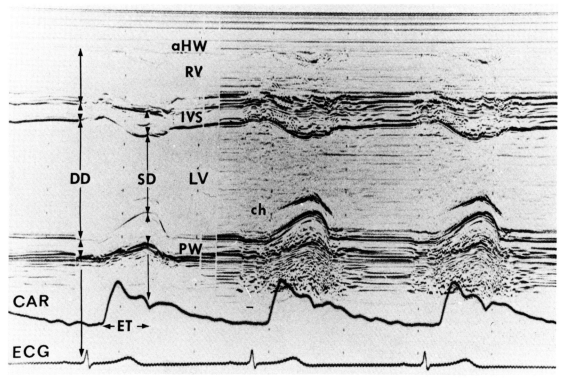

Fig. 4.33 Left ventricular (LV) echocardiogram. The end-diastolic (DD) right ventricular (RV) and LV dimensions are measured at the onset of the QRS complex of the simultaneously recorded electrocardiogram (ECG). End-systolic dimensions are measured at the moment of aortic valve closure (aortic component of second heart sound on simultaneously recorded phonocardiogram) or just before the incisura of the carotid artery tracing (CAR) as indicated here. It must be emphasised that the incisura has a delay of approximately 30 ms to aortic valve closure and the events on the echocardiogram. The interventricular septal (IVS) and posterior wall (PW) thickness are measured at the same moments in the cardiac cycle. Left ventricular ejection time (ET) is best measured from the CAR. aHW: anterior heart wall; ch: chordae tendineae.

Table 4.3 Differential diagnosis of patients with paradoxical motion of the interventricular septum on M-mode echocardiograms

Normal right ventricular dimension (< 30 mm)
1. Normal IVS thickening (> 30%)
   Postoperative patients
   Intraventricular conduction abnormalities (RV pacing, RV ectopy, Wolfe-Parkinson-White syndrome, congenital left bundle branch block)
   Constrictive pericarditis
   Acute RV volume overload
   (Idiopathic?)
2. Reduced IVS thickening (< 30%)
   Coronary artery disease
   Left bundle branch block
   Hypertrophic cardiomyopathy (increased septal thickness)

Increased right ventricular dimension (> 30 mm)
1. Normal IVS thickening (> 30%)
   RV volume overload
   Primary pulmonary hypertension
2. Reduced IVS thickening (< 30%)
   Coronary artery disease
   Dilated cardiomyopathy (rare)

conditions with obstruction to LV outflow and in hypertension. Absence of IVS echoes occurs in single ventricle and large VSD. Aortic-septal discontinuity is seen in tetralogy of Fallot, truncus and pseudotruncus arteriosus and pulmonary atresia.

The left ventricular posterior wall

By analogy with the IVS, the amplitude of motion, thickness and thickening can be analysed. The diagnostic potential of isolated LVPW analysis is considerably less than of the IVS. Increased amplitude of motion is seen in patients with LV volume overload and in conditions where the LVPW has to compensate for IVS abnormalities such as paradoxical motion, cardiomyopathic ASH and coronary artery disease. Decreased motion and thinning is seen with coronary artery disease

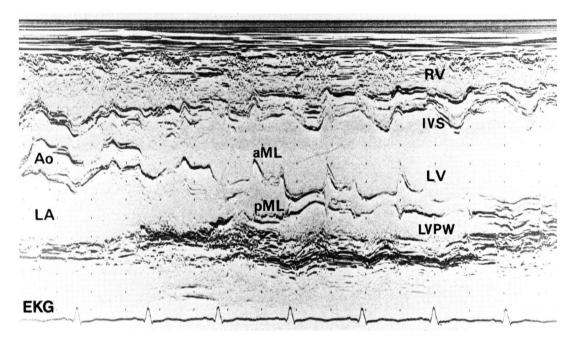

Fig. 4.34 M-scan obtained from a patient with an inferolateral myocardial infarction. The left ventricular posterior wall (LVPW) is akinetic (the observed amplitude of motion is most likely total cardiac displacement) and the interventricular septum (IVS) demonstrates increased motion and contraction. The left ventricle (LV) is dilated. LA: left atrium; RV: right ventricle; RVOT: right ventricular outflow tract.

involving the inferno-lateral wall. A thickening LVPW occurs in conditions with LV pressure overload. Systolic thickening (normal value > 30 per cent) is decreased in coronary artery diseases and dilated cardiomyopathy.

Assessment of left ventricular function

The different aspects which independently or in combination are responsible for altered LV function and can be studied echocardiographically include abnormalities in LV cavity size and shape, extent and rate of wall movements, and abnormalities in LV contraction, relaxation and filling.

Probably the most useful and important echocardiographic measurement is the end-diastolic LV cavity diameter, which represents a quick and reliable assessment of LV size. This cavity dimension is measured at the onset of the QRS complex as the distance between the leading edge of the endocardial echoes of the IVS and LVPW from the LV echocardiogram (Fig. 4.33). In the adult, this dimension ranges from 39 to 56 mm. The limited area of the LV transected by the M-mode beam, however, represents a significant limitation in assessing LV size in the abnormally shaped or segmentally diseased LV. Since two-dimensional echocardiography permits visualisation of both the spatial geometry and motion pattern of the LV walls, the size, shape and wall motion are better studied with this method in patients suspected or known to have ischaemic heart disease or cardiomyopathy.[42, 49] Measurement of cyclic change of LV cavity diameter allows assessment of the extent and velocity of LV wall motion, which is routinely done by measuring the LV dimensions at both end-diastole and at end-ejection defined at the moment of aortic valve closure (the aortic component of second heart sound on a simultaneously recorded phonocardiogram). IVS and LVPW thickness are measured at the same moments within the cardiac cycle and further allow for a specific analysis of myocardial contraction of the LV areas along the sound beam axis. The accuracy of all these measurements has been well validated by comparison with angiographic data.

Since the M-mode echocardiogram is typically sampled at a repetition rate of 1000 s^{-1}, it is possible to study LV cavity dimension and wall motion continuously with considerably better resolution in time than that available from angiographic methods. In order to do this, the LV echocardiogram may be placed on a digitising tablet and a cursor run along each of the echoes to be measured. The position of the cursor is detected electronically and converted to a series of digital coordinates which are fed to a computer. This process can be applied to echoes from the right and left sides of the IVS and endocardial and epicardial surfaces of the LVPW. A continuous measure of LV dimension and wall thickness can thus be obtained. It is also possible to differentiate these traces with respect to time so that their mean and peak rates of change can be calculated (Fig. 4.35).[11, 30]

M-mode derived parameters for assessment of LV function can be divided into three categories:

1. Volume-dependent indices: ejection fraction and fractional shortening. Ejection fraction relates stroke volume to end-diastolic volume and is an index of the extent of overall LV fibre shortening and wall movement. It requires echocardiographic volume estimation by cubing the LV cavity dimension at both end-diastolic and end-ejection, however, and this is a method which is not sufficiently precise for clinical use because of the many assumptions that have to be made. Furthermore, calculation of volume is no more than a means of converting an echo measurement into a more familiar term. Thus, in view of the close relationship between LV volume and the LV cavity dimension, it seems reasonable to use this latter dimension as a correlate of volume. The percentage change in dimension or fractional shortening (FS) during systole provides the same information as the more familiar ejection fraction and therefore is the most direct information that can be got from the echocardiogram (FS in per cent = $[DD - SD] \times 100/DD$, where DD and SD are end-diastolic and end-ejection dimensions). Fractional shortening ranges between 25 and 42 per cent.

2. Velocity-dependent indices: mean and peak rates of dimensional shortening and lengthening. The rate of change of the minor axis circumference during systole is closely related to the mechanical properties of the myocardium and is a useful measure to distinguish between normal and abnormal LV function. The mean rate of circumferential shortening (mean Vcf) can be simply determined from the end-diastolic (DD) and end-ejection (SD) dimensions. Assuming that the LV is circular at its

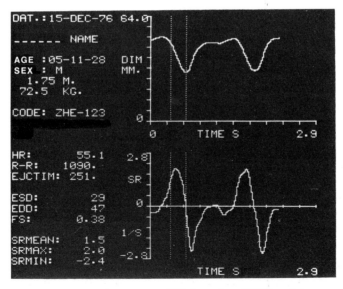

Fig. 4.35 Example of computer report showing instantaneous LV dimensions (DIM), and the normalised first derivative of the dimensional change (SR). These were calculated from the digitised left IVS endocardial and LVPW endocardial echoes entered into a computer via a digitising tablet. Vertical dotted lines show points of end-diastole (EDD) and end-ejection (ESD). Values of heart rate (HR), R–R interval, ejection time, fractional shortening (FS), mean and maximal shortening as well as maximal lengthening rates are calculated and listed.

minor axis, the difference between DD and SD divided by the duration of shortening or ejection time (ET) yields mean Vcf. In order to compare ventricles of different sizes it is convenient to normalise the value to DD. Mean Vcf ranges from 1.02 to 1.94 circumferences per second. Peak Vcf has been reported to be a more sensitive measure of LV function than mean Vcf. Peak Vcf as well as peak velocity of dimension increase require computer-aided analysis. The measurement of continuous rates of dimensional change in diastole allows the study of LV filling patterns. These are altered in mitral valve disease, hypertrophy and cardiomyopathy.

3. Wall thickening and relaxation indices. The measurement of fractional wall thickening (FWT in per cent) is a specific and sensitive index for the assessment of myocardial contraction of the LV segments explored by the sound beam. FWT is calculated using end-diastolic (DWT) and end-ejection (SWT) thickness measurements for both IVS and LVPW (FWT in per cent = [SWT − DWT] × 100/DWT). Normal values are greater than 30 per cent. Mean velocity of thickening is calculated by dividing the change of thickness by the contraction time (measured from the onset of contraction to its peak). Again, correction should be made for the mean velocity of wall thickening in the initial wall thickness or DWT. Peak rates of contraction and relaxation are obtained by digitising techniques and the validity of these parameters is now beginning to be studied. It appears that the peak rate of relaxation reflects the myocardial abnormality causing the reduction in peak rate of dimension increase in patients with hypertrophy and cardiomyopathy. In patients with LV inflow obstruction, the rate of relaxation is normal and the rate of dimension increase is reduced causing abnormal LV wall motion.

LV pressure-dimension loops can also be constructed with the use of computer techniques when the LV echocardiogram is recorded simultaneously with the LV pressure trace.[31] The area of the loop represents the actual work done by the myocardium on the circulation per unit LV area studied by the sound beam. The loop is distorted when the LV dimension changes during the isovolumic periods and is a sensitive method in demonstrating incoordinate LV contraction and relaxation. Since only

timing of the upstroke and downstroke rather than absolute pressure is required, incoordinate wall motion can also be detected by using the apexcardiogram instead of pressure, thus obviating cardiac catheterisation.[94]

Two-dimensional echocardiography yields real-time information on an entire cardiac cross-section. Thus, a larger portion of the LV at any point in time can be studied as compared to M-mode echocardiography. This would allow a more accurate assessment of LV volume and function since no major geometric assumptions have to be made and formulae largely tested in angiography can be applied. The recording and display media, however, do not lend themselves to rapid quantitative measurements. In addition endocardial boundaries are blurred and difficult to localise on stop-frame images and adequate recordings are not obtained in all patients. Usually, accurate measurements of the LV long-axis and surface areas from cardiac views are not possible because of large intra- and interobserver variability.[4] Nevertheless, extreme accuracy has been reported in experimental studies, proving that the models and formulae applied are satisfactory.[22, 101] The reported results in patients, however, have been less convincing.[8, 12, 27, 80] At present, subjective visual analysis of two-dimensional images allows a reasonable estimate of overall and regional LV function. Quantitation awaits further instrument improvements and better display techniques.

Pericardial effusion

Echocardiography has become the standard method for the diagnosis of pericardial effusion and has replaced previously employed techniques. The method also offers good insight into underlying or associated cardiac disorders. Therefore, assessment of pericardial effusion is a sufficient reason to have echocardiography available in any general hospital. In the presence of pericardial fluid, the potential pericardial space fills with fluid which is usually echo-free and the anterior and posterior heart walls become separate from the pericardium. Because posteriorly the pericardium is deflected from the back of the LA into the pulmonary veins, no potential space exists for the accumulation of pericardial fluid behind the LA.

It should be remembered, however, that there exists posteriorly a blind recess of pericardial cavity (oblique pericardial sinus) in which pericardial fluid exceptionally may accumulate.

The importance of an adequate examination technique cannot be overemphasised. Clear visualisation of the aHW and all components of LVPW using different gain settings is mandatory. The best approach is to make a good LV study and to look for an effusion behind the LV. Too high gain settings may produce ultrasonic reflections from the pericardial fluid itself, thereby 'filling-in' the echo-free space and preventing visualisation of the effusion. Another pitfall in the recording technique pertains to the direction of the ultrasound beam. When the transducer is tilted too far medially, an echo-free space behind the LV may be recorded on account of structures such as the coronary sinus, pulmonary veins or aorta. The earliest manifestations of the presence of a small pericardial effusion is a slight separation, through systole and part of diastole, of the LVPW epicardial echo from a relatively stationary pericardial echo (Fig. 4.36b). The lower limit of sensitivity of a technically

optimum properly interpreted echocardiogram is about 20 ml.[45]

Small degrees of effusion are better detected by M-mode echocardiography. With increasing accumulation of fluid, the posterior echo-free space becomes evident throughout the entire cardiac cycle (moderate degree of pericardial effusion) (Figs. 4.29 and 4.36c). With large pericardial effusion, the posterior space further widens and the fluid also appears anteriorly as an echo-free space between the aHW and the chest wall. In this situation, but not invariably, the heart may be observed to be moving freely in the pericardial sac ('swinging' heart syndrome) (Fig. 4.37). Abnormalities of IVS and valvular motion then occur frequently, but they are 'pseudo' in nature. Therefore dimensions should never be measured and motion patterns never interpreted in the presence of a large effusion. Recently, a miniature hand-held real-time scanner has been developed which permits the immediate diagnosis of the presence of a large pericardial effusion in emergency situations at the bedside (Fig. 4.38).[77]

In some patients with large effusion, an echo-free space posterior to the LA may also be found. In the absence of a large effusion, this finding usually indicates the presence of a pleural effusion. Two-dimensional echocardiography and especially parasternal short-axis views are helpful in differentiating pleural versus pericardial effusion and in recognising patients with both pleural and pericardial effusions. Pleural effusion is seen as an echo-free space posterior to the descending aorta which is in close anatomical relationship with the heart. In pericardial effusions the echo-free space separates the heart from the descending aorta. When both pericardial and pleural effusions are present, the aorta is insulated and the parietal pericardium, brought out by the pleural effusion, can be seen as a band of echoes enclosing the pericardial fluid and heart.[38] The distribution and localisation of effusions are better recognised with two-dimensional echocardiography.[55]

Pronounced phasic variation in dimension of the ventricles, consisting of an increase of RV dimension with a reciprocal decrease of the LV dimension during inspiration, suggest cardiac tamponade. These phasic variations disappear during held respiration and after pericardiocentesis. Fibrous

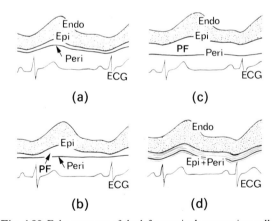

Fig. 4.36 Echo patterns of the left ventricular posterior wall and pericardium seen with increasing amounts of pericardial fluid (PF) and fibro-adhesive pericardial disease. (a) An echo-free space in systole and early diastole between the moving epi- and pericardium represents an amount of less than 20 ml of fluid, which may be the normal volume. (b) When the moving epicardium touches a stationary pericardial echo in diastole, a small pericardial effusion is present.
(c) Separation throughout the cardiac cycle, the classic pattern, indicates larger amounts of pericardial fluid. (d) A sustained small separation of the epi- and pericardial echoes showing equal motion is seen with fibro-adhesive pericardial disease. For further details see text.

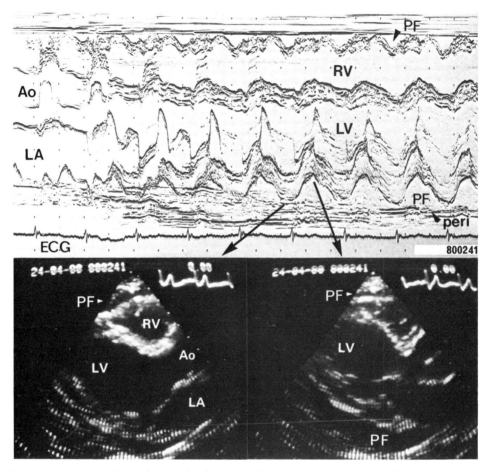

Fig. 4.37 M-mode scan and stop-frame photographs of parasternal long axis views of a patient with severe pericardial effusion and 'swinging' heart syndrome. Note anterior displacement of the total heart of the two-dimensional images during systole resulting in paradoxical motion of the septum. Ao: aorta; LA: left atrium; peri: pericardium; PF: pericardial fluid; LV and RV: left and right ventricles.

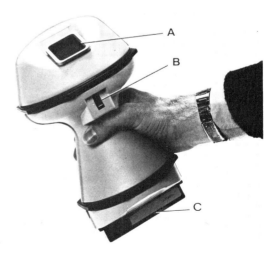

Fig. 4.38 Photograph of the hand-held ultrasound scanner showing the small display screen (A), the switch at thumb level, which controls the on/off function and the semiautomatic time-sensitive gain (B), and the linear-array transducer (C).[77]

material and adhesions result in higher reflectivity, sustained separation and virtually identical motion of the epi- and pericardial echoes (Fig. 4.36d). The pattern is commonly seen in postoperative patients and is easier to appreciate from M-mode recordings.[81]

The cardiomyopathies

Echocardiography provides useful clues in diagnosing and in differentiating between different forms of cardiomyopathies. Three districtive types of cardiomyopathy can be recognised—dilated, infiltrative (restrictive) and hypertrophic cardiomyopathy (Table 4.4). The findings in the dilated type of cardiomyopathy include an enlarged LV and LVOT, decreased amplitude of motion of the LV walls (low fractional shortening), and a mitral valve configuration consistent with dilated LV and elevated end-diastolic pressure (Fig. 4.39). In two-dimensional images a globular shaped heart with global hypokinesis is seen. In severe cases, all cardiac chambers are enlarged. This group of findings is not specific for primary dilated cardiomyopathy, but can also be found in patients with severe LV dysfunction of any aetiology. Diffuse wall motion abnormalities are in favour of primary dilated cardiomyopathy, whereas segmental abnormalities occur in coronary artery disease.[19] The history and other clinical data are helpful in further differentiating the latter two conditions. Symmetric hypertrophic or infiltrative cardiomyopathy is relatively uncommon. It is caused by amyloid infiltration of the myocardium, iron (hemachro-

matosis), eosinophilic leukaemia or glycogen. Symmetric LV hypertrophy with normal or decreased volume is a common finding (Fig. 4.40). These findings may also be encountered in patients with hypertension and aortic stenosis. In amyloidosis there is also a thickened RV wall which seems rather specific for this disease entity, and pericardial effusion.[88] Echocardiography is of value in distinguishing infiltrative cardiomyopathy from constrictive pericarditis of which the clinical picture is similar. These two entities have different therapeutic and prognostic implications, however, and are difficult to differentiate clinically and haemodynamically. Normal wall thickness, normal LVPW and paradoxical IVS motion are seen with constrictive pericarditis.

The asymmetric type of hypertrophic cardiomyopathy is characterised by an IVS which is thicker than the LVPW (ASH or asymmetrical septal hypertrophy) (Fig. 4.41). Echocardiography appears to be the easiest and most reliable method of detecting this anatomical marker of the disease.[43, 56] The septal thickening may be diffuse, or confined to its upper, middle or lower parts. IVS thickening and its localisation are more reliably defined with two-dimensional echo than with M-mode methods since the length and width of the abnormality can be studied from long and short axis planes respectively. Septal thickness should be at least 15 mm and the ratio between IVS thickness and the LVPW greater than 1.3. Using this echocardiographic marker, ASH has been shown to be transmitted as an autosomal dominant trait by studying first degree relatives of patients with

Table 4.4 Classification of cardiomyopathies with echocardiography

	Left ventricle Cavity dimension	Outflow tract	Wall thickness	Fractional shortening	Cavity shape	Mitral valve Systolic motion	Diastolic motion	Exclude
Dilated types	↑	↑	N	↓	Globular	Multi layering	Double diamond B-bump	CAD AS
Non-dilated types Symmetric hypertrophy	N or ↓	N	↑	N	Ellipse	N	E–F slope ↓ (B-bump)	BP ↑ Athletes CAD
Asymmetric hypertrophy	N or ↓	↓	ASH	N or ↑	Banana	N, SAM (IHSS	E–F slope ↓ (B-bump)	RVH

Abbreviations: AS: aortic stenosis; ASH: asymmetric hypertrophy of the interventricular septum; BP: blood pressure; CAD: coronary artery disease; IHSS: idiopathic hypertrophic subaortic stenosis; N: normal; SAM: systolic anterior motion.

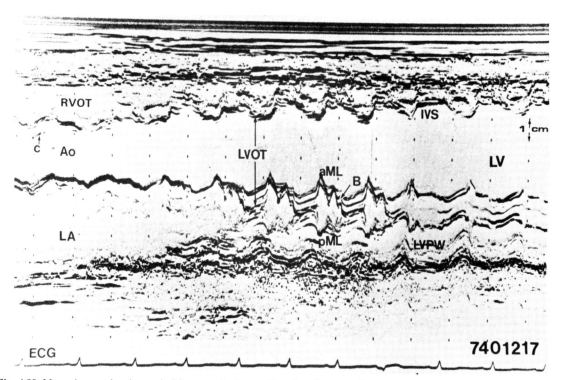

Fig. 4.39 M-mode scan showing typical features of primary dilated cardiomyopathy. The left ventricular (LV) cavity diameter, as well as the dimension of the LV outflow tract (LVOT), are increased. The 'double diamond-shaped' motion pattern and B-notch of the mitral valve (aML and pML) indicate LV dysfunction and an elevated end-diastolic pressure which is also reflected by the dilated left atrium (LA). Ao: aorta; C: aortic valve cusps; IVS: interventricular septum; LVPW: left ventricular posterior wall.

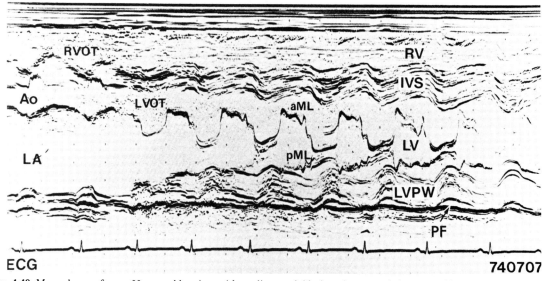

Fig. 4.40 M-mode scan from a 22-year-old patient with cardiac amyloidosis and symmetric hypertrophic cardiomyopathy. The thickness of both the interventricular septum (IVS) and the left ventricular posterior wall (LVPW) are increased (14 mm). The anterior mitral valve leaflet (aML) has a delayed closure (long AC interval and B-notch) indicating an elevated end-diastolic pressure in the left ventricle (LV). Note the small pericardial effusion (the patient was in congestive heart failure). The left atrium (LA) is dilated. Ao: aorta; PF: pericardial fluid; pML: posterior mitral valve leaflet; RV: right ventricle; RVOT: right ventricular outflow tract.

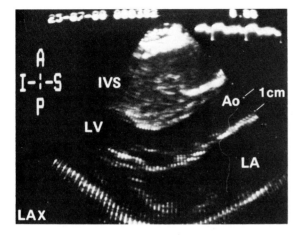

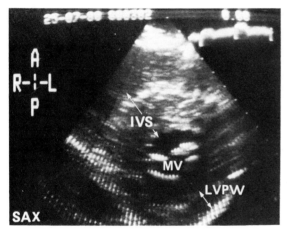

Fig. 4.41 Long axis (LAX) (top) and short axis (SAX) (bottom) views of a patient with hypertrophic obstructive cardiomyopathy. Note massive hypertrophy of the interventricular septum (IVS) as compared to the posterior wall (LVPW). Ao: aorta; IVS: interventricular septum; LA and LV: left atrium and left ventricle; MV: mitral valve. Orientation: A: anterior, P: posterior, R and L: right and left.

hypertrophic cardiomyopathy.[18, 21] The isolated echocardiographic demonstration of ASH may occasionally be found in other conditions (an IVS thickness of 20 mm or more, however, is rare in these instances) such as RV pressure overload, aortic stenosis, hypertension and coarctation of the aorta. The disease may manifest without or with outflow obstruction (IHSS or idiopathic hypertrophic subaortic stenosis) (Fig. 4.42). LV outflow obstruction results chiefly from systolic anterior movement (SAM) of the aML in the narrowed outflow tract. SAM may be observed intermit-

tently, it may increase in, for example, post-extrasystolic complexes, or be evoked by amylnitrate. SAM may also be observed in other conditions, and false patterns should be recognised (Table 4.5). In addition to abnormal motion of the

Table 4.5 Systolic anterior motion (SAM) of anterior mitral valve leaflet

True SAM
1. Diagnosis
    Independent motion of aML from LV posterior wall
    On baseline before aortic valve closure
2. Clinical conditions
    Hypertrophic obstructive cardiomyopathy
    Fixed aortic stenosis
    Aortic incompetence
    High output states

False patterns
    Beam width artifact
    Mitral valve prolapse syndrome
    Large pericardial effusion (swinging heart syndrome)

mitral valve leaflets in systole, many patients also have abnormal valve motion in diastole reflecting a decreased filling rate of the LV. Because of its noninvasive nature, echocardiography can be used repetitively to study the long term effects of drugs and surgery.[85] Three cases developing dilated LV and congestive heart failure after myotomy, myectomy and propranolol therapy have been documented.[15]

Coronary artery disease

There are significant problems with M-mode echocardiography in assessing the LV in patients with coronary artery disease since the technique does not allow examination of the entire ventricle. The small area of the LV transsected by the M-mode beam represents a significant limitation in assessing left ventricular structure and function in the segmentally diseased LV.[25] In the areas recorded, however, wall motion and thickening abnormalities (a specific marker of ischaemic myocardium) can be detected with a high degree of sensitivity[48, 95] (see page 127 and Figs. 4.34, 4.43 and 4.44).

Two-dimensional echocardiography adds significantly to the ability to examine areas of the LV not commonly seen with the M-mode technique such as the lateral and medial walls and the cardiac

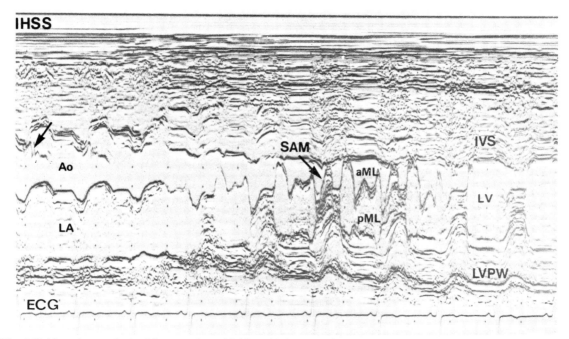

Fig. 4.42 M-mode scan obtained from a patient with idiopathic hypertrophic subaortic stenosis. The interventricular septum (IVS) is much thicker than the left ventricular posterior wall (LVPW). The anterior mitral valve leaflet demonstrates a typical systolic anterior motion (SAM) which further narrows the already narrowed outflow tract and causes obstruction to left ventricular ejection. There is a midsystolic closure of the aortic valve (arrow). Motion patterns of valves are easier recognised from M-mode recordings. Ao: aorta; aML and pML: anterior and posterior mitral valve leaflets; LA: left atrium; LV: left ventricle.

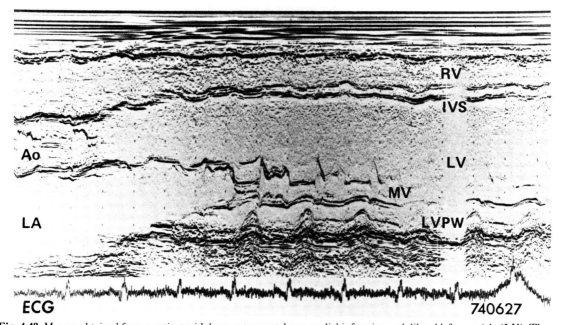

Fig. 4.43 M-scan obtained from a patient with large anteroseptal myocardial infarction and dilated left ventricle (LV). The interventricular septum (IVS) is akinetic and does not contract during systole. The left ventricular posterior wall (LVPW) is hypokinetic. The findings are consistent with ischaemic cardiomyopathy. The left atrium (LA) is dilated. Ao: aorta; MV: mitral valve; RV: right ventricle; RVOT: right ventricular outflow tract.

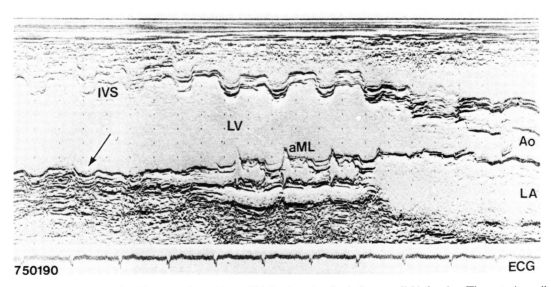

Fig. 4.44 M-mode scan obtained from a patient with an old inferolateral and apical myocardial infarction. The posterior wall of the left ventricle (LV) bulges posteriorly at the level of the mitral valve. The observed anterior movement of the posterior wall in that area is probably passive motion as a result of total cardiac displacement in systole which, therefore, might be akinetic. Scanning lower in the direction of the apex demonstrates paradoxic motion of the posterior wall (see arrow). Note the exaggerated amplitude of septal (IVS) motion (12 mm) with normal systolic thickening. The left ventricle (LV) is dilated (about 75 mm) and mitral valve closure is delayed and interrupted by a B-notch. These findings indicate an elevated end-diastolic LV pressure as does the dilated left atrium. Ao: aorta; aML: anterior mitral valve leaflet.

apex. Preliminary data suggested that two-dimensional echocardiography can be utilised to identify and quantify regional motion abnormalities in a way similar to LV cineangiocardiography.[49] The major limitation of the method remains the low success rate in obtaining good quality images, poor definition of the endocardium and the difficulty of studying the LV during exercise. Several approaches are now under investigation to estimate functional infarct size by studying the amount of muscle moving abnormally, wall thinning and regional myocardial dilatation. [23, 42] The assessment of global LV function remains subjective and semiquantitive (see page 130). While the left main coronary artery can often be visualised by cross-sectional echocardiography, [71, 99] obstructive lesions cannot be reliably predicted at present. The great potential of echocardiography in patients with coronary artery disease lies in the detection of complications of myocardial infarction such as pericardial effusion, rupture of the IVS,[16, 24] left ventricular thrombus[1, 59] (Fig. 4.45), right ventricular infarction[86] and papillary muscle rupture.[65] In patients with complications of chronic coronary artery disease, aneurysm formation is frequent. Two-dimensional echocardiography has proved to be sensitive in determining the site and extent of these aneurysms in a manner similar to angiocardiography especially when using apical views.[100] False aneurysms are also reliably detected echocardiographically.[29, 78]

Congenital heart disease

Echocardiography has many applications in the evaluation of congenital heart disease but its use is complex. A detailed discussion of echocardiography in congenital heart disease is not presented here, but only some general aspects pertinent to the diagnosis of common types of complex structural abnormalities. Many of the noncyanotic types of congenital heart disease, such as valvular lesions, left-to-right shunts and cardiomyopathies are discussed earlier in this chapter.

Transposition of the great arteries (TGA) represents the commonest form of complex cyanotic heart disease in newborns. Great vessel orientation is best assessed in the parasternal short axis view. In TGA, both great vessels are seen as tubular structures with the aorta being displaced

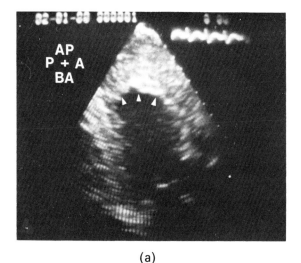

(a)

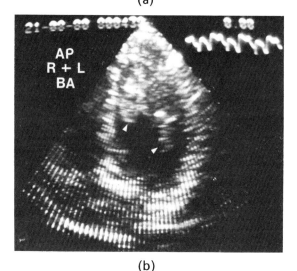

(b)

Fig. 4.45 (a) Apical long axis view of patient with an ischaemic aneurysm of the apical and antero-septal area and a large thrombus. Note distinct margin of thrombus (see arrowheads). (b) Apical four chamber view of patient with recent inferior myocardial infarction and systemic embolisation. There is a massive thrombus with parts loosely hanging in the left ventricular cavity (see arrowheads). This type of thrombus represents a significant risk for embolisation as compared to the more organised thrombus seen in (a).

anteriorly. The demonstration of a bifurcation of one of the great vessels proves it to be the pulmonary artery. Demonstration of the origin of the coronary arteries proves that particular artery to be the aorta. Contrast echocardiography allows the examiner to follow the blood flow through the right-sided heart cavities. When using parasternal and apical views, the presence of right-to-left shunting can be detected and its level determined.

The combination of parasternal and suprasternal notch views is extremely helpful in demonstrating the great vessel orientation because of the constant anatomical relationship of the aortic arch, right pulmonary artery and left atrium from the suprasternal position. By observing the pattern of opacification it is possible to demonstrate the level of right-to-left shunting and to determine which artery receives most of the systemic venous return.[68] Other conotruncal abnormalities which are readily detected are tetralogy of Fallot, where an enlarged RV with the Ao overriding the IVS is seen, and truncus arteriosus where a single large vessel is seen as a single circle. No great vessel can be seen originating from the LV in any view in the presence of double outlet RV.[40] The apical four-chamber view is particularly useful when complex lesions are present.[87] It uniquely allows for simultaneous visualisation of all four cardiac chambers, the IVS, the interatrial septum and both atrioventricular valves. Because of the configuration of the endocardial cushion, normally the tricuspid valve is situated to the patient's right and inserts more apically than does the mitral valve. With ventricular inversion, the AV valves are also inverted and the lower-inserted tricuspid valve is seen to the patient's left. Downward displacement of tricuspid valve insertion on the IVS is typical of Ebstein's anomaly.[74] In patients with atrial septal defect of the primum type, the lower part of interatrial septum is missing. Despite good visualisation of the interatrial septum, however, even the presence of a secundum type atrial septal defect is difficult to diagnose, since dropouts of a normal interatrial septum frequently occur. Features of RV volume overload should be present and contrast studies are diagnostic.[2, 28]

In patients with AV canal, there is absence of the endocardial cushion and therefore a common level of insertion[39] (Fig. 4.46). Overriding and straddling tricuspid or mitral valves,[51] tricuspid atresia,[83] and absence of IVS with a common ventricle can also be diagnosed.[82]

ACKNOWLEDGEMENTS

I am grateful to the following for permission to reproduce illustrations: Research Studies Press (Fig. 4.3),[76] and the Editor of the *Journal of Clinical Ultrasound* (Fig. 4.38).[77] I thank Machtelt Brussé for her much appreciated help in the preparation of the manuscript of this chapter.

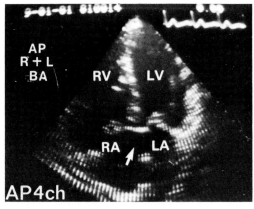

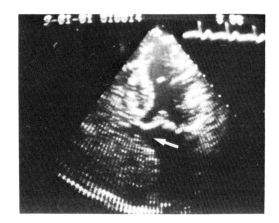

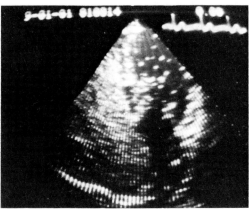

Fig. 4.46 Stop-frame images of apical four-chamber views obtained from a patient with an atrial septal defect (type primum) indicated by the arrow in the top left photograph. The bottom left photograph shows opacification of the right-sided cavities after a peripheral venous injection of 5 ml of dextrose 5 per cent in water. Echo-contrast appears in the left ventricle (LV) proving a right-to-left shunt. The top right photograph shows the negative contrast effect when noncontrast blood flows from the left atrium (LA) into the right atrium (RA) (see arrow). RV: right ventricle.

REFERENCES

1. Asinger A W, Mikell F L, Sharma B, Hodges M 1981 Observations on detecting left ventricular thrombus with two-dimensional echocardiography: emphasis on avoidance of false positive diagnosis. Am J Cardiol 47:145–156
2. Assad-Morell J C, Tajik A J, Giuliani E R 1974 Echocardiographic analysis of the interventricular septum. Prog Cardiovasc Dis 17:219–237
3. Baker D W 1978 The present role of Doppler techniques in cardiac diagnosis. Prog Cardiovasc Dis 21:79–91
4. Bastiaans O L, Meltzer R S, Vogel J A, Verbeek P W, Roelandt J 1981 Quantification from two-dimensional echocardiographic images. In: Rijsterborgh H (ed) Echocardiology. Nijhoff, The Hague, p 131–143
5. Bierman F Z, Williams R G 1979 Subxiphoid two-dimensional imaging of the interatrial septum in infants and neonates with congenital heart disease. Circulation 60:80–90
6. Bom N, Lancée C T, Zwieten G van, Kloster F E, Roelandt J 1973 Multiscan echocardiography. I. Technical description. Circulation 48:1066–1074
7. Bommer W, Neef J, Neumann A, Weinert L, Lee G, Mason D T, DeMaria A N 1978 Indicator-dilution curves obtained by photometric analysis of two-dimension echo contrast studies. Am J Cardiol 41:370
8. Bommer W, Weinart L, Neumann A, Neff J, Mason D T, DeMaria A 1979 Determination of right atrial and right ventricular size by two-dimensional echocardiography. Circulation 60:91–100
9. Brandestini M, Howard A, Eyer M, Stevenson J, Weiler T 1979 Visualisation of intracardiac defects by M/Q-mode echo-Doppler ultrasound. Circulation suppl II:59–60
10. Brodie B R, Grossman W, McLaevien L, Starek P J K, Craige E 1976 Diagnosis of prosthetic mitral valve malfunction with combined echo-phonocardiography. Circulation 53:93–100
11. Brower R W, Dorp W G van, Vogel J A, Roelandt J 1975 An improved method for the quantitative analysis of M-mode echocardiograms. Eur J Cardiol 3:171–179
12. Carr K W, Engler R L, Forsythe J R 1979 Measurement of left ventricular ejection fraction by mechanical cross-sectional echocardiography. Circulation 59:1196–1206
13. Cate F J ten, Dorp W G van, Hugenholtz P G, Roelandt J 1979 Fixed subaortic stenosis: value of echocardiography for diagnosis and differentiation between various types. Br Heart J 41:159–166
14. Cate F J ten, Kloster F E, Dorp W G van, Meester G I, Roelandt J 1974 Dimensions and volumes of left atrium and ventricle by single beam echocardiography. Br Heart J 36:737–746
15. Cate F J ten, Roelandt J 1979 Progression to left

ventricular dilation in patients with hypertrophic obstructive cardiomyopathy. Am Heart J 97:762–765

16. Chandraratna P A N, Balachandran P K, Shah P M, Hodges M 1975 Echocardiographic observations in ventricular septal rupture complicating acute myocardial infarction. Circulation 51:506–510

17. Chang S, Feigenbaum H 1973 Subxiphoid echocardiography. J Clin Ultrasound 1:14–20

18. Clark C E, Henry W L, Epstein S E 1973 Familial prevalence and genetic transmission of idiopathic hypertrophic subaortic stenosis. N Eng J Med 298:709– 714

19. Corya B C, Feigenbaum H, Rasmussten S, Black M J 1974 Echocardiographic features of congestive cardiomyopathy compared with normal subjects and patients with coronary artery disease. Circulation 49:1153–1159

20. DeMaria A N, Miller R R, Amsterdam E A, Markson W, Mason D T 1976 Mitral valve early diastolic closing velocity in the echocardiogram: relation to sequential diastolic flow and ventricular compliance. Am J Cardiol 37:693–700

21. Dorp W G van, Cate F J ten, Vletter W B, Dohmen N, Roelandt J 1976 Familial prevalence of asymmetric septal hypertrophy. Eur J Cardiol 4:349–357

22. Eaton L W, Maughan W L, Shoukas A A, Weiss J L 1979 Accurate volume determination in isolated ejecting canine left ventricle by two-dimensional echocardiography. Circulation 60:320–326

23. Eaton L W, Weiss J L, Bulkley B H, Garrison J B, Weisfeldt M L 1979 Regional cardiac dilation after acute myocardial infarction: recognition by two-dimensional echocardiography. New Engl J Med 300:57–62

24. Farcot J C, Boisante L, Rigaud H, Bardet J, Bourdarias J P 1980 Two-dimensional echocardiographic visualisation of ventricular septal rupture after acute anterior myocardial infarction. Am J Cardiol 45:370–377

25. Feigenbaum H 1975 Echocardiographic examination to the left ventricle. Circulation 51:1–7

26. Feigenbaum H 1981 Echocardiography. 3rd edn. Lea & Febiger, Philadelphia

27. Folland E D Parisi A F, Moynihan P F, Jones D R, Feldman C L, Tow D E 1979 Assessment of left ventricular ejection fraction and volumes by real-time, two-dimensional echocardiography: a comparison of cineangiographic and radionuclide techniques. Circulation 60: 760–766

28. Fraker T D, Harris P J, Behar V S, Kisslo J A 1979 Detection and exclusion of interatrial shunts by two-dimensional echocardiography and peripheral venous injection. Circulation 59:379–384

29. Gatewood R P, Nanda N C 1980 Differentiation of left ventricular pseudoaneurysm from true aneurysm with two-dimensional echocardiography. Am J Cardiol 46:869–878

30. Gibson D G, Brown D J 1973 Measurement of instantaneous left ventricular dimension and filling rate in man, using echocardiography. Br Heart J 35:559

31. Gibson D G, Brown D J 1976 Assessment of left ventricular systolic function from simultaneous echocardiographic and pressure measurements. Br Heart J 38:8–17

32. Gilbert B W, Haney R S, Crawford F, McClellan J, Gallis H A, Johnson M L, Kisslo J A 1977 Two-dimensional echocardiographic assessment of vegetative endocarditis. Circulation 55:346–353

33. Gilbert B W, Schatz R A, Ramm O T von, Behar V S, Kisslo J A 1976 Mitral valve prolapse: two-dimensional and angiocardiographic correlation. Circulation 54:716– 723

34. Goldberg B B 1971 Suprasternal ultrasonography. JAMA 215:245–250

35. Gramiak R, Shah P M, Kramer D H 1969 Ultrasound cardiography: contrast studies in anatomy and function. Radiology 92:939–948

36. Griffith J M, Henry W L 1974 A sector scanner for real time two-dimensional echocardiography. Circulation 49:1147–1152

37. Griffith J M, Henry W L 1978 An ultrasound system for combined cardiac imaging and blood flow measurement in man. Circulation 57:925–930

38. Haaz W S, Mintz G S, Kotler M N, Parry W, Segal B L 1980 Two-dimensional echocardiographic recognition of the descending thoracic aorta: value in differentiating pericardial from pleural effusions. Am J Cardiol 46:739– 743

39. Hagler D J, Tajik A J, Seward J B, Mair D D, Ritter D G 1979 Real-time wide-angle sector echocardiography: atrioventricular canal defects. Circulation 59:140–150

40. Hagler D J, Tajik A J, Seward J B, Mair D D, Ritter D G 1980 Wide-angle two-dimensional echocardiographic profiles of conotruncal abnormalities. Mayo Clin Proc 55:73–82

41. Hagler D J, Tajik A J, Seward J B, Mair D D, Ritter D G, Ritman E L 1978 Videodensitometric quantitation of left-to-right shunts with contrast sector echocardiography. Circulation suppl II:57–58

42. Heger J J, Weyman A E, Wann L S, Dillon J C, Feigenbaum H 1979 Cross-sectional echocardiography in acute myocardial infarction: detection and localization of regional left ventricular asynergy. Circulation 60:531–538

43. Henry W L, Clark C E, Epstein S E 1973 Assymetric septal hypertrophy (ASH): the unifying link in the IHSS disease spectrum. Circulation 47:827–832

44. Henry W H, DeMaria A, Gramiak R, et al 1980 Report of the American Society of Echocardiography committee on nomenclature and standards in two-dimensional echocardiography. Circulation 60:212–217

45. Horowitz M S, Schultz C S, Stinson E B, Harrison D C, Popp R L 1974 Sensitivity and specificity of echocardiographic diagnosis of pericardial effusion. Circulation 50:239–247

46. Johnson M L 1981 Echocardiography of prosthetic mitral valves. In: Vlieger M de et al (eds) Handbook of clinical ultrasound. Wiley, New York, p 479–486

47. Kerber R E, Kioschos J M, Lauer R M 1974 Use of an ultrasonic contrast method in the diagnosis of valvular regurgitation and intracardiac shunts. Am J Cardiol 34:722–729

48. Kerber R E, Marcus M L, Abboud F M 1977 Echocardiography in experimentally induced myocardial ischaemia. Am J Med 63:21–28

49. Kisslo J A, Robertson D, Gilbert B W, Ramm O T von, Behar V S 1977 A comparison of real-time two-dimensional echocardiography and cine-angiography in detecting left ventricular asynergy. Circulation 55:134– 141

50. Kronik G, Mösslacher H, Schmoliner R, Hutterer B 1980 Kontrastechokardiographie bei Patienten mit kleinen interatrialen Kurzschlussverbindungen (Offnes Foramen ovale). Wien Klin Wochenschrift 92:290–293

51. LaCorte M A, Fellows K E, Williams R G 1976 Overriding tricuspid valve: echocardiographic and angiographic features. Am J Cardiol 37:911–919

52. Lieppe W, Behar V S, Scallion R, Kisslo J A 1978 Detection of tricuspid regurgitation with two-dimensional echocardiography and peripheral vein injections. Circulation 57:128–132

53. Ligtvoet C M, Ridder J, Lancée C T, Hagemeijer F, Vletter W B 1977 A dynamically focused multiscan system. In: Bom N (ed) Echocardiology. Nijhoff, The Hague, p 313–324

54. Markiewicz W, Stoner J, London E, Hunt S A, Popp R L 1976 Mitral valve prolapse in one hundred presumably healthy young females. Circulation 53:464–473

55. Martin R P, Rakowski H, French J, Popp R L 1978 Localisation of pericardial effusion with wide angle phased array echocardiography. Am J Cardiol 42:904–912

56. Martin R P, Rakowski H, French J W, Popp R L 1979 Idiopathic hypertrophic subaortic stenosis viewed by wide-angle phased array echocardiography. Circulation 59:1206–1217

57. Martin R P, Rakowski H, Kleiman J H, Beaver W, London E, Popp R L 1979 Reliability and reproducibility of two-dimensional echocardiographic measurement of the stenotic mitral valve orifice area. Am J Cardiol 43:560–568

58. Mathey D G, Decoodt P R, Allen H N, Swan H F C 1976 The determinants of onset of mitral valve prolapse in the systolic click-late systolic murmur syndrome. Circulation 53:872–878

59. Meltzer R S, Guthaner D, Rakowski H, Popp R L, Martin R P 1979 Diagnosis of left ventricular thrombi by two-dimensional echocardiography. Br Heart J 42:261–265

60. Meltzer R S, Hoogenhuyze D van, Serruys P W, Halebos M, Roelandt J 1981 Diagnosis of tricuspid regurgitation by contrast echocardiography. Circulation 63:1093–1099

61. Meltzer R S, Martin R P, Robbins B S, Popp R L 1980 Mitral annular calcification: clinical and echocardiographic features. Acta Cardiol 35:189–202

62. Meltzer R S, Meltzer C, Roelandt J 1980 Sector scanning views in echocardiography: a systematic approach. Eur Heart J 1:379–394

63. Meltzer R S, Serruys P W, McGhie J, Verbaan N, Roelandt J 1980 Pulmonary wedge injections yielding left-sided echocardiographic contrast. Br Heart J 4:390–394

64. Meltzer R S, Tickner E G, Sahines T P, Popp R L 1980 The source of ultrasonic contrast effect. J Clin Ultrasound 8:121–127

65. Mintz G S, Kottler M N, Segal B L, Parry W R 1978 Two-dimensional echocardiographic recognition of ruptured chordae tendinae. Circulation 57:244–250

66. Mintz G S, Kotler M N, Segal B L, Parry W R 1979 Comparison of two-dimensional and M-mode echocardiography in the evaluation of patients with infective endocarditis. Am J Cardiol 43:738–744

67. Mintz G S, Kotler M N, Segal B L, Parry W R 1979 Two-dimensional echocardiographic evaluation of patients with mitral insufficiency. Am J Cardiol 44:670–678

68. Mortera C, Hunter S, Tynan M 1979 Contrast echocardiography and the suprasternal approach in infants and children. Europ J Cardiol 19:437–454

69. Nanda N C, Gramiak R, Robinson T 1975 Echocardiographic evaluation of pulmonary hypertension. Circulation 50:575–581

70. Nichol P M, Gilbert B W, Kisslo J A 1977 Two-dimensional echocardiographic assessment of mitral stenosis. Circulation 55:120–128

71. Ogawa S, Chen C C, Hubbard F E, Pauletto F J, Mardelli T J, Morganroth J, Dreifus L S, Akaishi M, Nakamura Y 1980 A new approach to visualize the left main coronary artery using apical cross-sectional echocardiography. Am J Cardiol 45:301–304

72. Phillips D J, Blackshear W M, Baker D W, Strandness D E 1978 Ultrasound duplex scanning in peripheral vascular disease. Radiology/Nuclear Medicine Jan–Feb:6

73. Popp R L, Fowles R, Coltart J, Martin R P 1979 Cardiac anatomy viewed systematically with two-dimensional echocardiography. Chest 75:579–585

74. Ports T A, Silverman N H, Schiller N B 1978 Two-dimensional echocardiographic assessment of Ebstein's anomaly. Circulation 58:336–343

75. Ramm O T von, Thurstone F L 1976 Cardiac imaging using a phased array ultrasound system I. System design. Circulation 53:258–262

76. Roelandt J 1977 Practical echocardiology. Research Studies Press, Forest Grove

77. Roelandt J, Bom N, Hugenholtz P G 1980 The ultrasound cardioscope: a hand-held scanner for real-time cardiac imaging. J Clin Ultrasound 8:221–225

78. Roelandt J, Brand M van den, Vletter W B, Nauta J, Hugenholtz P G 1975 Echocardiographic diagnosis of pseudoaneurysm of the left ventricle. Circulation 52:466–472

79. Roelandt J, Dorp W G van, Bom N, Laird J D, Hugenholtz P G 1976 Resolution problems in echocardiology: a source of interpretation errors. Am J Cardiol 37:256–262

80. Schiller N B, Acquatella H, Ports T A, Drew D, Goerke J, Ringerts H, Silverman N H, Brundage B, Botvink E H, Boswell R, Carlsson E, Parmley W W. 1979 Left ventricular volume from paired biplane two-dimensional echocardiography. Circulation 60:547–555

81. Schnittger I, Bowden R E, Abrams J, Popp R L 1978 Echocardiography: pericardial thickening and constrictive percarditis. Am J Cardiol 42:388–395

82. Seward J B, Tajik A J, Hagler D J, Ritter D G 1977 Contrast echocardiography in single or common ventricle, Circulation 55:513–519

83. Seward J B, Tajik A J, Hagler D J, Ritter D G 1978 Echocardiographic spectrum of tricuspid atresia. Mayo Clin Proc 53:100–112

84. Shapira J N, Martin R P, Fowles R E 1979 Two-dimensional echocardiographic assessment of patients with bioprosthetic valves. Am J Cardiol 43:510–519

85. Shapira J N, Stemple D R, Martin R P, Rakowski H, Stinson E B, Popp R L 1978 Single and two-dimensional echocardiographic visualisation of the effects of septal myectomy in idiopathic hypertrophic subaortic stenosis. Circulation 58:850–860

86. Sharpe N S, Botvinick E H, Shames D M, Schiller N B, Massie M B, Chatterjee K, Parmlee W W 1978 The noninvasive diagnosis of right ventricular infarction. Circulation 57:483–490

87. Silverman N H, Schiller N B 1978 Apex echocardiography. A two-dimensional technique for

evaluating congenital heart disease. Circulation 57:503–511

88. Siqueira-Filho A G, Cunha C L P, Tajik A J, Seward J B, Schattenberg T T, Giuliani E R 1981 M-mode and two-dimensional echocardiographic features cardiac amyloidosis. Circulation 63:188–196

89. Stevenson G, Brandestini M, Weiler T, Howard A, Eyer M 1981 Digital multigate Doppler with color echo and Doppler display-diagnosis of atrial and ventricular septal defects. Circulation suppl II:59–69

90. Tajik A J, Seward J B, Hagler D J, Mair D D, Lie J T 1978 Two-dimensional real time ultrasonic imaging of the heart and great vessels. Technique, image orientation, structure identification and validation. Mayo Clinic Proc 53:271–303

91. Taylor K J W 1974 Current status of toxicity investigations. J Clin Ultrasound 2:149–156

92. Tickner E G, Rasor N S 1978 Noninvasive assessment of pulmonary hypertension using bubble ultrasonic resonance pressure (BURP) method. NIH report HR-62917-2A. National Institutes of Health, Bethesda

93. Ulrich W D 1974 Ultrasound dosage for experimental use on human beings. IEEE Trans Biomed Engng BME-21:48–51

94. Venco A, Gibson D G, Brown D J 1977 Relation between the apexcardiogram and changes in left ventricular pressure and dimension. Br Heart J 39:117–125

95. Wann L S, Paris J V, Childress R H, Dillon J C, Weyman A E, Feigenbaum H 1979 Exercise cross-sectional echocardiography in ischemic heart disease. Circulation 60:1300–1308

96. Weiss J L, Weisfeldt M L, Mason S J, Garrison J B, Livengood S V, Fortuin N J 1979 Evidence of Frank-Starling effect in man during severe semi-supine exercise. Circulation 59:655–661

97. Weyman A C, Dillon J C, Feigenbaum H, Chang S 1974 Echocardiographic patterns of pulmonic valve motion in valvular pulmonic stenosis. Am J Cardiol 34:644–651

98. Weyman A E, Feigenbaum H, Dillon J S, Chang S 1975 Cross-sectional echocardiography in assessing the severity of valvular aortic stenosis. Circulation 52:828–834

99. Weyman A E, Feigenbaum H, Dillon J C, Johnston K W, Eggleton R C 1976 Non-invasive visualization of the left coronary artery by cross-sectional echocardiography. Circulation 54:169–174

100. Weyman A E, Peskoe S M, Williams E S, Dillon J C, Feigenbaum H 1976 Detection of left ventricular aneurysms by cross-sectional echocardiography. Circulation 54:936–944

101. Wyatt H L, Heng M K, Meerbaum S, Davidson R, Corday E 1978 Evaluation of models for quantifying ventricular size by two-dimensional echocardiography. Am J Cardiol 41:369

5

Arteries and veins

J. P. Woodcock and R. N. Baird

INTRODUCTION

Ultrasonic investigation now plays an important role in the diagnosis of extracranial carotid disease and in the diagnosis of occlusive arterial disease of the lower limb. Although blood vessels were visualised ultrasonically as early as 1957, [12] it is only since the development of Doppler ultrasound techniques that any major advance has been made in the diagnosis and management of patients with systemic arterial disease. The major applications of Doppler ultrasound are based on the study of blood flow and it is necessary to describe the type of equipment now available in order to appreciate the clinical information obtainable. There are basically four types of equipment which are used for studies of blood vessels and blood flow. As explained in Chapter 1, these are the simple Doppler flowmeter, the Doppler imaging system, the real-time two-dimensional B-scanner and the duplex scanner.

INSTRUMENTATION

Doppler flowmeter

When an ultrasonic wave is reflected or scattered by a moving object, the signal returning to the receiving transducer is shifted in frequency. If the ultrasound beam is inclined at an angle γ to the direction of the flow of blood, then (according to Equation 1.8) the Doppler shift frequency f_D is given by:

$$f_D = 2v \, (\cos \gamma) \, f/c \qquad (5.1)$$

where f is the transmitted frequency, c is the speed of sound in blood and v is the speed of the blood. It can be seen that the Doppler shift, for a given transmitted frequency, is proportional to the velocity of the blood, and the angle of inclination γ.

The basic Doppler instrument is a continuous wave device, which means that the emitting transducer is continuously excited. The scattered wave is received by a second transducer and the instantaneous frequency difference between the emitted and received waves is measured. This Doppler-shift frequency varies both in time, over the cardiac cycle, and in position, across the lumen of the blood vessel. An example of such an instrument is shown in Figure 1.24.

The Doppler-shift signal contains information about the way blood moves in a blood vessel. Two questions which must be asked are, firstly, 'What useful information is contained within this signal?' and, secondly, 'Is it possible to study physiological and pathological processes using Doppler-shift flowmeters?' The first of these questions can be answered relatively easily by a consideration of the physical properties of blood flowing in a vessel. The blood cells scatter the incident ultrasound waves and, because the cells are moving, a Doppler frequency shift is detected at the receiving transducer. Consequently the information contained within this signal relates to the distribution of blood cell velocities within the blood vessel and how this distribution varies with time over the cardiac cycle. These parameters themselves are affected by the diameter and elastic properties of the blood vessel, and the impedance generated by the arterial system. Therefore, in theory, it should be possible to attempt to provide some answers to the second question about the possibility of using Doppler ultrasound to study physiological and pathological changes in the circulation.

The continuous wave flowmeter obtains blood flow information from any blood vessel which intersects the ultrasonic beam from the transmitter, but no depth information is available. If the transmitting transducer is pulsed, as explained on page 28, it is possible to determine from what part of the blood vessel the received Doppler signal comes. It is also possible, in practice, to produce a pulsed Doppler flowmeter with a number of range gates, each capable of processing a Doppler signal. (see page 9). In this way the velocity profile across the vessel may be measured and the flow estimated.

Doppler imaging systems

The simplest Doppler imaging system consists of a continuous-wave Doppler flow detector, a storage monitor to display the image, and the probe position resolver (see page 28). The position resolver is mechanically linked to the Doppler probe and delivers signals to the storage monitor. When a Doppler shift signal is detected, a bright spot, corresponding to the probe position, is stored on the monitor. As the probe is moved over the skin overlying the blood vessel an image of the projection of the blood vessel on the skin surface is generated on the screen. The limitation of this simple continuous wave system is that only one projection can be obtained. In order to study the three-dimensional flow image it is necessary to pulse the transmitted ultrasonic signal, and time-gate the received signal.

Pulsed Doppler imaging systems require a position computer which combines the probe position signals, and the signals corresponding to the position of the detected flow point in front of the transducer. It is then possible to produce images in three orthogonal planes of the internal lumen of the blood vessel. A block diagram of a multi-channel Doppler imaging system is shown in Figure 5.1. Three orthogonal projections of the Doppler data from a carotid artery are shown in Figure 5.2.

Real-time two-dimensional B-scanners

The principles of instruments for real-time two-dimensional B-scanning are described on page 20. For imaging arteries and veins, fast mechanical scanners are generally preferred at present. They

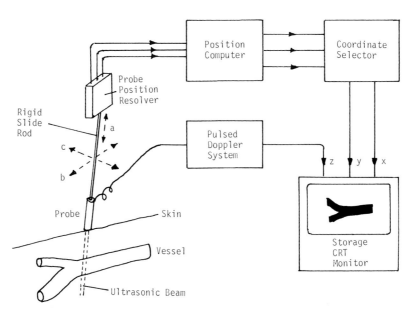

Fig. 5.1 Block diagram of multigate Doppler imaging system. The position of the probe on the surface of the patient and the direction of the ultrasonic beam are measured by the probe position resolver in terms of voltages corresponding to the displacements a, b and c. These voltages are fed to the position computer and the coordinate selector to produce x and y deflection voltages corresponding to the desired scan plane for display on the storage cathode ray tube (CRT) monitor. The presence or absence of a Doppler signal at the depth of interest controls the brightness (z-modulation) of the CRT.

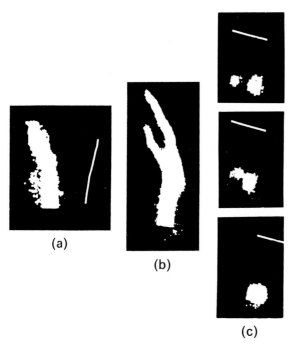

(a)

(b)

(c)

Fig. 5.2 Three orthogonal projections of the carotid bifurcation produced using a range-gated Doppler imaging system. (a) Lateral projection. (b) Antero-posterior projection. (c) Cross sections: top, above bifurcation; middle, at bifurcation; bottom, below bifurcation. The thin lines in (a) and (c) correspond to the skin surface.

give images with higher line densities than are easily obtainable with array scanners, and this is important when the presence of small plaques and other abnormalities is suspected.

Duplex scanners

Duplex scanners are a combination of a real-time B-scanner and a direction-resolving, pulsed, range-gated Doppler flowmeter (see page 28). Instruments such as these produce a real-time B-scan image of the area of interest, and the sample volume of the Doppler flowmeter can be placed at a particular point of interest within the image plane.

A typical system operates at 5 MHz and consists of three transducers mounted on a wheel which rotates continuously, sweeping out new image frames at a rate of 30 per second. The image can be stored, and one of the transducers used as a gated pulsed Doppler transducer to sample the blood

velocity at a particular point in the vessel. The position of the Doppler beam is indicated by a white line on the display screen, and the position of the sample volume is shown as a white spot on this line. The position of the sample volume and the direction of the Doppler ultrasound beam are adjusted using an arm connected by means of a servomechanism to the transducer.

CLINICAL APPLICATIONS

The area where Doppler ultrasound plays a major diagnostic role are in investigations of atherosclerosis of the carotid arteries, aneurysms of the abdominal aorta, and in occlusive disease of the lower limb arteries. There is a limited role in the detection of deep vein thrombosis. As previously mentioned, there are three ways by which information can be obtained using ultrasound techniques. Firstly, the variation in blood velocity over the cardiac cycle may be measured. Secondly, the Doppler image of the blood vessel may be studied. Thirdly, the blood vessel may be visualised by real-time two-dimensional scanning. In studies of the lower limb arteries, there is a fourth technique, that of systolic blood pressure measurement, which plays a major role in diagnosis.

Lower limb

Blood pressure measurement

Two major techniques have been developed, using Doppler ultrasound, to measure systolic blood pressure. In the first, the movement of the arterial wall is detected and, in the second, the Doppler signal is obtained from blood moving under the sphygmomanometer cuff.

Ultrasound kinetoarteriography. Two transducers are incorporated into an ordinary sphygmomanometer cuff which is placed around the limb in such a way that the transducers are over the artery.[26] The cuff is inflated and slowly deflated until the movement of the artery wall is first detected as the artery opens. A comparison[23] of the results of this method with intra-arterial pressure measurements gave the very good correlation coefficient of 0.98.

Doppler sphygmomanometry. This method, in

essence, replaces the conventional stethoscope with a Doppler blood flow detector.[33] The technique is similar to the conventional method for measuring blood pressure except that diastolic pressure cannot be determined in this way. The systolic pressure has, however, been found to be a useful index of the severity of arterial disease in the lower limb. An important point to note is that in a normal supine limb the resting ankle systolic pressure is the same or higher than the brachial systolic pressure. Pressure measurements in vascular disease now play an important role in the investigation of the severity of disease.[7] These useful measurements have become deservedly popular because they are made easily, quickly and reliably and the rechargable, portable instruments are robust and inexpensive.

In general, pressure information is used in clinical diagnosis either as the absolute value at a particular site, or as the ratio or the difference between two sites. The term 'pressure index' is defined as the ratio of the ankle systolic pressure to the brachial systolic pressure. In normal individuals this index is greater than unity and in patients with occlusive arterial disease it is always less than unity. The results of a study of the correlation between the severity of the disease, assessed using the pressure index, and the patients' symptoms, are shown in Figure 5.3. The difference between symptoms of increasing severity is highly significant ($P < 0.01$). The question which thus arises is whether the absolute value of blood pressure at a particular site, or the pressure difference between two sites, or the pressure index, is most sensitive to the presence of arterial disease. The absolute value of ankle systolic pressure varies widely in patients with intermittent claudication because of the wide variability of central blood pressures encountered in patients with lower limb ischaemia. Arm systolic pressures range commonly from 110–220 mmHg, necessitating the use of the popular pressure index, or the perhaps more logical pressure difference. In

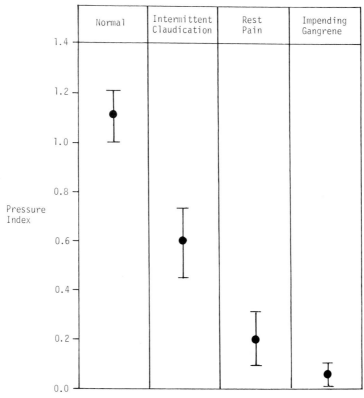

Fig. 5.3 Pressure index as an indicator of the severity of arterial disease of the lower limb.[31] ($P < 0.01$.)

a study[7] comparing the correlation between the maximum walking time before the onset of claudication with the pressure difference at rest and with the pressure index in a group of patients, the regression coefficients were -0.69 and 0.66 respectively. Another study[13] compared absolute values of ankle systolic pressure with pressure index and pressure difference and found that ankle systolic pressure measurements could only distinguish between complete arterial occlusion and a normal artery. Using either pressure differences or pressure index it was possible to distinguish occlusions and major stenoses from normal arteries. Despite their sensitivity in the case of foot ischaemia, ankle pressure measurements are unhelpful in localising the level of the dominant arterial lesion, whether the aortoiliac or femoropopliteal arterial segment is primarily affected. This problem is mentioned again in the next section.

Occlusive disease of lower limb arteries

Exercise-related pain. Ankle pressure measurements are helpful if the symptoms of intermittent claudication are atypical or do not fit with the appearance of the extremity and the presence of palpable arterial pulses. For example, the exercise-related symptoms may not be relieved promptly by rest, or there may be sensory changes of numbness or tingling and a history of low back pain suggestive of nerve root entrapment. If neurogenic claudication is suspected, ankle pressure measurements can help to decide on the appropriate invasive investigation (e.g., whether to suggest myelography or arteriography).

The limb arterial pulses may be only weakly palpable throughout at rest, suggesting incomplete stenosis of the iliac artery, or the pulses may be present at rest only to fade after exercise, as in entrapment of the popliteal artery by a muscular band. If the resting arterial pressures are nearly normal, occult arterial narrowing can be revealed by repeating the measurements after augmenting blood flow by exercise on a treadmill. During exercise, blood flow to the legs increases by up to ten times the resting value, and pathophysiological principles are similar to exercise provocation tests in angina pectoris and in respiratory diseases. The fall in ankle pressure and the recovery time — the time taken for the ankle pressure to be restored to its resting value — are sensitive indicators of arterial occlusions.

Improvement in foot perfusion following arterial reconstruction. Upon completion of a vascular anastomosis, a sterile Doppler pencil probe applied intraoperatively to the reconstruction and distal vessels can be used to confirm, through the evidence of undamped blood-velocity signals, that the reconstruction is working well. Obstruction of the anastomosis by thrombus, atheroma or by the intimal layer of the arterial wall can also sometimes be detected by inspection and palpation of the operative area. Other monitoring techniques include operative arteriography, electromagnetic flow metering and pulse volume recordings.

The risk of closure of arterial reconstructions is greatest within hours of operation, when the distal pulses may be difficult to palpate because of peripheral vasoconstriction. If, as is not infrequently the case, the peripheral circulation is 'shut down', the Doppler ankle pressure measurement at this stage is often reassuring. In one study[32] of more than 500 cases, successful arterial reconstruction in single segment disease (aortoiliac or superficial femoral) was followed by an immediate increase in pressure index to a value of approximately 1.0, or within six hours postoperatively. The patency of femoropopliteal and femorotibial grafts can be easily confirmed by insonating the graft with a Doppler pencil probe applied to the medial aspect of the thigh between the sutured surgical incisions.

As time passes the patency of arterial reconstructions can be put at risk by the development of a neointimal hyperplasia composed of fibrinplatelet aggregates and smooth muscle cells. The graft-artery anastomosis is particularly susceptible. Graft function can be restored by timely surgical intervention. These changes are sometimes detected clinically as a worsening of symptoms and confirmed by a fall-off in Doppler ankle pressures.

Ultrasound imaging techniques have been used in sequential studies of implanted arterial grafts.[3] Using a pulsed Doppler imaging system, the internal diameters of 21 arterial grafts, 14 aorto-iliofemoral and 7 axillobifemoral, were compared with those of 12 iliopopliteal grafts. All grafts at implantation were 10 mm internal diameter. The iliopopliteal grafts were reduced to 7.6 ± s.d. 0.4 mm at the inguinal level and to 6.3 ± s.d. 0.4 mm just above the popliteal anastomosis. No distal narrowing was found in any of the 14 aortofemoral grafts. Twenty-five internal diameter measurements of 14 iliopopliteal grafts were made up to 5 years after implantation. Sequential measurements in time were made on several grafts and popliteal narrowing was found to be progressive with time ($r = -0.54, P < 0.01$). The inguinal internal diameter of the axillobifemoral grafts was not reduced. No comparison has yet been made between the accuracies of Doppler imaging, duplex imaging and contrast arteriography in the lower limb. It is extremely unlikely, however, that these comparisons would differ considerably from similar studies of the carotid artery, reported later in this chapter. A direct comparison of the two ultrasonic systems in imaging a dacron graft in the lower limb is shown in Figure 5.4. Both images show a wide bore,

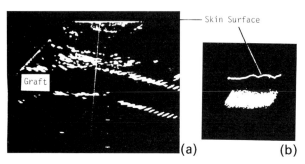

Fig. 5.4 Longitudinal ultrasound scans of dacron graft in lower limb artery. (a) Duplex scan. (b) Pulsed Doppler scan.

patent graft. Using this approach, it may be possible to estimate the growth of neointimal hyperplasia.[8]

Level of amputation. Measurement of the Doppler systolic pressure helps to select the level of amputation at which the blood supply to the skin is sufficiently good to ensure that healing occurs. The choice of amputation level usually rests between below or above the knee, the former being preferred but with a poorer blood supply. If a below-knee amputation fails to heal, the patient's already long hospital stay may be doubled together with increased hospital mortality, reinforcing the need for care in selecting the appropriate level.

If Doppler signals cannot be detected at the ankle, or if the calf systolic pressure is less than 40 mmHg, the skin below the knee may not heal and an above-knee amputation is indicated. An adequate blood supply is assured if the ankle pressure is more than 70 mmHg and there are reports of good results of below-knee amputations with ankle pressures in the range of 40–70 mmHg. For example, determination of the distal thigh pressure is very useful in determining the level of amputation.[32] In 31 major amputations for ischaemia, the presence of distal thigh pressures of 50 mmHg or greater successfully predicted healing of a below-knee stump in 91 per cent of patients. In the presence of failed femoropopliteal grafts, below-knee amputation was feasible in 16 out of 19 patients with a thigh pressure above 50 mmHg, suggesting that a failed femoropopliteal graft should not necessarily lead to above-knee amputation. Moreover, if the ankle systolic pressure at rest was greater than 40 mmHg, then skin ulceration healed.

Blood-velocity/time waveform over the cardiac cycle. The changes which occur in the shape of the blood-velocity/time waveform over the cardiac cycle, proximal to, distal to, and over an occlusion of the profunda femoris artery were first described qualitatively in 1967.[24] Since then, various attempts have been made to quantify waveform shape in occlusive disease of the lower limb. In 1972, it was shown how, by measuring the time delay between flow occurring at two points along the limb, combined with the calculation of the pulsatility index (PI) of the waveforms at each site and the damping factor (Λ), it is possible to differentiate between an occluded artery, generalised narrowing and localised plaques.[30] Pulsatility index was originally defined in terms of Fourier analysis, but the calculation was subsequently simplified by taking the PI at a particular site to be given by the equation:[11]

$$PI = (\text{peak-to-peak amplitude})/(\text{mean amplitude}) \quad (5.2)$$

using the measurements indicated in Figure 5.5. The damping factor Λ along a segment of artery is given by the equation:

$$\Lambda = (\text{PI at proximal site})/(\text{PI at distal site}) \quad (5.3)$$

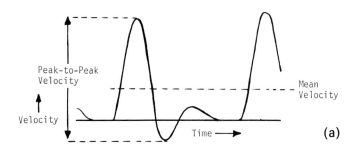

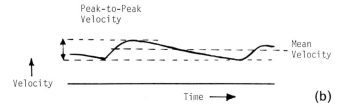

Fig. 5.5 Maximum blood-velocity/time waveforms from lower limb artery. (a) Normal. (b) Occlusive disease.

Correct differential diagnosis in 89 per cent of diseased lower limbs has been claimed using this modified technique.[11]

More recent developments[22, 21] derive a mathematical description of the blood-velocity/time waveforms in the lower limb. With this technique it is possible to derive coefficients from actual waveforms recorded from patients. These coefficients correlate with physiological variables such as vessel lumen size, distal impedance and vessel elasticity. This type of approach allows a physiological interpretation of a change in waveform shape. The blood-velocity/time waveform is digitised and the finite discrete Fourier transform is calculated. This allows three coefficients ω_0, γ and δ to be derived which relate to the physiological characteristics of the artery*. It has been shown that ω_0 is related to the elastic properties of the major conduit arteries, γ to the distal impedance

---

* The mathematics involves the evaluation of the Laplace transform $H(s)$ by means of a curve-fitting procedure:

$$H(s) = 1/[(s^2 + 2\delta\omega_0 s + \omega_0^2)(s + \gamma)]$$

The curve fitting is carried out on a small digital computer which calculates the values of the coefficients of the equation which give the best fit to the original frequency distribution. This relationship is usually expressed graphically by plotting the positions of the poles on the Argand diagram.

and δ to the radius of the iliofemoral arterial segment. The term δ is called the 'Laplace transform damping' (LTD). This LTD term is very sensitive to the presence of occlusive arterial disease.

Diagnosis of iliac artery stenosis

The sensitivities and specificities of pulsatility index (PI) and Laplace transform damping (LTD) in diagnosing stenoses of the iliac arteries have been compared using the receiver operator characteristic (ROC).[2] An estimate of the true-positive and false-positive rate in detecting stenotic lesions of greater than 50 per cent was obtained using stepwise threshold criteria of PI and LTD. The sensitivity and specificity of the two ultrasound tests were thus compared whilst avoiding reference to a single arbitrary value. LTD values correlated well with the severity of iliac artery stenosis ($r = -0.75$). There was a significant difference between minor stenoses (< 50 per cent diameter, LTD $= 0.5 \pm$ s.d. 0.15) and major stenoses (> 50 per cent, LTD $= 0.78 \pm$ s.d. 0.16) ($P < 0.02$). There was a highly significant difference between the normal volunteers and patients with minor iliac diseases ($P < 0.001$). LTD was not affected by an increased distal impedance and there was good correlation ($r = -0.73$) with iliac stenosis in the presence of an occlusion of the superficial femoral artery. Mean PI values for both minor (7.4 ± s.d. 4.2) and major (3.6 ± s.d. 1.6) stenoses were within the normal range. There was a good correlation with iliac stenosis if the distal vessels were patent ($r = 0.75$), but if the SFA was occluded, PI did not correlate with the presence or extent of the iliac disease ($r = 0.51$).

The ROC curves are shown in Figure 5.6. The LTD curve is 'better' than the PI, since it passes closer to perfect sensitivity and specificity. At an LTD value of 0.6, sensitivity was 85 per cent and specificity 84 per cent. PI at the lower limit of normal was insensitive in detecting iliac disease.

Doppler imaging

The iliac, common, superficial and profunda femoris arteries,[16, 17] the popliteal artery,[10] and the femoral artery bifurcation,[3] can all be imaged by Doppler scanning. In one study,[3] lateral, cross-sectional and antero-posterior scans were obtained of the common femoral, superficial femoral and profunda femoris arteries, and blood-velocity/time waveforms were recorded from the centre stream in each artery. Pulsatility indices and damping factors were then calculated. In 12 limbs in which

the profunda femoris origin was later seen to be normal at operation, the common femoral to profunda femoris damping factor was invariably less than 1.40 (mean 0.96 ± s.d. 0.05). In 13 limbs of normal patients without clinical evidence of atherosclerosis, the damping factor was less than 1.34 (mean 0.92 ± s.d. 0.05). In 8 limbs, at operation the profunda femoris was found to be completely occluded or severely stenosed (> 50 per cent). In these limbs the damping factor was always greater than 1.50 (mean 1.74 ± s.d. 0.08). Single plane arteriography of 7 limbs correctly identified all 3 complete occlusions, but only 1 of 4 severe stenoses.

Aneurysms of the abdominal aorta

The abdominal aorta can be outlined from the level of the diaphragm to its bifurcation and the common iliac arteries can be visualised to the level of the brim of the pelvis, by real-time B-scanning at an ultrasonic frequency around 3 MHz. Accurate measurements of the transverse outer diameter of aneurysms can be obtained; in the normal aorta, this dimension is in the range 12–18 mm. Enlarging aneurysms and those greater than 50 mm diameter are usually replaced by arterial dacron grafts because of the risk of rupture. When an aneurysm is managed conservatively, change in its size can

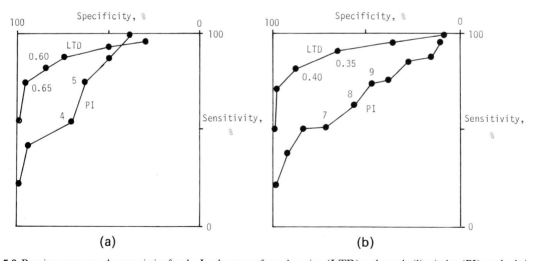

(a)   (b)

Fig. 5.6 Receiver operator characteristics for the Laplace transform damping (LTD) and pusalatility index (PI) methods in the assessment of arterial disease of the lower limb. (a) The accuracy with which LTD and PI can distinguish stenoses of less than 50 per cent from those of more than 50 per cent. (b) The accuracy with which LTD and PI can distinguish between normal vessels and those with less than 50 per cent stenosis.

be monitored by repeat scanning every six months or every year and the possibility of operation reconsidered. An example of a transverse scan of an aortic aneurysm is shown in Figure 5.7. Longitudinal scans only help to define the extent of an aneurysm. Other diagnostic methods include

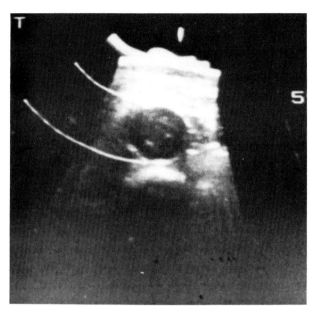

Fig. 5.7 Abdominal aortic aneurysm shown in transverse section. The diameter, measured by the electronic calipers, is 59.9 mm.

X-ray CT scans, which show aneurysms well, and X-ray contrast aortagrams, which define the important surgical relationship between the upper margin of the aneurysm and the renal arteries and which reveal the presence of distal arterial occlusive disease.

Carotid arteries

Clinical considerations

Atherosclerosis of the carotid bifurcation is an important cause of cerebral ischaemia and infarction. Premonitory signs of an impending stroke include transient ischaemic attacks or 'ministrokes', and a transient curtain-like loss of vision known as *amaurosis fugax*. A murmur may be discovered in the neck on auscultation of the carotid arteries.

The origin of the internal carotid artery is susceptible to localised atherosclerosis as a major stenosis resulting in cerebral underperfusion. Alternatively the atherosclerotic area ulcerates and releases fibrin-platelet thrombi and cholesterol microemboli which can lead to both transient and permanent neurological changes. Atherosclerosis at the carotid bifurcation is implicated as a cause of symptoms in less than half of the patient population affected by strokes and uncontrolled hypertension is an important risk factor. In those with carotid stenosis or ulceration, carotid endarterectomy can reduce the risk of stroke and X-ray contrast carotid angiography is the definitive preoperative investigation. In straight-forward, severe and worrying cases, angiography is proceeded to directly. In those with milder or atypical symptoms and those in whom evidence of carotid disease is felt desirable prior to angiography there are several useful noninvasive tests. Those not involving ultrasound include the indirect study of oculoplethysmography in which suction cups are applied to the sclera of the eyes to measure ophthalmic artery pressure and demonstrate underperfusion of the internal carotid artery territory distal to carotid stenosis. Direct tests in those with murmurs at the carotid bifurcation include analysis of the sounds of turbulence caused by flow disturbance as in carotid phonoangiography with spectral analysis of the recorded sounds. The ultrasound tests described in this section are the direct waveform analysis and imaging techniques which identify atheroma at the carotid bifurcation and the indirect periorbital Doppler techniques.

Blood-velocity/time waveform over the cardiac cycle

Typical maximum blood-velocity/time waveforms from the normal carotid arteries are shown diagrammatically in Figure 5.8. There is a relatively higher diastolic flow component in the internal carotid artery than in the common or the external.

The most consistent numerical index of changes with age and disease in the maximum frequency envelope recorded from the carotid arteries is the ratio A/B of the two peaks in systole.[5] If the A/B ratio in either the common carotid or in the ipsilateral supraorbital artery is less than 1.05 then there is an 88 per cent probability of disease at the carotid bifurcation. When the ratio is greater than

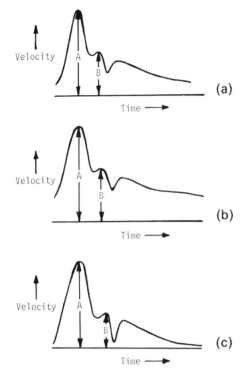

Fig. 5.8 Diagrammatic representation of the maximum blood-velocity/time waveforms detected from normal carotid arteries. (a) Common carotid artery. (b) Internal carotid artery. (c) External carotid artery. The amplitudes A and B correspond to the two systolic velocity maxima.

1.05 then there is an 80 per cent probability of a normal bifurcation.

Pattern recognition techniques such as principal component analysis (PCA) have been used successfully to determine important diagnostic features in the maximum frequency envelope.[15] It is interesting to note that PCA selected the A/B ratio as the characteristic of greatest significance, and further, that it is only necessary to measure this ratio from the common carotid artery. The technique is as sensitive to the presence of occlusive disease using just the common carotid signal, as using both the common carotid and the technically more difficult supraorbital artery.

Spectral analysis of the Doppler-shifted signals recorded from the carotid arteries is a sensitive indicator of the presence of occlusive disease.[18] In one study,[6] 66 per cent of vessels with angiographically demonstrable mild irregularity of the wall had minimal spectral broadening. In the range 10–49 per cent diameter stenosis, 78 per cent exhibited spectral broadening, while all the high grade stenoses also showed spectral broadening. Examples of this spectral broadening are shown in Figure 5.9.

Extra information can be gained about the carotid circulation, from peri-orbital Doppler

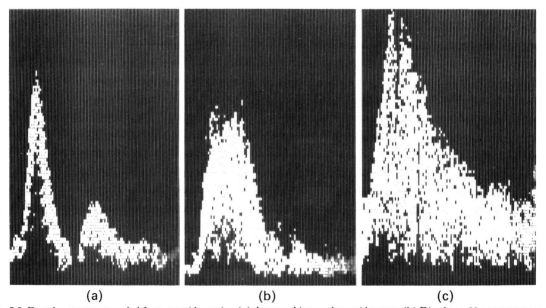

| (a) | (b) | (c) |

Fig. 5.9 Doppler spectra recorded from carotid arteries. (a) A normal internal carotid artery. (b) Distal to a 30 per cent stenosis. (c) Distal to a 75 per cent stenosis. Note that the 'window sign' (absence of low frequency signals at peak systole) disappears as the degree of stenosis increases.

signals and various compression manoeuvres. These indirect investigations determine flow direction in the supraorbital or frontal artery, and by a series of compression tests, attempt to determine any collateral circulation present. These tests have been well reviewed elsewhere[4] and if applied correctly have a sensitivity of about 98 per cent in the detection of carotid lesions of greater than 50 per cent of the vessel diameter.

The predictive value of this type of examination in the selection or exclusion of patients for endarterectomy is 39 per cent; in other words, only 39 per cent of patients presenting with abnormal periorbital signals are candidates for carotid endarterectomy.

Doppler scanning

The processes by which Doppler scanning systems, operating either in continuous wave or pulsed modes, produce functional flow images are explained earlier in this chapter. In comparisons of pulsed Doppler imaging with X-ray contrast arteriography, the sensitivity of the Doppler method ranged from 75 per cent[6] to 88 per cent[14] in detecting all grades of stenosis. Summarising the results[29] of comparisons of continuous wave Doppler scanning with arteriography,[19, 22, 27] it can be concluded that, for stenoses of less than 50 per cent, pulsed systems are more sensitive than continuous wave. This is probably a result of the three-dimensional imaging capability of the pulsed type. It is sometimes difficult to image small stenoses in the anterior-posterior plane, but they are usually detected in the lateral display, as illustrated in Figure 5.10. This is one of the major

reasons for choosing the pulsed Doppler imaging system over the continuous wave.

Duplex scanning

In a comparison of duplex scanning with X-ray contrast arteriography, the overall sensitivity of the duplex system in assessing the degree of stenosis of the carotid arteries was 84 per cent.[6] In another study of 46 patients with a history of transient ischaemic attacks, the overall sensitivity and specificity of duplex scanning was approximately 90 per cent.[14]

Real-time ultrasound scans and corresponding Doppler images are shown in Figure 5.11. It can be seen that the two images are complementary.

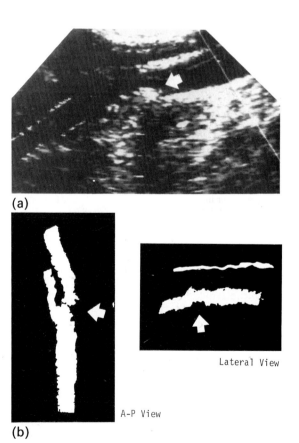

(a)

(b)

Lateral View

A-P View

Fig. 5.11 Scans of a carotid bifurcation. (a) Real-time image of longitudinal section; the arrow indicates a plaque. (b) Pulsed Doppler images; the arrows indicate narrowing of the lumen by the plaque.

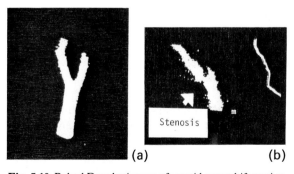

Stenosis

(a)   (b)

Fig. 5.10 Pulsed Doppler images of carotid artery bifurcation. (a) Antero-posterior projection. (b) Lateral projection.

Summary

The Doppler shift signals from the carotid arteries contain clinically useful diagnostic information. The variation of the Doppler signals over the cardiac cycle can be quantified either as some index obtained from the maximum frequency envelope, such as the A/B ratio, or as the degree of spectral broadening. Periorbital investigations show that for stenoses of greater than 50 per cent of the lumen the improved compression test detects approximately 99 per cent of angiographically demonstrable disease. Both the duplex systems and the pulsed Doppler imaging systems have sensitivities around 90 per cent for the detection of arterial diseases around the carotid bifurcation. These results are summarised in Table 5.1. It can be seen that

Table 5.1 Relative success rates of various Doppler ultrasound methods in detecting minor and major stenoses of the carotid arteries.[28]

Investigation	Percentage of diagnoses at difference degrees of stenosis and complete occlusion that were correct		
	< 50	> 50	Occlusion
Abnormal flow direction	—	—	85
Temporal artery occlusion test	9	64	64
Improved occlusion test	—	98	98
Spectral broadening	80	96	—
Duplex scanning	74	95	75*
CW Doppler imaging	23	84	87
Pulsed Doppler imaging	45	80	92

* Only small numbers in this category.

spectral broadening is the most sensitive technique for the detection of minor stenosis. If such a technique is used in conjunction with either the duplex of the pulsed Doppler imaging system, the sensitivity of the combined techniques to small degrees of stenosis should be improved.

Veins

Doppler shift signals from veins in the upper and lower limbs can be classified as 'spontaneous' or 'augmented'.[1] Spontaneous signals occur without external stimuli such as manual compression of a limb or muscle contraction. Augmented signals occur in the vein due to various types of manual compression. Investigation of the venous circulation involves the use of both types of signal, and these can play a major role in the clinical management of patients with deep vein thrombosis, incompetent venous valves and incompetent perforating veins.

Manual compression of the lower limb distal to the measurement site empties the underlying normal veins and forces the blood towards the heart, thus augmenting the venous signal. Proximal compression causes cessation of flow in the normal due to the presence of venous valves. If no competent valves are present, however, some reflux occurs as a result of this compression.

In principle, the detection of venous obstruction such as deep vein thrombus depends on the occurrence of the corresponding abnormal Doppler signals during compression testing.[25] The overall accuracy of the Doppler method ranges from 44 to 96 per cent, with an average of 88 per cent. False positives may be as high as 50 per cent, with an average of 12 per cent; the corresponding incidences of false negatives is 52 per cent and 13 per cent. Most errors occur when the thrombosis is below the knee. Doppler imaging can give supporting evidence.[9]

CONCLUSIONS

Doppler ultrasound plays an important role in the assessment of patients with arterial disease and in the postoperative phase of patient care. The optimum configuration of equipment is either the pulsed Doppler imaging system or the duplex scanner in combination with spectral analysis. The Doppler imaging systems are limited to the investigation of the larger peripheral blood vessels and so the most versatile configuration must be the combination of a real-time ultrasound scanner with a pulsed Doppler flowmeter and spectral analysis. At this stage it is important to be clear about what else has emerged from discussion. The basic hypotheses are firstly, that every blood vessel has a characteristic Doppler-shift signature and secondly, that this signature is modified by the presence of disease. Present developments include techniques for the measurement of volume flow in specific vessels. These will be particularly useful in studying ipsilateral and contralateral carotid compensation after stroke. They will also find application in the assessment of balloon angioplasty particularly of the iliac and femoral arteries.

ACKNOWLEDGEMENT

Figure 5.7 was kindly provided by F. G. M. Ross.

REFERENCES

1. Atkinson P, Woodcock J P 1982 Doppler ultrasound and its use in clinical measurement. Academic Press, London, p 235–244
2. Baird R N, Bird D R, Clifford P C, Lusby R J, Skidmore R, Woodcock J P 1980 Upstream stenosis diagnosed by Doppler signals from the femoral artery. Arch Surg 115:1316–1322
3. Baird R N, Lusby R J, Giddings A E B, Bird D R, Skidmore R, Woodcock J P 1979 Pulsed Doppler imaging in lower limb arterial surgery. Surgery 86:818–825
4. Barnes R W, Russell H E, Bone G E, Slaymaker E E 1977 Doppler cerebrovascular examination: improved results with refinements in technique. Stroke 8:468–471
5. Baskett J J, Beasley M G, Murphy G J, Hyams D E, Gosling R G 1977 Screening for carotid junction disease by spectral analysis of Doppler signals. Cardiovasc Res 11:147–155
6. Blackshear W M, Phillips D J, Thiele B L, Hirsch J H, Chikos P M, Marinelli M R, Ward K J, Strandness D E 1979 Detection of carotid occlusive disease by ultrasonic imaging and pulsed Doppler spectrum analysis. Surgery 86:698–706
7. Carter S A 1979 Role of pressure measurements in vascular disease. In: Bernstein E F (ed) Noninvasive diagnostic techniques in vascular disease. Mosby, St Louis, p 261–287
8. Clifford P C, Skidmore R, Bird D R, Lusby R J, Baird R N, Woodcock J P, Wells P N T 1980 Pulsed Doppler and real-time duplex imaging of dacron arterial grafts. Ultrasonic Imaging 2:381–390
9. Day T K, Fish P J, Kakkar V V 1976 Detection of deep vein thrombosis by Doppler angiography. Br Med J 1:618–20
10. Fish P J 1972 Visualising blood vessels by ultrasound. In: Roberts V C (ed) Blood flow measurement. Sector, London, p 29–32
11. Gosling R G, King D H 1975 Ultrasonic angiography. In: Harcus A W, Adamson L (eds) Arteries and veins. Churchill Livingstone, Edinburgh, p 61–98
12. Howry D H 1957 Techniques used in ultrasonic visualization of soft tissues. In: Kelly E (ed) Ultrasound in biology and medicine. American Institute of Biological Sciences, Washington, p 49–63
13. Johnston K W, Kakkar V V 1976 Rate of risk of arterial pressure as an index of peripheral arterial disease. In: Woodcock J P (ed) Clinical blood flow measurement. Pitman Medical, Tunbridge Wells, p 38–41
14. Lusby R J, Woodcock J P, Skidmore R, Jeans W D, Hope D T, Baird R N 1981 Carotid artery disease: a prospective evaluation of pulsed Doppler imaging. Ultrasound Med Biol 7:365–370
15. Martin T R P, Barber D C, Sheriff S B, Prichard D R 1980 Objective feature extraction applied to the diagnosis of carotid artery disease using a Doppler ultrasound technique. Clin Phys Physiol Meas 1:71–81
16. Mozersky D J, Hokanson D E, Baker D W, Sumner D S, Strandness D E 1971 Ultrasonic arteriography. Arch Surg 103:663–667
17. Mozersky D J, Hokanson D E, Sumner D S, Strandness D E 1972 Ultrasonic visualization of the arterial lumen. Surgery 72:253–259
18. Reinertson J E, Barnes R W 1976 Carotid flow velocity scanning: diagnostic value in Doppler ultrasonic arteriography. Clin Res 24:594a
19. Shoumacher R D, Bloch S 1978 Cerebrovascular evaluation: assessment of Doppler scanning in carotid arteries, ophthalmic Doppler flow and cervical bruits. Stroke 9:563–566
20. Skidmore R, Woodcock J P 1980 Physiological interpretation of Doppler shift waveforms. I. Theoretical considerations. Ultrasound Med Biol 6:7–10
21. Skidmore R, Woodcock J P 1980 Physiological interpretation of Doppler shift waveforms. II. Validation of the Laplace transform method for characterisation of the common femoral blood-velocity/time waveform. Ultrasound Med Biol 6:219–225
22. Spencer M P, Brockenbrough E C, Davies D L, Reid J M 1977 Cerebrovascular evaluation using Doppler CW ultrasound. In: White D N, Brown R (eds) Ultrasound in medicine, Vol 3B, Plenum, New York p 1291–1310
23. Stegall H F, Kardon M B, Kemmerer W T 1968 Indirect measurement of arterial blood pressure by Doppler ultrasonic sphygmomanometry. J Appl Physiol 25:793–797
24. Strandness D E, Schultz R D, Sumner D S, Rushmer R F 1967 Ultrasonic flow detection. A useful technic in the evaluation of peripheral vascular disease. Am J Surg 113:311–319
25. Sumner D S 1979 Diagnosis of venous thrombosis by ultrasound. In: Bergan J J, Yao S T (eds) Venous problems. Chicago: Year Book Medical, 159–186
26. Ware R W 1965 New approaches to indirect measurement of human blood pressure. Proceedings of the Third National Biomedical Instrumentation Symposium, Dallas; ISA BM-65
27. White D N, Curry G R 1978 A comparison of 424 carotid bifurcations examined by angiography and the Doppler echoflow. In: White D N, Lyons E A (eds) Ultrasound in medicine, Vol 4, Plenum, New York, p 363–376
28. Woodcock J P 1980 Doppler ultrasound in clinical diagnosis. Br Med Bull 36:243–248
29. Woodcock J P 1981 Special ultrasonic methods for the assessment and imaging of systemic arterial disease. Br J Anaesth 53:719–730
30. Woodcock J P, Gosling R G, Fitzgerald D E 1972 A new non-invasive technique for assessment of superficial femoral artery obstruction. Br J Surg 59:226–231
31. Yao S T 1970 Haemodynamic studies in peripheral arterial disease. Br J Surg 57:761–766
32. Yao S T 1979 Surgical use of pressure studies in peripheral arterial disease. In: Bernstein E F (ed) Noninvasive diagnostic techniques in vascular disease. Mosby, St Louis, p 281–293
33. Yao S T, Hobbs J, Irvine N T 1968 Pulse examination by an ultrasonic method. Br Med J 4:555–557

6

The infant brain

P. S. Warren, W. J. Garrett and George Kossoff

INTRODUCTION

A detailed demonstration of the newborn and infant brain can be obtained in almost any plane of section with all types of B-mode ultrasound apparatus, including real-time equipment. Early grey-scale studies were directed principally at the diagnosis of hydrocephalus,[3] but subsequent refinements in instrumentation have enabled more subtle diagnoses to be made. The recent realisation that the incidence of intracranial haemorrhage is very high in babies born under 1500 g weight[5] has directed much interest to echoencephalography in the newborn. In this age group, particularly in premature babies, there is enhanced differentiation of brain structures echographically compared to older children which is thought to be due to the thinner skull vault and wider sutures, the presence of relatively more cerebrospinal fluid in the cerebral subarachnoid space, and the greater water content of the brain.

The attenuating and refracting properties of the growing skull interfere increasingly with the transmission of ultrasound and beyond the age of eighteen months it is difficult to obtain an elegant demonstration of brain anatomy routinely. The width of the bodies of the lateral ventricles, however, can be studied well into childhood as the lateral walls of the bodies of the lateral ventricles can usually be demonstrated even in the absence of other intracranial detail.[4]

TECHNIQUES

The best results are obtained when the examination is performed with the child relaxed after feeding.

The use of a water-bath static scanner affording a rapid, no-touch technique removes the need for additional sedation in neonates and most infants and produces a panorama of anatomy in almost any plane of section.

Sector and linear array real-time scanners also produce an excellent demonstration of the brain particularly when the anterior fontanelle is used as an acoustic window,[2] and have the advantage of being able to be taken to the bedside, a particularly useful feature when a sick premature baby is to be assessed. When contact static and real-time machines are used to scan children over three to four months of age it is often difficult to obtain strictly sequential sections parallel to a particular plane, for detailed retrospective analysis, without some form of sedation as the irritation of the transducer may cause the child to move.

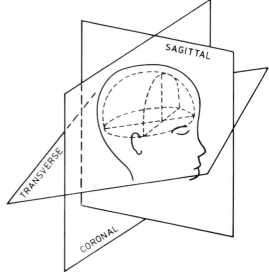

Fig. 6.1 Image planes used in the triplanar approach for echoencephalography.

Hypothermia, a potential hazard with premature babies, is obviated by wrapping the child in a metallised ethylene film ('space blanket') and using an overhead radiant heater. When using water bath equipment the water temperature should be set to 37°C.

The brain is routinely examined in three planes, transverse (either parallel to the canthomeatal line or with a slight tilt to it), coronal and sagittal (Fig. 6.1). To enable the various intracranial structures to be interrogated by the axial resolution of the beam, three approaches are used. These are through the side of the head (lateral), through the occipital region and through the crown (vertical) with two approaches for each plane (Fig. 6.2):

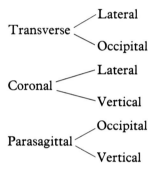

The supratentorial compartment is well served by the transverse and coronal planes from the lateral approach, with the occipital and vertical approaches coming into their own in the demonstration of the posterior fossa contents (Figs. 6.3–6.8). Retrospective analysis is facilitated by having available echograms taken at right angles with each approach and with the line of rotation marked (Fig. 6.9).

When using a real-time scanner to interrogate the brain through the anterior fontanelle, rocking the transducer in the coronal and sagittal planes produces a continuous sweep of oblique 'coronal' and oblique 'parasagittal' sections, from which selected sections can be recorded.

ANATOMY

The complete ventricular system with its contained choroid plexus can be well demonstrated. The characteristic change in the shape of the anterior horns and lateral ventricular bodies is well seen in serial coronal sections whilst the inferior horns are often best demonstrated in parasagittal sections. The atrium or trigone of the lateral ventricle, where the bodies, posterior and inferior horns meet behind the thalamus, is not restrained by the presence of adjacent nuclei and it is this region, together with the posterior horn, which generally becomes fuller than other parts of the ventricular system in transverse sections in early to moderate hydrocephalus.

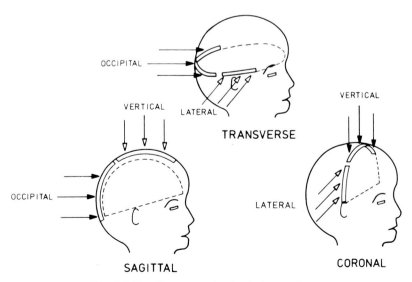

Fig. 6.2 Coupling approaches for the image planes.

The shape of the third ventricle depends on the extent to which the medial nuclei of the thalami bulge into the lumen and a suprapineal recess is often demonstrated extending from its postero-superior aspect. The floor of the ventricle is usually best seen in sagittal sections.

The cavum in the septum pellucidum (which should not be mistaken for the third ventricle) is a constant finding in babies born under 2000 g weight and in 80 per cent of neonates born at term. It may be seen to extend posteriorly to form a cavum vergae and in most cases can be shown to disappear on serial studies.

With careful scanning the fourth ventricle can be demonstrated in the neonatal period in both sagittal and horizontal sections through the occiput. Occasionally, in lateral coronal sections, the rhomboid shape of the floor of the fourth ventricle may be seen.

All three approaches provide a demonstration of the cerebral convolutions, corpus callosum and basal nuclei. The caudate nucleus indenting the lateral wall of the body of the lateral ventricle, the putamen and globus pallidus and the anterior and posterior limbs of the internal capsule can be seen in coronal and horizontal sections. Some of the thalamic nuclei can be resolved, with the dorsal medial nucleus often prominent in sagittal echograms.

A characteristic landmark in all planes is the low

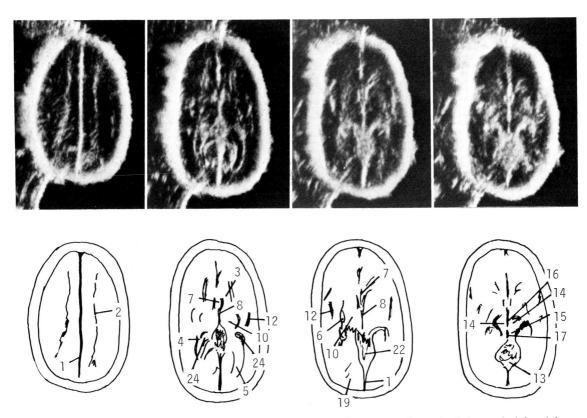

Fig. 6.3 Transverse (lateral) sections. The images are shown in sequence, the most superior section being on the left and the most inferior, on the right. The anatomical structures shown in Figs. 6.3–6.8 inclusive are identified according to the following key: 1. falx cerebri; 2. lateral wall of body of lateral ventricle; 3. anterior horn of lateral ventricle; 4. atrium of lateral ventricle; 5. posterior horn of lateral ventricle; 6. inferior horn of lateral ventricle; 7. cavum in septum pellucidum; 8. third ventricle; 9. fourth ventricle; 10. thalamus; 11. dorsal medial nucleus of thalamus; 12. sulcus over insula; 13. tentorium cerebelli; 14. cerebral peduncles; 15. brain stem; 16. basilar artery; 17. cerebral aqueduct; 18. pituitary fossa; 19. occipital lobe; 20. cerebellar hemisphere; 21. cerebellum (central part); 22. subarachnoid space; 23. corona radiata; 24. choroid plexus; 25. orbit; 26. clivus; 27. pons; 28. head of caudate nucleus; 29. anterior limb of internal capsule; 30. temporal lobe; 31. choroid fissure (ambient wings); 32. frontal lobe; 33. putamen and globus pallidus; 34. cingulate gyrus.

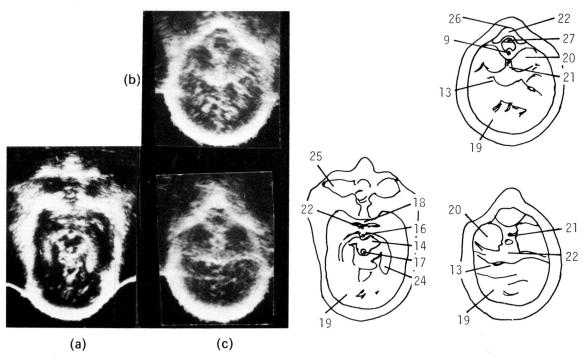

Fig. 6.4 Transverse (occipital) sections. (a) Section through peduncles and orbits. (b) and (c) Different brain sectioned at the level of the pons and the medulla respectively. For key, see Figure 6.3.

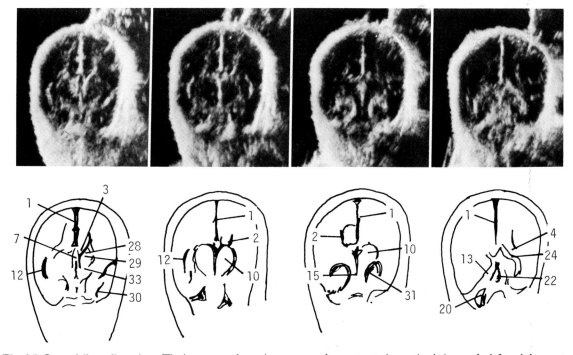

Fig. 6.5 Coronal (lateral) sections. The images are shown in sequence, the most anterior section being on the left and the most posterior, on the right. For key, see Figure 6.3.

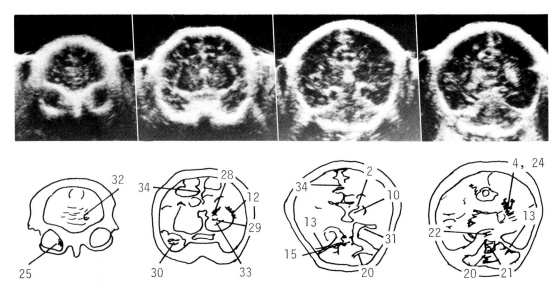

Fig. 6.6 Coronal (vertical) sections. The images are shown in sequence, the most anterior section being on the left and the most posterior, on the right. For key, see Figure 6.3.

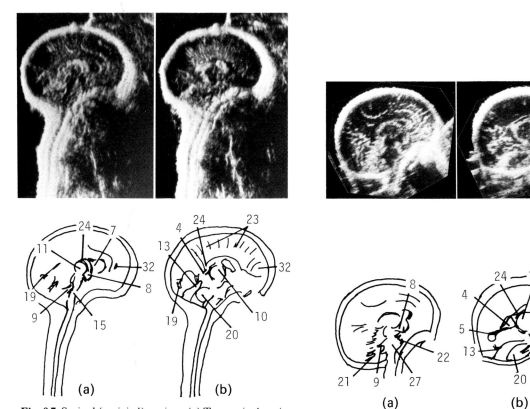

Fig. 6.7 Sagittal (occipital) sections. (a) True sagittal section including superior part of spine. (b) Slightly tilted parasagittal section demonstrating corona radiata. For key, see Figure 6.3.

Fig. 6.8 Sagittal (vertical) sections. (a) True sagittal section. (b) Parasagittal section through lateral ventricle. For key, see Figure 6.3.

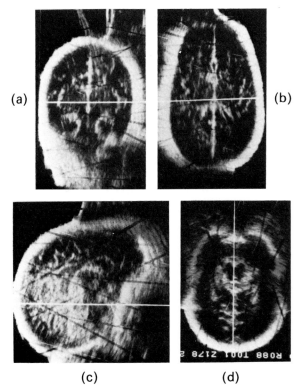

(a)   (b)

(c)   (d)

Fig. 6.9 Markers showing axes of rotation. Lateral approach in (a) Coronal and (b) Transverse scans. Occipital approach in (c) Sagittal and (b) Transverse scans. (See Figs. 6.1 and 6.2 for explanatory diagrams.) The acquisition of two sections at right angles to each other greatly facilitates the recognition of structures in retrospective analysis.

echo amplitude brain stem. The pons and cerebral peduncles are always well demonstrated and with careful scanning the colliculi can be defined.

The central structures of the cerebellum return high amplitude echoes which in the sagittal plane have the appearance of a snowball behind the brain stem. Laterally, however, the cerebellar hemispheres return echoes of low to moderate amplitude surrounded by a rim of high amplitude echoes which probably originate from the sulci on the surface of the hemispheres.

Although the cerebral subarachnoid space contains cerebrospinal fluid, it returns echoes of high amplitude corresponding to the numerous fine trabeculae and blood vessels crossing it. The subarachnoid space is typically best seen echographically in the posterior fossa, but may also be well demonstrated in the interhemispheric region behind the third ventricle.

MEASUREMENT

The lateral ventricular ratio (LVR) may be used for assessing ventricular size. The LVR is the width of the body of the lateral ventricle divided by half the greatest distance between the inner tables of the parietal bones (the ratio a/b in Fig. 6.10). The normal range is 0.24–0.36, but the upper part of the range (0.34–0.36) may include some cases of early hydrocephalus.[4] Displacements of the midline and extracerebral liquid collections usually do not affect the LVR.

The atrium of the lateral ventricle, which typically expands early in hydrocephalus, has a dimension c (Fig. 6.10) which is usually less than

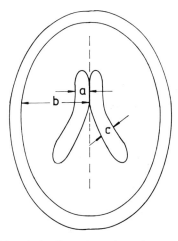

Fig. 6.10 Ventricular dimensions, shown in transverse section.

10 mm in transverse diameter. The width of the third ventricle is variable, but a recent series shows that it measures less than 5 mm in 92.5 per cent of patients whose LVR was 0.24–0.33.[4]

ABNORMALITIES

Hydrocephalus

Early hydrocephalus is best demonstrated in coronal sections as a fullness of the lateral ventricles with rounding of the usually sharp inferior and lateral angles and widening of the atria. Choroid plexus tissue almost fills the normal atrium, but with increasing ventricular volume echo-free cerebrospinal fluid can be seen between the atrial walls. As the hydrocephalus progresses the LVR increases beyond the normal range. In premature babies the absence of a cavum in the septum pellucidum may be a helpful sign, suggesting that it has been obliterated by increased pressure. In severe hydrocephalus (when the LVR is not so meaningful) measurements of the cortical thickness in serial studies are of value. The cause of the hydrocephalus may be demonstrated (by e.g., visualisation of intracranial haemorrhage or space-occupying lesion), but usually the site of an obstruction can only be inferred by the relative sizes and shapes of the lateral, third and fourth ventricles (Fig. 6.11).

Intracranial haemorrhage

Intraventricular haemorrhage (IVH) which occurs primarily in immature babies has been shown at autopsy to arise from the germinal matrix overlying the heads of the caudate nuclei and from the choroid plexus. A major IVH fills the ventricular system with blood and returns echoes of high amplitude. Fresh bleeding into the ventricular system has been observed with real-time equipment, when choroid plexus tissue was needled during in-utero decompression of hydrocephalus. A cascade of high amplitude echoes was seen arising from the traumatised choroid plexus. A clot may form at the site of the bleed and the underlying brain tissue may swell. In early cases it is possible to demonstrate a fluid level between blood and CSF. Subsequently mobile blood clots (often with an echogenic rim and low level echo centre) may be seen. Hydrocephalus usually follows the bleed and is associated with prominent choroid plexus tissue and a thickened high amplitude echo line around the boundaries of the ventricles indicating an ependymal reaction (Fig. 6.12).

Pulsatile flow changes in the anterior cerebral arteries can be monitored through the anterior fontanelle by Doppler ultrasound techniques.[1] Changes in the pulsatility of the waveforms, which can be expressed quantitatively, may give an indication of impending intraventricular and intracerebral haemorrhage.

Intracerebral haemorrhage (Fig. 6.13) typically produces a reasonably well defined area of moderate-to-high amplitude echoes often with a subse-

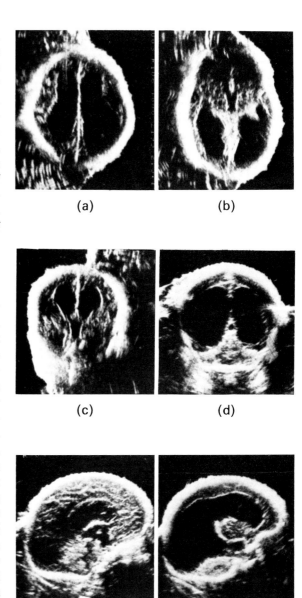

(a)   (b)

(c)   (d)

(e)   (f)

Fig. 6.11 Hydocephalus due to aqueduct stenosis. The child had a rapidly-increasing head size in the first month of life but was clinically normal and thriving. (a) and (b) Lateral transverse sections. (c) Lateral coronal section at the level of the foramen of Monro. (d) Vertical coronal section showing depression of the tentorium cerebelli. (e) Sagittal section showing normal-sized fourth ventricle. (f) Parasagittal section. The lateral ventricular ratio (LVR) is 0.51 on both sides, and the atria measure 46 mm on the left side and 40 mm on the right. The caudate nuclei are flattened and the third ventricle measures 15 mm in transverse diameter.

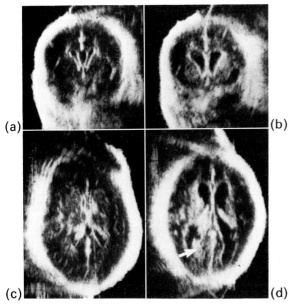

(a)

(b)

(c)

(d)

Fig. 6.12 Intraventricular haemorrhage in one of twins aged two weeks born at 32 weeks amenorrhoea. Coronal (lateral) sections in (a) Normal twin and (b) Twin with IVH. Transverse (lateral) sections in (c) Normal twin and (d) Twin with IVH. The twin with IVH has enlarged lateral ventricles, thickened ependymal lining of the ventricles and blood clot (arrow) in the right posterior horn. The cavum in the septum pellucidum is squashed; it subsequently reappeared as the hydrocephalus resolved.

quent development of a low echo amplitude centre. The resolution of the haemorrhage can be monitored by serial studies.

Extracerebral haemorrhage is characterised by an echo-free area which separates the brain from the skull and falx. The diagnosis of minor collections should be made cautiously in the neonate, particularly in premature babies, as the peripheral parts of the cerebral hemispheres normally are very cellular and have a high water content and therefore appear relatively echo-free (Fig. 6.14).

CONGENITAL ABNORMALITIES AND TUMOURS

Many congenital anomalies and space-occupying lesions can be demonstrated in high-resolution echograms. They may manifest themselves by their effects on the size and shape of the ventricular system or by asymmetry in the cerebral echo

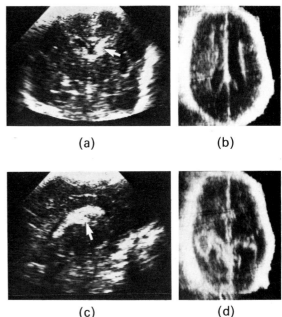

(a)

(b)

(c)

(d)

Fig. 6.13 Intracerebral haemorrhage. (a) Coronal and (b) Parasaggital real-time sector scans through the anterior fontanelle of a premature baby born at 32 weeks amenorrhoea, showing left-sided subependymal bleed (arrows). No IVH was demonstrated but follow-up studies revealed a subsequent subependymal bleed on the right side and a recurrent bleed on the left. (c) and (d) Transverse (lateral) sections in a three week old child born at 28 weeks amenorrhoea, showing well defined haemorrhage with a rim of high amplitude echoes around a lower echo amplitude centre and the hydrocephalus consequent on an associated IVH. This lesion was shown to resolve on serial studies over six weeks.

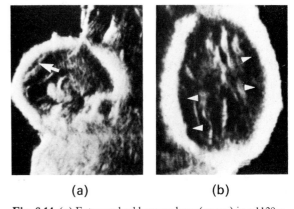

(a)

(b)

Fig. 6.14 (a) Extracerebral haemorrhage (arrow) in a 1120 g premature baby born at 30 weeks amenorrhoea. (b) The relatively echo-free peripheral cerebral cortex (arrowheads) is demonstrated in a normal premature baby born at 34 weeks amenorrhoea; it should not be confused with an abnormal liquid collection.

patterns (Fig. 6.15). Meticulous care with gain settings and time-gain compensation is necessary if subtle echographic signs are not to be lost.

CONCLUSIONS

Advances in ultrasound technology are meeting the new clinical demands raised by the survival of increasing numbers of high risk infants as a result of improved obstetrical and neonatal care. Ultrasound systems can now dependably demonstrate normal and abnormal intracranial structures in any plane of section. Echoencephalography enables high-risk and post-operative infants to be evaluated safely, repeatedly and rapidly at relatively low cost.

ACKNOWLEDGEMENT

The scans in Fig. 6.13a and 6.13b were kindly provided by Albert Lam.

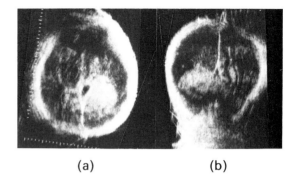

(a)   (b)

Fig. 6.15 Cerebral tumour in a six month old boy who presented with a large head but no neurological symptoms or signs. (a) Transverse (lateral) and (b) Coronal (lateral) sections showing the relatively poorly defined abnormal high amplitude echo area in the left parieto-occipital region which has squashed the left lateral ventricle and which can be seen to extend through the tentorial hiatus into the posterior fossa. A very vascular tumour was found at craniotomy.

REFERENCES

1. Bada H S, Hajjar W, Chua D, Sumner D S 1979 Noninvasive diagnosis of neonatal asphyxia and intraventricular hemorrhage by Doppler ultrasound. J Pediat 95:775–779.
2. Ben-Ora A, Eddy L, Hatch G, Solida B 1980 The anterior fontanelle as an acoustic window to the neonatal ventricular system. J Clin Ultrasound 8:65–67
3. Garrett W J, Kossoff G, Jones R F C 1975 Ultrasonic cross-sectional visualisation of hydrocephalus in infants. Neuroradiology 8:279–288
4. Garrett W J, Kossoff G, Warren P S 1980 Cerebral ventricular size in children — a two-dimensional ultrasonic study. Radiology 136:711–715
5. Papile L, Burstein J, Burstein R, Koffler M 1978 Incidence and evolution: a study of infants with birth weights less than 1500 g. J Pediat 92:529–534

7

The eye

R. L. Dallow

INTRODUCTION

Ultrasonic techniques are essential for clinical evaluation of eyes with opaque media due to corneal scarring, cataract, haemorrhage, or inflammation, for assessment of suspected intraocular tumours, and for orbital soft tissue analysis of tumours or inflammatory-congestive changes. Ultrasound findings complement other radiographic and computed tomographic studies and influence directly the course of medical and surgical treatment of many disorders of the globe and surrounding orbital tissues. Additionally, ultrasound is used for accurate biometry of the eye. Measurements with accuracies of 0.01 mm are needed for the calculation of prosthetic intraocular lens implants inserted during cataract surgery.

Ophthalmic applications of diagnostic ultrasound were first described in 1956 by Mundt and Hughes[11] using A-scan techniques. One-dimensional time-amplitude ultrasound (A-mode) diagnostic features were subsequently further elaborated.[3, 9, 10, 12, 13] Two-dimensional ultrasonic imaging (B-scan) was introduced in 1958 by Baum and Greenwood.[1] The method was developed,[4, 7, 14] and a practical water immersion system with simultaneous displays of A-scan and two-dimensional B-scan was produced.[6] Several large series of clinicopathologic correlations documented the validity of these techniques. Accuracy of 85 to 95 per cent correct diagnosis is now assumed in ophthalmic ultrasonography. A recent publication of multiple authorship summarises the status of all of the currently available diagnostic ultrasound techniques applicable to ophthalmology.[8]

TECHNICAL FACTORS

For the eye, as in other medical diagnostic ultrasound applications, a pulse-echo technique is used with A-mode, B-scan, D-scan (an isometric display, described later in this chapter) and M-mode display systems. Echoes relate closely to anatomical boundaries of the eye and orbit structures. B-scan images serve as two-dimensional sections through the eye for topographic analysis of the location, size, and shape of lesions. A-mode is used to characterise tissue types more fully by depicting the entire range of echo amplitudes, providing a more complete grey scale than do B-scan images. D-scan (isometric) accentuates the surface aspect of contours and the texture of echoes within lesions. M-mode is used for physiological studies of lens changes and choroidal pulsations, and occasionally for magnet tests of foreign bodies.

Unique features of the eye are its small size (24 mm diameter), cystic composition, and fine anatomical details measuring less than 0.1 mm. The bony orbit is about 40 mm deep and filled with heterogeneous fat, muscles, vessels and nerves. The requirements of high resolution (0.1 mm), shallow penetration (40 mm), as well as the cystic character of an exposed organ permit use of high frequency ultrasound transducers. Commonly used frequencies for eye ultrasound are 8 to 10 MHz, although 15 or 20 MHz are used for higher resolution and biometry. A 5 MHz transducer may be used occasionally for deeper penetration to the orbital apex or for higher sensitivity to weaker echoes. The transducers generally are weakly focused with a 1 mm beam width between 10 and 30 mm from the face.

Two-dimensional B-scan images may be produced as automated real time sector scans or compound storage static scans. Grey scale images are produced with an automated real time sector scanner with the motorised transducer encased in a small fluid filled compartment to achieve transducer stand-off from the eye. The smooth anterior surface of this compartment is placed directly on the eyelids or the topically anaesthetised cornea. Most such sector scanners have sweep rates of 20 to 30 frames per second. A-scan is utilised for tissue characterisation instead of relying on B-scan grey scale information. B-scan eye images may be enhanced for complete outlining using high contrast imaging on a storage oscilloscope and compound scanning patterns. This requires a water immersion scanning system. The eye is surrounded by a saline-filled plastic bag with a large aperture over the eye. The transducer is submerged in the saline about 1 cm above the eye and manipulated manually on a cantilevered carriage. Bistable ultrasound units with a hand operated transducer moved across the eye in a combined linear and sector pattern produce echoes from all aspects and contours of the eye. Thus, there are two methods of two-dimensional B-scan ultrasound examination of the eye and two corresponding equipment assemblies. The first is the *contact method* (Fig. 7.1a) with an automated, real time sector scanner. The second is the *immersion method* (Fig. 7.1b) with

a hand operated transducer and selective scanning patterns, transducer frequencies, and signal processing. The more flexible and more sophisticated immersion system produces the best results of B-scan imaging, whilst the contact system is simpler to operate and thus more popular.

A-scan for application to the eye has the same features as those described for two-dimensional B-scan. A-scan can be performed with a separate direct contact method or in combination with two-dimensional B-scan either by contact or immersion techniques. The examiner reduces receiver sensitivity to characterise echo intensities on a qualitative scale, usually using a highly reflective scleral coat of the eye as a reference echo. Some controversy exists over the type of amplifier system best suited for A-scan machinery. Linear, logarithmic, and S-shaped amplitude response amplifiers are available, each tending to emphasise a different part of the echo amplitude spectrum. If one amplifier system is chosen as a standard, then quantitative A-scan analysis is possible, but only applicable to that one system. Until now, however, standardisation has not been feasible on a wide scale because of equipment diversity. Commercially available ultrasound units with combinations of these features for A-scan and B-scan individually or combined are marketed by several different manufacturers specifically for eye applications. In general, ultrasound units used for abdominal or

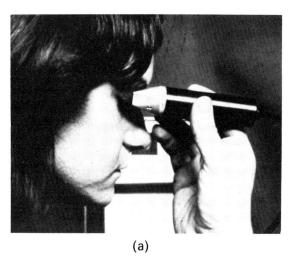

(a)

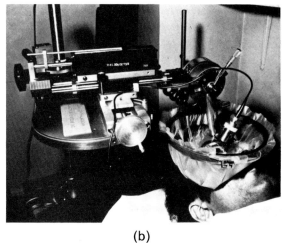

(b)

Fig 7.1 Techniques for ophthalmic ultrasound examination (a) Contact A-scan and two-dimensional B-scan probe with motorised sector real-time capability. (b) Immersion system for hand operated real-time sector scan or storage static compound scan, with interchangeable transducers.

other body areas have not been adaptable to eye studies, as they lack high resolution, compound scanning, and the necessary mechanical flexibility.

Eye and orbit images produced by the various ultrasound systems described are depicted in Figure 7.2. Other techniques have been devised to

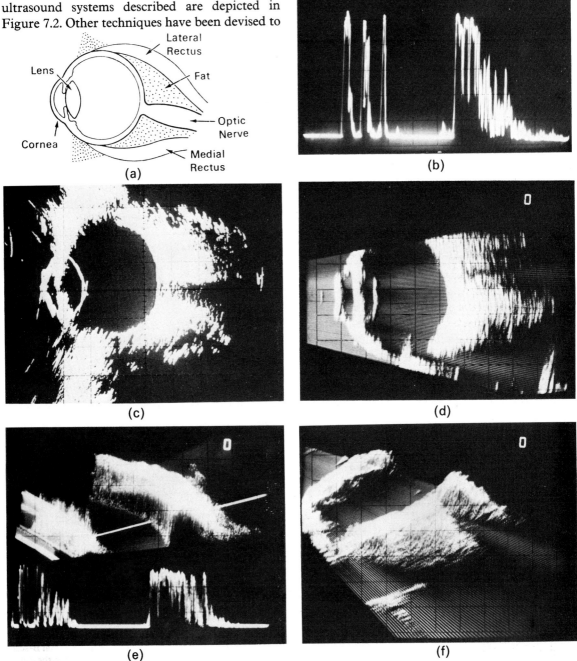

Fig 7.2 Normal eye. (a) Diagram of major eye and orbit components, shown in horizontal (transverse) section. (b) A-scan along the central axis. (c) Contrast enchanced (bistable) two-dimensional compound B-scan. (d) Grey scale two-dimensional sector B-scan. (e) Isometric D-scan demonstrating retinal surface; the bright vector line identifies the location of the displayed A-scan. (f) Isometric D-scan with image rotated to allow the optic nerve to be viewed from behind the eye.

accentuate features of B-scan images, notably colour coded and isometric D-scans. The latter shows the scan tilted on edge so that the A-scans can be viewed superimposed on the two-dimensional B-scan image. The most promising development in ophthalmic ultrasound appears to be computer aided signal analysis of spectral components of the returned echoes. It is hoped that this technique will lead to more precise tissue differentiation in the future.

OCULAR DIAGNOSIS

A normal eye has a smooth round contour with several internal reflecting surfaces (cornea, lens, iris, and the retina, choroid, and scleral layers of the globe wall). Each of these structures produces high amplitude, sharply defined echoes on A-mode and smooth contours on two-dimensional B-scan images. As its equator, the globe wall is incompletely depicted because the ultrasound beam is oblique to this surface. Equatorial regions can be filled in by having the patient look from side to side so the beam becomes perpendicular to the surfaces. Internally, the globe is filled with aqueous anteriorly and with a vitreous gel posteriorly, through which sound is transmitted readily with little attenuation and no internal echoes. The crystalline lens of the eye is a compactly organised tissue with no internal reflections unless cataract is forming.

Retinal and choroidal detachment

Normally the retina is appositional with the choroid along the inner globe wall. Fluid or haemorrhage leaking beneath the retina causes it to detach and become elevated within the vitreous cavity. Even when fully 'detached', the retina retains its anatomical adhesion to the optic nerve head posteriorly and to the ora serrata at the anterior periphery. Ultrasonically, detached retina appears as a complete, smooth, highly reflecting membrane within the vitreous (Fig. 7.3). This membrane moves freely when seen with real time scanning as the patient turns his eye.

There are many variations seen with retinal detachment. It may show only a localised area of elevation, instead of a total detachment. Long-

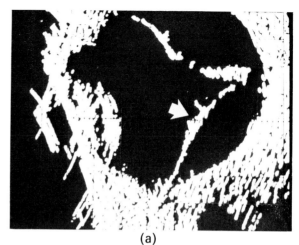

(a)

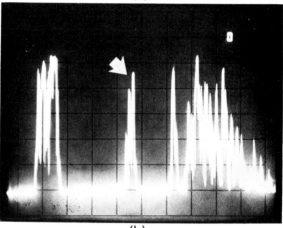

(b)

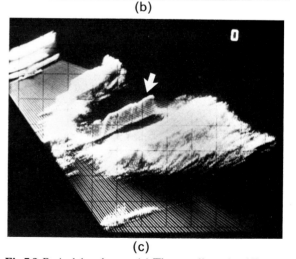

(c)

Fig 7.3 Retinal detachment. (a) The two-dimensional B-scan shows a smooth membrane (arrow) emanating from the optic nerve and extending to the globe periphery. (b) The A-scan shows a high amplitude echo (arrow) in the mid-vitreous cavity. (c) The isometric D-scan shows a sheet of uniform high amplitude echoes (arrow).

standing retinal detachment becomes organised and immobile, resulting in the late stage appearance of a tight funnel configuration with connecting dense vitreous membranes and debris. A common retinal detachment pattern seen in diabetic, hypertensive, or trauma related vitreous haemorrhage or fibroproliferation is the appearance of retina 'tented up' in a peaked configuration with connecting vitreous membranes pulling it up (Fig. 7.4a).

Another major membrane configuration within the vitreous cavity is choroidal detachment, which appears as a smooth convex elevated membrane located at the equatorial region of the eye, usually circumferentially on both sides of the two-dimen-

sional B-scan image of the eye (Fig. 7.4b). Because choroidal detachment is thicker than retina and inserts anteriorly near the lens and posteriorly near the equator, it is always distinguishable ultrasonically from retinal detachment. The clinical implications of retinal and choroidal detachments are quite different.

Vitreous opacities

Haemorrhage or inflammation occurring within the eye often spreads diffusely throughout the vitreous cavity, producing scattered low amplitude echoes. Denser debris of cellular aggregates may become confluent in some areas and layered along

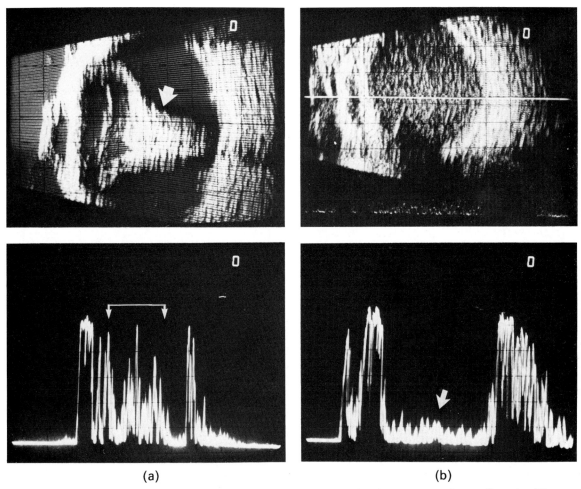

(a) (b)

Fig 7.4 Vitreous cavity abnormalities. (a) Total organised inoperable retinal detachment, seen on the two-dimensional B-scan as a funnel-like pattern (arrow) of moderate- and high-amplitude echoes from the disorganised fibrous tissue (arrows on A-scan). (b) Intraocular haemorrhage giving rise to diffuse low amplitude echoes; indicated by arrow on the A-scan.

vitreous membranes. Compared to retinal detachment, vitreous membranes appear somewhat incomplete with irregular surfaces and only mid-density echoes. These echoes tend to disappear on two-dimensional B-scan as gain is lowered, whereas retinal detachment echoes persist at much lower gain settings. The density, location, multiplicity, and configuration of the debris and membranes may detach the retina partially or totally by producing a progressively tightening traction within the eye.

The vitreous gel is firmly attached to the globe wall at the same locations as retina (the optic nerve head and the ora serrata region). Hence, dense vitreous membranes sometimes simulate the appearance of retinal detachment, making their differential diagnosis impossible at times. Serial ultrasound examinations done over several weeks or months show the evolving patterns of vitreo-retinal pathology and serve as the primary guidelines for determining when vitrectomy and retinal operations are appropriate.

Foreign bodies

The eye is vulnerable to direct trauma from blunt as well as sharp objects. Blunt injury produces internal changes such as retinal detachment or vireous haemorrhage. Propelled small foreign bodies (such as metal chips or glass fragments) may cause penetrating injury with the object retained within the eye, possibly with internal haemorrhage obscuring its location. Surgical removal of intraocular foreign bodies is usually advisable if they can be located precisely. Ultrasound serves this purpose admirably.

All foreign materials within the eye are detectable with ultrasound, but some significant limitations exist. Metal, glass, plastic, and stone are prominent reflectors, producing higher amplitude echoes than any normal structure except bone. Wood and vegetable matter reflect only intermediate amplitude echoes and may be quite difficult to identify. Ultrasound indicates not only the presence of foreign bodies, but also their relationships to the globe wall and the presence of associated disorders such as retinal detachment.

Foreign body detection is aided by the ultrasonic artifact of acoustical shadowing when the foreign body is strongly reflective. Lowering the gain of the receiver demonstrates persistence of a strong foreign body echo beyond that of all other echoes (Fig. 7.5). The degree of foreign body response to a magnetic field can be graphically portrayed by A-scan or M-scan. At the time of surgical removal of a foreign body, ultrasound can be of major importance in precise localisation and extraction during the operative procedure.

Foreign bodies do present several problems to ultrasound diagnosis. They are often quite small and linear in shape. The ultrasound beam must be precisely perpendicular to the flat side of such

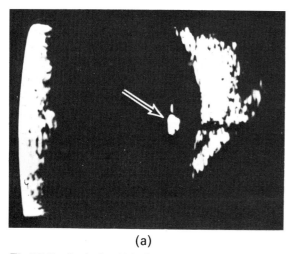

(a)

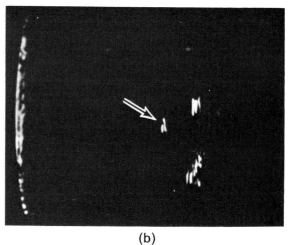

(b)

Fig 7.5 Foreign body within the eye. (a) High amplitude localised echo (arrow) with acoustic shadowing. (b) The echo remains (arrow) with reduced receiver gain.

foreign bodies in order to produce a detectable echo. Thus, some foreign bodies may be missed on ultrasound scanning because the beam orientation is not optimally perpendicular. It is sometimes easier first to identify the presence and general location of foreign material by plain radiography or computed tomography, and then to utilise ultrasound for precise localisation of it in reference to other ocular structures and for assessment of other soft tissue abnormalities, such as retinal detachment.

Intraocular tumours

The most common intraocular tumours in adults are malignant melanoma, metastatic tumours, and haemangioma. In children, retinoblastoma is more likely. Other mass lesions may simulate tumour clinically, such as subretinal haemorrhage, parasites, and inflammatory foci. Tumour underlying retinal detachment or vitreous haemorrhage may be undetectable clinically. Ultrasonography is capable of detecting all mass lesions larger than 1.0 mm in elevation and of characterising them into cystic, solid, or angiomatous types. Some difficulty exists in differentiating fibroproliferative masses from true neoplasms, but monitoring of growth of a lesion on serial ultrasound examinations over time usually reveals the presence of neoplasm. Ultrasound assumes a crucial role when opacities of the lens, vitreous, or retina obscure the view for clinical examination of posterior portions of the eye, as it is the sole means presently available of identifying underlying eye pathology including tumours.

The usual two-dimensional B-scan ultrasonic appearance of malignant melanoma is a mass along the inner globe contour with a highly reflective, smooth, convex contour (Fig. 7.6). The internal tissue A-scan pattern is one of gradual decay attenuation of closely spaced echoes throughout the mass. These echoes derive from cellular aggregates, fibrous septae, and vessels within the mass. Of course, this appearance can vary. Most melanomas consist of compact masses of similar cells, but occasionally a necrotic centre or a fungating growth pattern alters the ultrasound findings. A polypoid or 'collar button' appearance may be produced. Other types of tumours do not

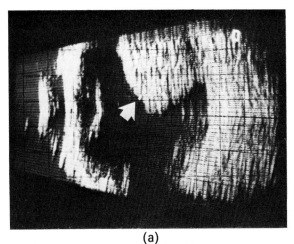

(a)

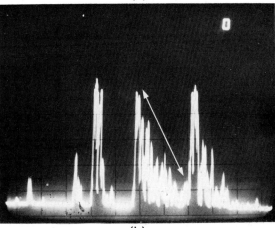

(b)

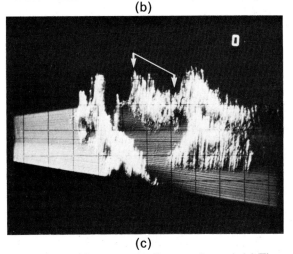

(c)

Fig 7.6 Intraocular tumour (malignant melanoma). (a) The two-dimensional B-scan shows a well-defined, rounded mass of confluent echoes (arrow). (b) The A-scan shows that the amplitudes of these echoes have a characteristic exponential decay (arrows). (c) The isometric D-scan provides a better appreciation of the internal tumour texture of echoes.

generally show this configuration. While melanomas expand primarily inward within the eye, they also frequently produce an apparent shallow 'excavation' posteriorly by replacement of the choroidal coat of the eye. This appears as a concave indentation of the normal globe contour posterior to the tumour and is highly characteristic of malignant melanoma. A-scan can be critical in differentiating lesions, especially when the two-dimensional B-scan image is equivocal. A localised vitreous haemorrhage may appear compact enough on two-dimensional B-scan to resemble a tumour mass. A-scan amplitudes show a high lead echo for tumour, with a gradual decay of amplitudes, whereas amplitudes from a haemorrhage are much lower, more closely arranged, and scintillate because of slight motion of red cell aggregates producing the echoes. Tumours may be measured by ultrasound to document size and growth using a high resolution transducer (15 or 20 MHz) for accuracy of better than 0.1 mm in tumour size.

Metastatic tumours in the eye have the same locations as do melanomas, being within the uveal tract, most frequently in the posterior pole of the eye. They may come from any source, including tumours of the breast, gastrointestinal system, lung, prostate, and so on. In contrast to melanomas, metastatic tumours tend to have a flat configuration on two-dimensional B-scan, appearing as low, undulating masses elevated only about 1.0 mm, but having a broad base diameter. Metastatic tumours may be multiple, whereas melanoma is nearly always solitary. A-scan patterns differ little from melanoma patterns.

Haemangiomas also arise within the choroid component of the uveal tract. This benign and generally non-progressive tumour has a very low elevation and a broad base dimension. The tumour composition of multiple small vascular channels and blood-filled spaces produces a characteristic A-scan appearance of sustained high amplitude echoes throughout the mass with a very regular spacing between them. If the lesion is large enough, it can usually be differentiated from a solid tumour by ultrasonic criteria.

Retinoblastoma is the most common intraocular malignancy in children, but it is often masked by opacities such as cataract, haemorrhage, or vitreous membranes. It is often seen initially in a fairly advanced state with a large fungating lesion having necrotic foci, calcium deposits, and associated vitreous debris and retinal detachment. The ultrasound pattern is quite variable because of these multiple elements. It always shows an irregular pattern of high amplitude echoes and membranes. The detection of multiple, 'foreign body' type of very high echoes reflected from the calcium flecks within the tumour is very suggestive of retinoblastoma. This sign, however, is not always present.

Other non-neoplastic conditions may simulate the appearance of intraocular tumours, both clinically and ultrasonically. One particularly difficult lesion to differentiate is an organised subretinal haemorrhage (disciform maculopathy), which is a common abnormality in elderly patients. It appears on two-dimensional B-scan as a small elevated mass in the posterior pole of the eye having the same size, shape, and configuration as a small melanoma. The A-scan pattern may also be indistinguishable from melanoma. Serial ultrasound examinations over time are necessary to demonstrate that a haemorrhage is static or regressing rather than showing the increase in size that implies neoplasm. Many other lesions can resemble tumours on ultrasound scans, including vitreous haemorrhage, retinal and choroidal detachment, ciliary body cysts, brawny scleritis, lymphoid hyperplasia, melanocytoma, dislocated cataract, foreign bodies, parasites and papillo-oedema. Most of these disorders give sufficient evidence on ultrasonography to be differentiated from neoplasms. The reliability of ultrasonic intraocular tumour differentiation has been documented by several investigators with the consensus of better than 90 per cent accuracy regardless of the specific techniques and equipment employed.

ORBITAL DIAGNOSIS

The soft tissues within the bony orbit have the potential for development of a large variety of tumours and inflammatory-congestive disorders. All may have similar clinical presentation of a protruding eye with no clinical evidence of the responsible disease. The most common causes of exophthalmos (protruding eye) are, in fact, inflammatory diseases (thyroid related Graves' disease, pseudotumours and cellulitis) rather than neoplasms. The incidence of specific tumours varies in different series and for different age groups. A rough descending order of orbital tumour frequency is as follows: neurogenic tumours, cysts (mucocele, dermoid), haemangioma, lymphoma, metastatic tumours, lacrimal epithelial tumours, and secondary tumours invading from the adjacent sinuses. Some arteriovenous anomalies can simulate orbital tumours, including varices and shunts. Radiography, computed tomography, and ultrasound are essential tests for defining orbital abnormalities and for guiding surgical approaches to tumours.

Normal orbits produce a consistent picture on two-dimensional B-scan ultrasonography. The eye portion of the scan shows clear delineation as a rounded cyst (Fig. 7.2). The retrobulbar pattern is derived primarily from the large fat pad, which has a triangular shape and which is bounded anteriorly by the globe concavity and on the sides by the extraocular muscles extending from the globe equator toward the apex of the orbit at the optic canal. The fat is quite heterogeneous, being composed of fat globules, fibrous septae, and nerves. High amplitude echoes are produced throughout the fatty tissue complex, giving it a 'filled in' appearance on the two-dimensional B-scan image and a decaying high amplitude pattern on the A-scan. The extraocular muscles and optic nerve are more compact, well organised, homogeneous tissues compared to the fat pad. These structures appear as relatively echo-free areas in contrast to adjacent fat. Hence a two-dimensional B-scan section at the level of the optic nerve produces a 'W-shaped' area of echoes posterior to the globe, with muscle and optic nerve seen in negative contrast. The bony orbital wall is represented only by a few low amplitude echoes, partly because the beam is not perpendicular to its surface and because the sound undergoes a shearing effect when it strikes bone.

Orbital evaluation with ultrasound is a proven, reliable, and often unique detector of orbital pathology. Because of the convenience of examination and lack of morbidity of this test, ultrasonography is often the first procedure indicated when orbit pathology is suspected. Not only can it direct further studies, but it can properly direct the choice and route of indicated surgery. Other abnormalities may be classified by ultrasonic criteria in four major categories: tumours, inflammatory-congestive changes, structural abnormalities, and foreign bodies. Each category is further subdivided into several specific types of lesions.

Tumours of the orbit

Orbital tumours, either malignant or benign, are classified on two-dimensional B-scan ultrasonography in terms of location, size, contour, and ability to transmit ultrasonic waves. A-scan criteria add additional information on the internal structures of tumours. Four general tumour types are easily identifiable. These are cystic, solid, angiomatous, and infiltrative.

Cystic tumours seen on two-dimensional B-scan have a smoothly rounded contour that is sharply defined from adjacent structures and which often causes some distortion of normal tissues by compression (Fig. 7.7). Cystic masses demonstrate good sound transmission with clear definition of the posterior wall of the lesion and of tissues posterior to the tumour. Internally, cysts are generally devoid of echoes because there are no significant tissue interfaces within the lesion. Cysts located anteriorly in the orbit generally represent dermoid or epithelial cysts, while those more posterior and along the bony orbital walls are frequently mucoceles.

Solid tumours also demonstrate well defined contours ultrasonically that contrast with adjacent tissues. Tumour contour may be smoothly rounded or somewhat irregular (Fig. 7.8). In contrast to cysts, a solid tumour produces significant sound attenuation so that penetration is poor and the posterior margin of the tumour may not be well defined. Tissue interfaces within the tumour produce multiple low- to mid-amplitude echoes within its substance, indicating a moderately heterogeneous tissue. A rounded outline with complete contour (encapsulated) and solid internal tissue characteristics may represent glioma or meningioma if located within the retrobulbar fat, or lacrimal gland neoplasm if located along the temporal bony wall. If the tumour has solid characteristics ultrasonically but an irregular contour (non-encapsulated), this suggests an infiltra-

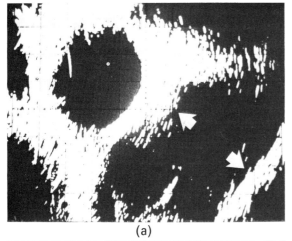

(a)

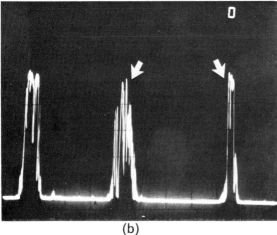

(b)

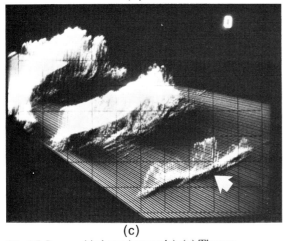

(c)

Fig 7.7 Large orbital cyst (mucocele). (a) The two-dimensional B-scan shows a sharply demarcated ovoid mass with smooth walls (arrows). (b) The A-scan shows strong interface echoes (arrows) with no internal tissue echoes. (c) The isometric D-scan with image rotation displays the smooth posterior wall (arrow) of the cyst.

tive neoplasm such as lymphoma, sarcoma, or metastatic tumour. Tumours extending into the orbit from the intracranial space or from adjacent sinuses usually have a very flat configuration that follows the contour of the orbital walls.

Angiomatous tumours produce very high amplitude echoes throughout the mass presumably because of the fluid-filled character of these lesions with multiple vessel walls. Sound transmission is quite good and the posterior extents of the lesions are well defined. Anterior contours of haemangiomas are sometimes difficult to define because the heterogeneous tissue pattern seems to merge with the retrobulbar fat pattern. Lowering the gain of the receiver permits differentiation between these two highly reflective tissues. The ultrasonic pattern of haemangiomas is in marked contrast to the cystic and solid tumour patterns already described.

Inflammatory-congestive orbital changes

Inflammatory and congestive processes of the orbit involve structures normally present, sometimes causing only subtle changes of the usual ultrasonic patterns, whereas tumours impose new contours intruding on the orbital pattern. Inflammatory ultrasound findings may be diffuse or localised to a particular area of tissue. A wide range of diseases produce orbital inflammation, including infections, lymphoid or granulomatous processes, or secondary passive congestion from arteriovenous anomalies. The aetiology can only be inferred from ultrasonography, since the same findings may be present in several different pathological disorders. Of major importance is the fact that inflammatory changes are clearly distinguishable from the findings associated with tumours.

Cellulitis and pseudotumour (idiopathic orbital inflammation) produce diffuse orbital ultrasound findings. The fat pad is generally the most involved tissue, with an abnormally diffuse mottled texture identified ultrasonically by widening of spaces between echoes. These changes probably result from interstitial oedema separating fat globules and connective tissue abnormally. A granuloma or abscess area may be identifiable ultrasonically as a focal mottled area, which is usually more evident after the area has become partially walled off from adjacent tissues. Oedema in the posterior space

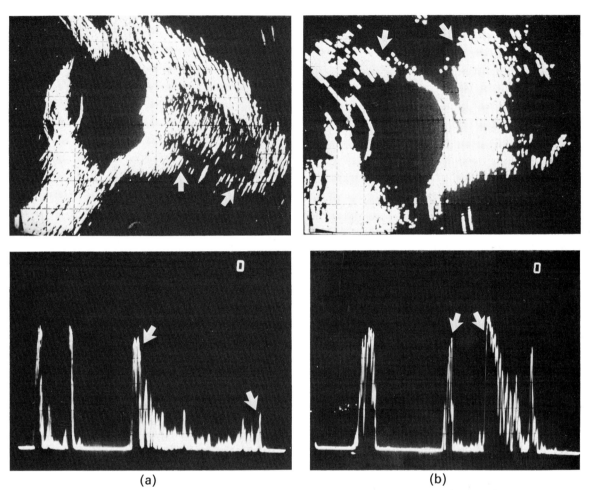

(a)   (b)

Fig 7.8 (a) Solid circumscribed orbital tumour (optic nerve glioma). The two-dimensional B-scan shows well-defined rounded contours (arrows). The A-scan shows that these echoes are significantly attenuated with increasing depth (arrows). (b) Infiltrative type of orbital tumour (lymphoma). The two-dimensional B-scan demonstrates irregular contour without complete boundaries (arrows). The A-scan shows low amplitude internal echoes with good sound transmission (arrows); this characteristic is variable with tumour type and size.

enveloping the globe and around the optic nerve sheath is particularly characteristic of pseudotumour. Dysthyroid exophthalmos (orbitopathy of Graves' disease) causes diffuse orbital oedema and inflammation with particular involvement of the extraocular muscles (Fig. 7.9). Accentuation and enlargement of these muscles is seen ultrasonically in nearly all patients with this disorder. On two-dimensional B-scan images this appears as a generalised widening of the echo-free area between the retrobulbar fat and the bony orbital wall. Muscle contours are much better defined than normally, and abnormal echoes may be seen within the disorganised and infiltrated muscle tissue. The degree of muscle enlargement varies considerably from subtle to very marked. Bilateral orbital changes are often evident in Graves' disease, even when exophthalmos appears unilateral clinically. Passive congestion of the orbit secondary to arteriovenous anomalies may produce similar ultrasound findings. Diffuse haemorrhage in the orbit, either spontaneous or traumatic, is also indistinguishable from the inflammatory-congestive ultrasonic changes already described.

Other orbital abnormalities

Structural abnormalities may account for apparent

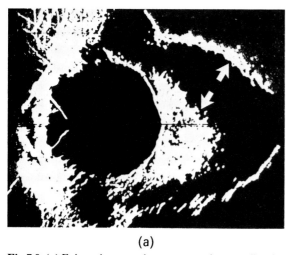
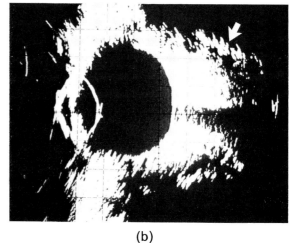

(a)   (b)

Fig 7.9 (a) Enlarged extraocular rectus muscles extending from the globe equator to the orbital apex (arrows) indicate a congestive or inflammatory orbital disease process. (b) The muscles of a normal orbit (arrow) contrast with the massive swelling that can occur.

exophthalmos. This may result from a shallow bony orbit which gives abnormal truncated contours of the orbital walls ultrasonically, or it may result from a unilaterally large globe with high myopia. The latter is identified by a large axial globe dimension seen on either A-scan or two-dimensional B-scan. This is usually a life-long non-progressive problem after the teenage years, but may occasionally need differentiation from orbital tumour or inflammatory disorders.

As described within the globe, foreign bodies of all types are detectable with ultrasonography. This becomes somewhat more difficult in the retrobulbar area because of the highly reflective heterogeneous fat. A small foreign body buried in the fat can be identified ultrasonically on a two-dimensional B-scan image by lowering the receiver sensitivity so that practically all other echoes disappear and only the spot from the foreign body persists. Small or irregular foreign bodies may escape detection with ultrasound. Even large foreign bodies with flat surfaces may be missed because there is less flexibility for getting the beam perpendicular to them in the deeper regions of the orbit, the bony orbital walls being a limiting factor in directing the transducer.

BIOMETRY

Precise measurement of eye dimensions to an accuracy of 0.01 mm is feasible with current ultrasound technology. Such accuracy requires a high frequency transducer (15 or 20 MHz) with weakly focused beam. Signal processing to truncate the returned echo enhances the apparent resolution. Electronic gating of the two echoes of interest gives much more accurate time measurement than visual analysis can do. Sound speed within each component of eye tissues has been determined experimentally, and must be applied to each segment of the time measurement of returned echoes when converting this to a distance measurement. Besides these electronic factors, the ultrasonic beam must be optimally aligned towards the eye with the maximum distance between the two echoes on the optical axis and perpendicular to their interfaces. Since there must be no compression of the cornea by the transducer, some type of water stand-off is necessary. Measurements require only A-scan, whereas diagnostic applications of ultrasound require A-scan and two-dimensional B-scan combined.

The primary clinical importance of eye biometry is calculation of the dioptric power for intraocular lens implants. A plastic prosthetic lens that is placed surgically into the eye at the time of cataract removal, must have the proper focusing power for that individual eye. A wide range of lens powers is used from +10.00 to +28.00 dioptres. Clinical assessment of the appropriate lens implant for an individual eye is derived from three parameters.

These are corneal curvature, axial eye length (by ultrasound), and the planned lens position in the eye. Lens power is then calculated using a nomogram or formula derived from physiological optical principles.[2]

SUMMARY

Ophthalmic ultrasonography has developed since its introduction in 1956 into an essential diagnostic aid. Reproducible ultrasonic criteria are now well established for identification of vitreoretinal disorders, foreign bodies, intraocular tumours, and orbital abnormalities including tumours, congestive-inflammatory processes, and structural anomalies. Accuracy of 85 to 95 per cent is obtainable by experienced examiners. Ultrasonic measurement of eyes also aids in cataract surgery for lens implant power determination.

The thrust of current investigation in ophthalmic ultrasound is being directed toward improved techniques for acoustic characterisation of tissues and the exploration of therapeutic applications in ocular disease.[5] Conventional ultrasound systems fail to utilise all the information contained in returned echoes. The simplified echo waveforms seen on oscilloscopes are designed to facilitate visual analysis. Computer spectral analysis of echoes has yielded far more detailed frequency components, thus making possible the identification of specific acoustic signatures for each type of ocular pathology. In relation to the therapeutic application of ultrasound, the destructive power of this energy is being used experimentally to produce filtering operations, to disperse vitreous haemorrhage, and to coagulate tumours in eyes. The fruits of these research efforts will be evident in the future.

REFERENCES

1. Baum G, Greenwood I 1958 The application of ultrasonic locating techniques to ophthalmology: part 2. Ultrasonic visualization of soft tissues. Arch Ophthal 60:263–279
2. Binkhorst R D 1975 The optical design of intraocular lens implants. Ophthal Surg 6:17–31
3. Bronson N R 1969 Quantitative ultrasonography. Arch Ophthal 81:460–472
4. Bronson N R, Fisher Y L, Pickering N C, Traynor E M 1980 Ophthalmic contact B-scan ultrasonography for the clinician. Williams & Wilkins, Baltimore
5. Chang S, Coleman D J, Dallow R L 1980 Trends in ophthalmic ultrasonography. In: Kurjak A (ed) Progress in medical ultrasound, 1. Excerpta Medica, Amsterdam, p 279–312
6. Coleman D J, Koenig W F, Katz L 1969 A hand operated ultrasound scan system for ophthalmic evaluation. Am J Ophthal 68:256–263
7. Coleman D J, Lizzi F L, Jack R L 1977 Ultrasonography of the eye and orbit. Lea & Febiger, Philadelphia
8. Dallow R L (ed) 1979 Ophthalmic ultrasonography: comparative techniques. Int Ophthal Clin 19(4):1–310
9. Gernett H, Franceschetti A 1967 Ultrasonic biometry of the eye. In: Oksala A, Gernet H (eds) Ultrasonics in ophthalmology. Karger, Basel, p 175–200
10. Jansson F 1963 Measurements of intraocular distances by ultrasound. Acta Ophthal 74 (suppl):1–51
11. Mundt G H, Hughes W F 1956 Ultrasonics in ocular diagnosis. Am J Ophthal 41:488–498
12. Oksala A, Lehtinen A 1957 Diagnostic value of ultrasonics in ophthalmology. Ophthalmologica 134:387–395
13. Ossoinig K C 1972 Clinical echo-ophthalmology. In: Blodi F (ed) Current concepts in ophthalmology, vol III. Mosby, St Louis, p 101–130
14. Purnell E W 1966 Ultrasound in ophthalmological diagnosis. In: Grossman C, Homes J H, Joyner C, Purnell E W (eds) Diagnostic ultrasound. Plenum, New York, p 95–109

8

Superficial structures

Catherine Cole-Beuglet

INTRODUCTION

Early investigators recognised the potential for ultrasonic evaluation of superficially located structures, but were limited in their imaging because the only transducers were low frequency nonfocused types. The resolution of these transducers was not sufficient for reliable diagnostic examinations. The development of higher frequency (5, 7.5 and 10 MHz) transducers, focused within a few centimetres of the crystal face, has allowed improved two-dimensional B-scan imaging of such superficial organs as the thyroid gland, and the mammary gland and the testicles. These transducers are frequently enclosed within a water bath which allows for ideal imaging of structures immediately below the skin surface. Real-time instruments that have the ability to image superficial structures are often referred to as 'small parts scanners'. Different types of equipment fall into this category including ophthalmic scanners, water path enclosed sector scanners, electronically focused linear arrays and duplex scanners which combine real-time imaging with Doppler flow recording capabilities (see Chapter 1).

NECK

In the early 1970s, contact two-dimensional B-scans of the neck using high frequency (5 MHz) narrow diameter transducers resulted in improved ultrasonic imaging of the thyroid and enlarged parathyroid gland.[4] The recent development of transducers of 5, 7.5 and 10 MHz frequency, focused within a few centimetres of the crystal face, allows resolution of abnormalities to within a few millimetres. Obtaining contact scans of the neck with these small diameter transducers attached to the articulated arms of a static B-scanner has been used routinely in many facilities. No matter what type of equipment is used, the patient is usually placed in the supine position with a bolus under the shoulder and the neck hyperextended. Free standing open water baths placed over the neck, with the ability of the transducers to be moved freely within the fluid, can also be used successfully. The use of a water bath eliminates the strong 'opening bang' echoes obtained when a transducer is in direct contact with the skin. Special water delayed self-contained transducer units can also be used successfully to obtain high resolution images.

Specialised small parts real-time scanners have been developed utilising high frequency transducers enclosed in a water bath.[14] The transducer is mechanically oscillated to produce images in real-time. These specialised small parts imagers have resolutions in the axial and lateral directions down to 1 mm. With this type of equipment, thyroid lesions as small as 2 or 3 mm in diameter can be detected even prior to the development of clinically palpable nodules. It is also possible occasionally to image normal parathyroid glands, as well as to detect enlargement, most often associated with hyperparathyroidism.[30]

Mass lesions in the neck are imaged on ultrasound B-scans to ascertain whether they are cystic, solid or complex (a combination of the two). In the region of the thyroid gland, fluid-filled areas represent thyroid cysts in 11 to 20 per cent of patients referred for ultrasound evaluation of a cold nodule following radionuclide uptake examination.[3, 12]

Simple cysts image as well defined smooth margined echofree areas and are usually unilateral in location (Fig. 8.1). In a small percentage of patients, the predominantly echo-free area contains linear or punctate echogenic foci which may represent haemorrhage or cellular debris within a thyroid cyst (Fig. 8.2). In these cases, there is often a history of recent development of a mass lesion in the region of the thyroid or rapid enlargement of a pre-existing nodule, or both.

A midline mass that extends from the junction of the chin and the neck to the suprasternal notch

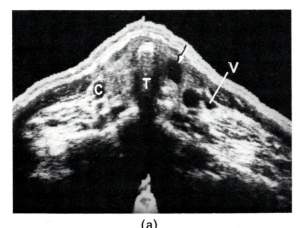

(a)

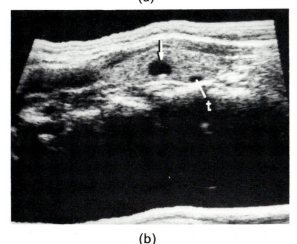

(b)

Fig. 8.1 Thyroid cyst. 59-year-old female with a left mid-thyroid 'cold nodule'. (a) Transverse 7.5 MHz B-scan at the level of the cold nodule in the left lobe. The well defined echo-free region measures 3 × 3 mm (arrow). Note enhanced sound transmission posterior to the fluid. T, trachea; C, common carotid artery; V, jugular vein. (b) Sagittal B-scan of the left lobe. Oval echo-free area within a normal-sized thyroid lobe. t, thyroid artery.

may represent a thyroglossal duct remnant.[3] This abnormality usually images as a well-delineated echo-free mass with occasional internally located linear or punctate echogenic foci. Positioning the patient in a decubitus or upright posture and rescanning usually demonstrates movement of these echogenic foci to a dependent position as would be expected with debris or clot. If the cyst is large, it may project to one side of the midline, deviating the trachea to the opposite side (Fig. 8.3). These may present clinically at any age.

Lateral neck fluid-containing mass lesions may represent branchial cleft cysts. Cystic masses occurring at the angle of the jaw usually are arising from the parotid or submandibular salivary glands, depending on their exact anatomical location and contiguity with the associated gland. Enlarged lymph nodes can also be imaged and occur most frequently in the lateral portions of the neck. The enlargement is most often due to lymphoma or metastatic disease. On occasion, they may undergo cystic degeneration and image as cyst-like masses (Fig. 8.4). Often by turning up the gain, weak internal echoes can be identified establishing their predominantly solid nature.

With the use of higher frequency transducers, masses may be identified within normal sized thyroid glands with negative radionuclide scans. When these are both solid and solitary, a management dilemma can occur, especially if there is a history of irradiation to the neck in childhood, since an increased incidence of thyroid cancer has been reported.[45] Fine needle aspiration biopsy has recently become the technique employed for analysis of the histopathology of these solid areas of abnormality.[13, 50] When multiple nodules or areas of solid architectural disruptions are imaged, there is an increased likelihood that they are benign in nature.[43]

Adenomas are the most common nodules that occur within the thyroid gland. If solitary, an adenoma usually presents clinically as a small palpable nodule within one lobe or the isthmus of the gland. Radionuclide scans usually demonstrate areas of decreased uptake when compared to the surrounding normal thyroid gland. At the time of the ultrasound examination, a physical examination of the neck is helpful in confirming the site and size of the mass lesion. Since benign thyroid

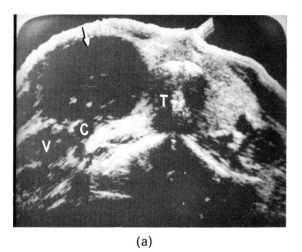

(a)

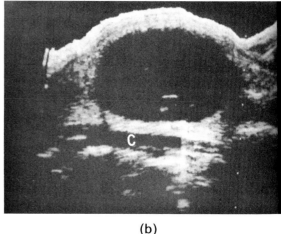

(b)

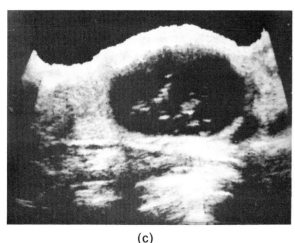

(c)

Fig. 8.2 Thyroid cyst with internal haemorrhage. Sudden development of a palpable right neck mass in a 38-year-old female. Large cold nodule in right lobe on radionuclide scan. (a) Transverse 5 MHz B-scan over the lower neck shows a 40 × 30 mm predominantly echo-free mass (arrow) distorting the right thyroid lobe laterally. The mass extends anterior to the common carotid artery (c) and the jugular vein (v). Echogenic internal foci represent haemorrhagic debris. Normal thyroid isthmus and left lobe. T, trachea. (b) Sagittal B-scan over the right thyroid mass shows a 45 mm echo-free area anterior to the carotid artery (C). (c) Sagittal B-scan over right thyroid medially with increased gain. Note the internal linear echogenic surfaces, the augmented transmission of sound distal to the fluid and the normal thyroid superior to the mass.

disease may present as bilateral or multiple adnormalities, ultrasound can be used effectively to delineate the location of the imaged cold nodules. Adenomas may have the same echotexture as the surrounding thyroid gland, or they may image as areas of decreased or increased echogenicity when compared to the adjacent thyroid tissue.[40] These lesions expand slowly and, as a result, at the periphery of the nodules, there is often recorded a 1 to 2 mm echolucent region, apparently due to compressed normal surrounding thyroid tissue

(Fig. 8.5). It has been reported that when such an echolucent line is seen, this indicates that it is a slowly growing benign lesion.[16] Several cases of malignant nodules that had a 'halo' echolucent ring have been reported, however, and thus it is not a specific finding with benign lesions.[37]

Adenomas are prone to internal haemorrhage and necrosis. On ultrasound, irregular shaped non-marginated echofree areas can be recorded within these solid tissue nodules (Fig. 8.5). Similar echo-free areas may also be imaged in regions of colloid

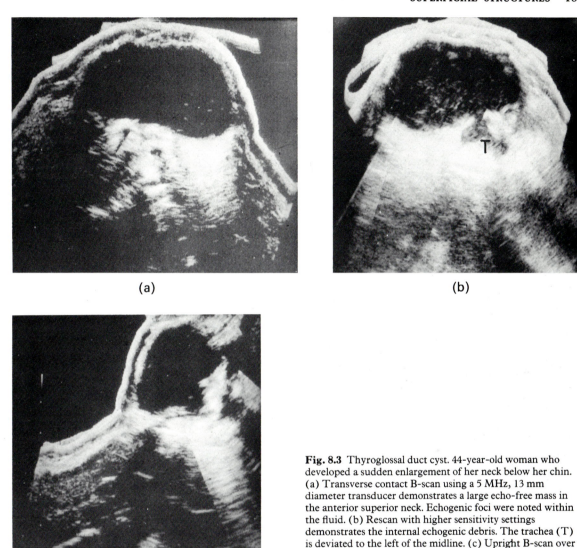

(a)

(b)

(c)

Fig. 8.3 Thyroglossal duct cyst. 44-year-old woman who developed a sudden enlargement of her neck below her chin. (a) Transverse contact B-scan using a 5 MHz, 13 mm diameter transducer demonstrates a large echo-free mass in the anterior superior neck. Echogenic foci were noted within the fluid. (b) Rescan with higher sensitivity settings demonstrates the internal echogenic debris. The trachea (T) is deviated to the left of the midline. (c) Upright B-scan over the mass demonstrates settling of the debris. The normal thyroid gland is imaged inferiorly on the left.

degeneration. Colloid is a gelatinous substance and frequently contains low level scattered echoes within it, cellular debris or proteinaceous material.

Solitary cold nodules represent carcinomas in 10 to 24 per cent of patients examined.[38, 43, 45] Carcinoma of the thyroid is usually a slow growing tumour which remains within the gland for many years. It may extend regionally or metastise to the lateral neck lymph nodes or to both. On an ultrasound B-scan, malignant tissue tends to be less echogenic than normal thyroid tissue and often

demonstrates acoustic attenuation.[7, 30, 40] The one exception is pure follicular carcinoma which can resemble a follicular adenoma on an ultrasound B-scan. These lesions can have the same echo-texture as the normal thyroid. If an echolucent capsule is imaged, the complete margin is examined for areas of disruption which may be helpful in distinguishing it from a benign adenoma.[30] The majority of thyroid cancers are mixed papillary and follicular lesions and these image as complex, predominantly solid masses. Margination between the lesions and

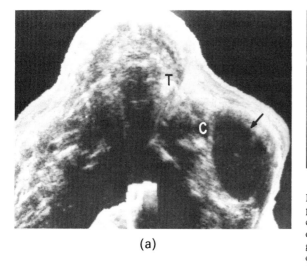

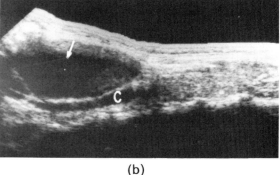

(a)

(b)

Fig. 8.4 Lymphoma. 33-year-old schoolteacher. Large palpable mass below the left jaw laterally. (a) Transverse contact B-scan of the neck demonstrates a large, weakly echogenic mass (arrow) in the left neck lateral to the thyroid gland (T) and the common carotid artery (C). (b) Sagittal oblique B-scan over the area demonstrates an oblong smooth-margined area (arrow) containing scattered weak internal echoes superior to the left thyroid lobe and anterior to the carotid artery (C).

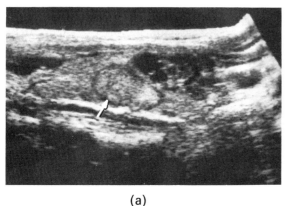

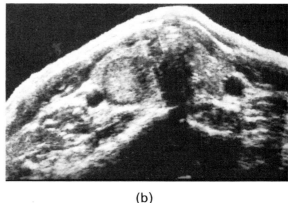

(a)

(b)

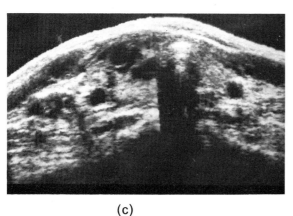

(c)

Fig. 8.5 Thyroid adenomas, one with an echo-poor margin (a capsule) and a second one with areas of haemorrhagic degeneration. 55-year-old male with a small palpable nodule in the mid portion of the right thyroid. Radionuclide ^{131}I scan imaged a cold area. 7.5 MHz water path contact B-scans. (a) Sagittal scan over right thyroid lobe demonstrates two distinct lesions. On the left superiorly is an oval 12×20 mm echogenic mass (arrow) containing variable strength echoes in a pattern similar to the adjacent thyroid tissue. Note the 1 mm echo-poor margin, the 'capsule'. Inferiorly, on the right there is an irregular-shaped, complex area in the anterior thyroid. Round echo-free regions with echogenic septations represent haemorrhage within a solid mass. (b) Transverse B-scan superior thyroid gland. Round 15 mm echogenic mass shows a narrow, weakly echogenic margin, a capsule. (c) Transverse B-scan inferior thyroid gland. Complex area, 20 mm in the anterior portion of the right lobe represents a benign adenoma with internal haemorrhage.

the normal thyroid parenchyma is not usually as distinct as it is with benign lesions. Needle aspiration biopsy is often required to make the diagnosis of thyroid cancer.[13, 50]

Calcification may occur in papillary carcinomas, occasionally imaged as punctate psammoma type on an X-ray of the neck. These small calcific deposits may be imaged ultrasonically, especially with high frequency transducers, as echo-dense reflections associated with distal acoustic shadowing. Large coarse calcifications can occur within the thyroid gland in such benign conditions as multinodular goitres, post-inflammatory changes and post-haemorrhagic necrosis. The location of the calcifications may have some significance, with

benign coarse calcifications tending to occur within the centre of benign adenomas (Fig. 8.6), and with peripheral fine calcifications occurring within carcinomas.

Generalised enlargement of the thyroid gland is usually associated with inflammatory conditions. Acute thyroiditis images as diffuse enlargement of both lobes and the isthmus. The echo-texture of the lobes is weaker than the echo-texture of the normal thyroid (Fig. 8.7). As the thyroiditis resolves, the gland may show a decrease in size and, at the same time, a brighter echo-texture due to

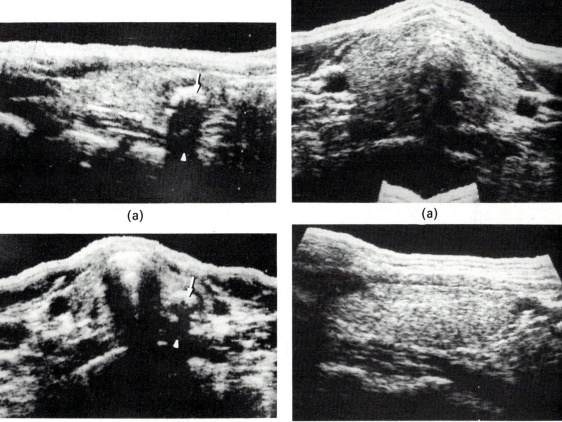

(a)

(b)

(a)

(b)

Fig. 8.6 Benign adenoma containing dystrophic calcification centrally. (a) Sagittal B-scan through the left 'cold' nodule shows a 16 mm oval shaped mass containing a curvilinear bright echofocus centrally (arrow). This represents coarse calcification which exhibits strong acoustic attenuation (arrowhead). (b) Transverse B-scan shows the anterior surface of the calcification (arrow) and the distal shadow (arrowhead).

Fig. 8.7 Diffuse thyroid enlargement. 44-year-old female who noticed swelling of her lower neck over several months. (a) Transverse B-scan over the thyroid isthmus. Diffusely enlarged thyroid and isthmus with increased echogenic glandular texture anteriorly and decreased echoes posteriorly due to attenuation of the sound beam by the enlarged gland. (b) Saggittal B-scan over the left lobe. Note the round contour of the inferior enlarged lobe.

deposition of fibrous tissue and, in a few cases, calcification.

Metastatic disease to the thyroid can occur from primary carcinomas anywhere in the body (Fig. 8.8). The appearances ultrasonically have no specific pattern that can differentiate them from primary tumours.[2]

With the recent development of small parts scanners containing high frequency transducers in the range of 7 to 10 MHz, it has become possible to image the parathyroid glands.[14] The normal parathyroid gland measures up to 5 mm in long axis and is most frequently located along the posterior margin of the thyroid gland. The gland may be separated from the posterior lobe of the thyroid by a thin fibrous tissue capsule or it may be located within the posterior thyroid capsule. The location of the glands (most commonly four to six in number) is variable and, not infrequently, some of these are found in other regions of the neck or in the substernal position. Localisation of the parathyroid glands has been attempted with radionuclide scanning, arteriography or CT scanning of the neck and mediastinum.[10, 46] Success rates in localising enlarged glands range from 30 to 75 per cent.[8, 11, 12, 46] Patients who present with symptoms of hyperparathyroidism may harbour a hyperplastic gland, an adenoma or, rarely, a carcinoma of the parathyroid.[17, 24] Occasionally, patients present with multiple endocrine neoplasia syndrome for which the treatment of choice is total parathyroidectomy. If primary hyperparathyroid-

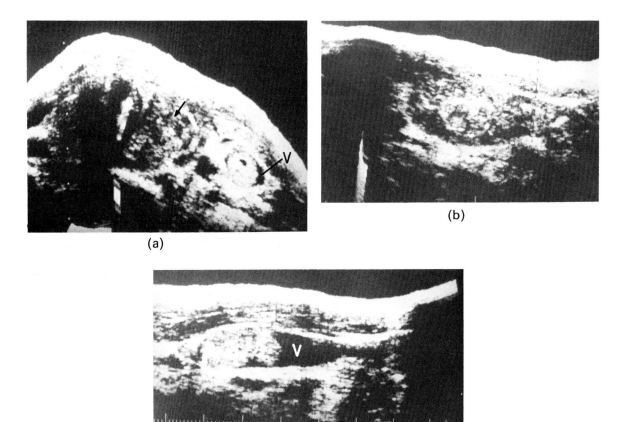

(a)

(b)

(c)

Fig. 8.8 Metastatic hypernephroma to left lobe of thyroid and internal jugular vein. Male with palpable mass in the left lower neck. Past history of nephrectomy for renal cell carcinoma 13 years previously. (a) Transverse B-scan over lower neck. Enlarged left lobe of thyroid with an abnormal echopattern (arrow). Echogenic mass within the left jugular vein (V). (b) Sagittal B-scan of the left thyroid lobe. Solid mass with an inhomogeneous echopattern. (c) Sagittal B-scan of the left jugular vein. Lobulated echogenic mass within the vein (V) represents tumour extension from the thyroid gland presumably along the inferior thyroid vein into the jugular vein.

ism occurs in the absence of a definite history of polyendocrine or familial disease, only the glands that are definitely enlarged should be removed to avoid the risk of permanent hypoparathyroidism.

The ability to image the soft tissues of the neck, using high frequency small parts scanners, has resulted in requests from clinicians for ultrasonic localisation of enlarged parathyroid glands.[1] The normal gland cannot be routinely imaged even with 10 MHz water delay systems.[30] In one series, enlarged glands over 5 mm in diameter were identified in 85 per cent of the 25 patients with primary hyperparathyroidism.[42] Using ultrasound, the gland can most often be localised in relation to the minor neurovascular bundle, along the posterior medial aspect of the thyroid and anterior to the longus colli muscle. The neurovascular bundles can be identified on both the transverse and oblique sagittal scans at the angle of the tracheal œsophageal groove. The left inferior parathyroid gland must be differentiated from the œsophagus which occasionally images posterior to the lower portion of the left thyroid lobe. This can be misdiagnosed as an enlarged parathyroid. The œsophagus is identified on a sagittal scan as an echofree tubular structure, often containing an echodense central area which represents the apposed mucosal surfaces (Fig. 8.9). The patient can be requested to swallow water while the examiner monitors this region, using real-time B-scan equipment to record the distention that occurs during the passage of the water bolus.

The internal echotexture of an enlarged parathyroid is weaker than the overlying thyroid echotexture. Hyperplasia is the commonest aetiology for an enlarged parathyroid and this may be confined to a single gland (Fig. 8.10). Sonic imaging of a single enlarged gland greatly assists the surgeon in preoperative localisation, and thus often eliminates the need for bilateral neck exploration. Parathyroid adenomas, which are benign tumours, present an ultrasonic appearance of low internal echogenicity, as is the case in hyperplasia (Fig. 8.11). Carcinoma of the parathyroid glands is extremely rare. From experience with thyroid carcinoma, an enlarged gland with a greater degree of echo attenuation might be expected to be seen by ultrasound.

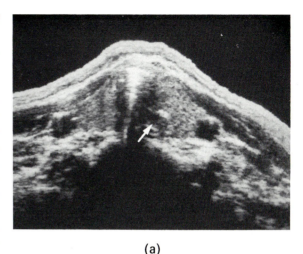

(a)

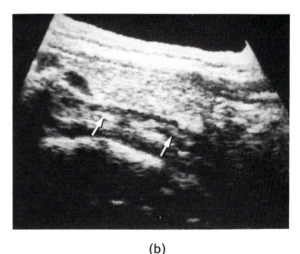

(b)

Fig. 8.9 Normal thyroid and œsophagus. 48-year-old female with symptoms of hyperparathyroidism. No palpable neck masses. (a) Transverse 7.5 MHz B-scan over mid-thyroid. Normal size, contour and echo-texture of the lobes. Note complex area in the angle of the trachea and left posterior lobe (arrow). (b) Saggittal B-scan over left thyroid lobe. Normal thyroid texture, oval weakly-echogenic area superiorly represents a vessel. Tubular weakly echogenic area (arrows) posterior to the lobe with central linear echogenic foci represents the œsophagus.

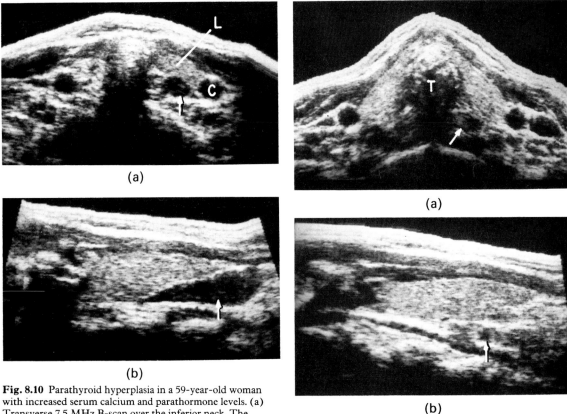

(a)

(b)

(a)

(b)

Fig. 8.10 Parathyroid hyperplasia in a 59-year-old woman with increased serum calcium and parathormone levels. (a) Transverse 7.5 MHz B-scan over the inferior neck. The thyroid gland shows a normal size and uniform echo reflective pattern. Note the 3 mm nodule (arrow) containing weak internal echoes along the posterior border of the left (L) thyroid lobe, medial to the carotid artery (C). (b) Sagittal B-scan over the left thyroid lobe. The teardrop-shaped solid mass (arrow) posterior to the inferior thyroid margin represents parathyroid hyperplasia. (Inferior to the right.)

Fig. 8.11 A 48-year-old patient with hypercalcaemia. No palpable neck masses. Left parathyroid adenoma, chief cell type. (a) Transverse 7.5 MHz B-scan inferior thyroid. Normal thyroid lobes on each side of the trachea (T). Mass (arrow) posterior to the left lobe contains weak internal echoes and measures 6 mm. (b) Sagittal B-scan of the left thyroid lobe. Lobulated, weakly-echogenic mass (arrow) along the inferior posterior margin of the thyroid represents a parathyroid adenoma on biopsy.

BREAST

The term 'ultrasound mammography' refers to the examination of the total breast parenchyma utilising ultrasound imaging. This has become possible with the technical and commercial development of dedicated breast scanners and high frequency superficial structure two-dimensional B-scanners. Previously, ultrasound examination of the breast had included evaluation of palpable breast masses using contact B-scan, static and real-time techniques over the area of palpable abnormalities.[5,48] The sonic evaluation of the internal composition of palpable masses was a useful adjunct for the referring clinician and mammographer to deter-

mine if a suspicious area represented a fluid-filled cyst or a solid mass lesion. It remained extremely difficult technically to evaluate the total breast parenchyma with contact B-scanning due to the mobility of this gland on the chest wall.

Various investigators throughout the world developed different methods of immobilising the breast. Wells in the United Kingdom constructed a water tank with the transducer mounted on a rotating arm capable of simple or compound scanning.[51] The patient, in a prone position, immersed her breasts within the open water bath. In Japan, Wagai and Kobayashi developed enclosed water path systems.[27, 49] One contained a 5 MHz transducer which could be mechanically driven at preset increments to obtain sequential transverse B-scans of approximately one quadrant of the breast. With this system, the patient, in a supine position, had the enclosed water path placed over a quadrant of the breast, and the size and weight of the instrumentation provided adequate immobilisation of the breast against the anterior chest wall. In Australia, Kossoff and associates developed an automated static water path scanner for general obstetrical and abdominal use, and a second smaller unit for breast imaging.[20, 29]

In the United States, two types of water path scanners have been developed. One type has transducers mounted at the base of a water bath and the patient, in a prone position, immerses the breast. The transducers are mechanically moved to obtain B-scan images of the gland for static or real-time viewing.[6] The second type has transducers mounted within an enclosed water path which is placed over the breast of the patient in a supine position.[15, 25] Dedicated breast scanners are now commercially available allowing B-scan imaging of the entire mammary gland and having rapid playback and viewing capabilities.

Indications for ultrasound mammography

The indications for ultrasound mammography described in this section were formulated during an evaluation of 2000 ultrasound mammograms performed at Thomas Jefferson University Hospital using water path techniques.

Patients with a palpable mass

When a mass lesion is imaged on an ultrasound B-scan, the area of the mass lesion is compared with the surrounding glandular tissue. Mass lesions are designated as fluid-filled or solid. The breast mass diagnostic ultrasound B-scan criteria established in the literature are used to differentiate fluid-filled and solid mass lesions (Table 8.1).[21, 26] Fluid-filled areas may be single or multiple (Figs. 8.12 and 8.13). When multiple variable sized fluid collections are noted throughout the breast parenchyma, in many cases palpable abnormalities are not appreciated clinically. When a patient with a

Table 8.1 Ultrasound two-dimensional B-scan analysis of breast masses

Classification	Characteristics	
I	Contour	smooth irregular
II	Shape	round oval lobulated
IIIA	Internal echo pattern	echo free contains echo (uniform or nonuniform)
IIIB	Internal echo strength	strong intermediate weak
IV	Boundary echoes (anterior and posterior)	strong intermediate weak absent
V	Attenuation	minimal (enhancement) intermediate great (shadow)

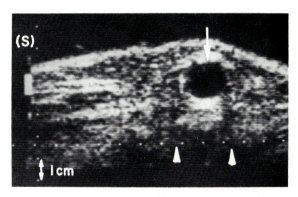

Fig. 8.12 Simple cyst. 39-year-old woman with a palpable mass in the upper outer quadrant of the left breast. Sagittal B-scan under compression over the mass shows a 24 mm echo-free area (arrow) with smooth, well-defined margins and enhanced sound transmission distal to the area. Note the lateral wall shadows (arrowheads). (S, superior.)

known history of fibrocystic disease develops a palpable mass, ultrasound mammography delineates the characteristics of the mass, and designates the area as fluid-filled or solid in nature. Occasionally, septations are noted within a fluid-filled area. Two adjacent cysts may give the appearance of a larger cyst with an internal septation. Carcinoma

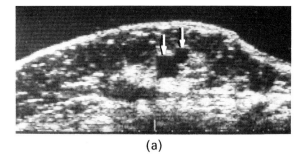

(a)

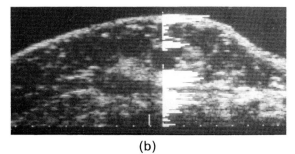

(b)

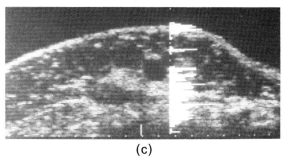

(c)

Fig. 8.13 A 41-year-old woman with a multicystic breast. (a) Sagittal B-scan 5 mm lateral to the left nipple. Two adjacent mass lesions (arrows) with well-defined margins containing no internal echoes. (b) A-mode vertical histogram through the larger mass confirms the absence of internal echoes. The amplitudes of the echoes distal to the mass are greater than those proximal to the mass demonstrating acoustic enhancement. (c) A-mode vertical histogram through the smaller mass demonstrates no internal echoes and distal acoustic enhancement.

within a cyst, arising from the internal lining epithelium, is a rare finding. The image might be expected to show a papillary solid growth projecting into the fluid-filled space from the internal wall. In the majority of cases, when such a cyst is aspirated, the clinician obtains haemorrhagic fluid rather than the clear serous fluid normally obtained from a simple cyst.

When a palpable mass images as a solid lesion, the ultrasound B-scan criteria are applied to determine if the solid is malignant or benign-

appearing. From a retrospective review of 117 confirmed breast carcinomas of all histological types, the majority of malignant-appearing solids demonstrate an irregular contour (Fig. 8.14). A solid breast mass is of weaker echo amplitude than the surrounding breast parenchyma. Acoustic attenuation is evident in over two-thirds of the cases. Infiltrating duct carcinoma comprises the majority of infiltrating breast carcinomas in the United States, and these tumours elicit a desmoplastic reaction as they grow in a stellate fashion into the breast parenchyma. For this reason, they are called 'scirrhous' carcinomas. On a B-scan, they image as irregular margined variable-shaped masses containing weak, nonuniform internal echoes. Acoustic attenuation images as a central shadow distal to the mass (Fig. 8.15).

A small fraction of breast carcinomas are termed 'cellular', as opposed to the previously-described 'acellular' carcinomas. Medullary carcinoma is an example of an acellular lesion. These are bulky tumours and growth from the periphery occurs in a smooth fashion compared to the tentacle stellate peripheral growth from an infiltrating duct carcinoma. When these malignancies are imaged on ultrasound, they have smooth or slightly lobulated margins and contain weak internal echoes. The distribution of the echoes within the mass may be inhomogenous. These mass lesions do not exhibit as great an acoustic attenuation as do the infiltrating duct carcinomas. Thus, echoes are usually recorded distal to these mass lesions (Fig. 8.16).

In contrast, when a palpable mass images as a smooth margined solid, usually with a round or oval shape, and demonstrates either intermediate or no significant acoustic attenuation, it is likely that this represents a benign tumour. The majority of benign-appearing solid mass lesions represent fibroadenomas. Generally, a younger patient presents to her clinician with a recently discovered mobile palpable mass in the breast. Occasionally, X-ray mammography images such a mass separate from the dense glandular tissue, normally present within the young breast. Ultrasound mammography images these as smooth margined, weakly echogenic masses within the strongly echogenic breast parenchyma (Fig. 8.17). During an evaluation of 100 biopsy-confirmed fibroadenomas, it was found that in six cases these benign-appearing

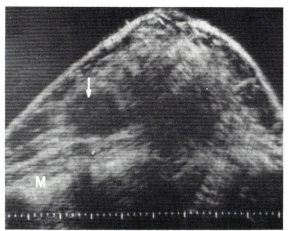

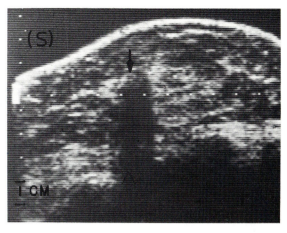

Fig. 8.14 Infiltrating duct carcinoma. Palpable mass in the superior left breast of a 42-year-old woman. Sagittal B-scan just lateral to the nipple shows a 30 mm irregular-margined, solid mass (arrow) within the superior breast parenchyma. The mass contains variable strength internal echoes.

Fig. 8.15 Infiltrating duct carcinoma in a post menopausal woman. Sagittal B-scan right lateral breast shows a 15 mm area of acoustic shadowing (arrow) that extends through the pectoralis major muscle (arrowhead). This represents great acoustic attenuation from the tumour. (S, superior.)

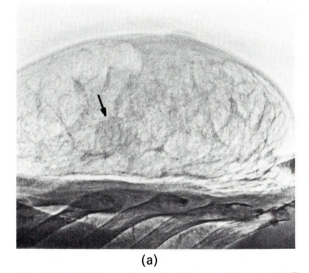

(a)

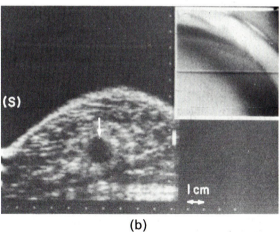

(b)

Fig. 8.16 Medullary carcinoma in a 52-year-old woman. (a) Dominant mass density (arrow) imaged on a mediolateral xeromammogram. (b) Sagittal B-scan under compression over the palpable mass shows a round 20 mm solid lesion (arrow) with weak internal echoes. (S, superior.) The inset photograph of the breast immersed in the scanner tank shows the scan section position.

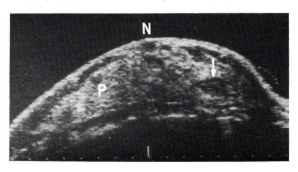

Fig. 8.17 Small fibroadenoma. 42-year-old female with a palpable mass inferior to the right nipple. Sagittal B-scan at the level of the right nipple (N). Mass (arrow) within the parenchyma (P) is oval shaped, contains weak internal echoes and does not demonstrate excess acoustic attenuation.

masses proved on biopsy to be infiltrating duct carcinoma. For this reason, the ultrasound B-scan criteria have been expanded to include all imaged solid mass lesions within the breast as potentially malignant. Thus, histopathology of a dominant solid mass lesion is usually confirmed by a fine needle or open excisional biopsy (Fig. 8.18).

Patients with dense glandular tissue demonstrated on X-ray

When a patient presents with a radiographic diagnosis of a dense breast, masses may not be delineated. Ultrasound mammography may be able to demonstrate lesions within the otherwise uniformly echogenic glandular tissue. Frequently, these patients are less than 50 years of age. Since the American College of Radiology does not recommend periodic filming of the breast to detect carcinoma in patients less than 50 years of age, ultrasound becomes the method of choice in imaging dense breasts (Fig. 8.19).[15]

Pregnant patients

During pregnancy and lactation, the proliferation of the terminal ducts within the mammary gland causes increased density on X-ray. These patients are not candidates for X-ray evaluation of the

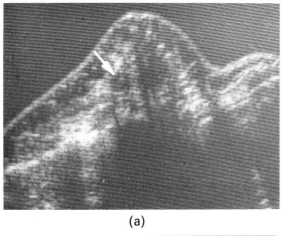

(a)

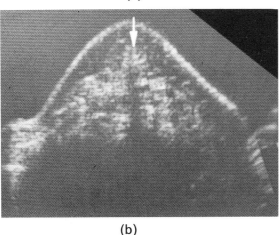

(b)

Fig. 8.18 Small infiltrating duct carcinoma in a 48-year-old woman. (a) Sagittal B-scan at the nipple level shows a 10 mm round mass containing variable strength internal echoes (arrow) within the strongly echogenic breast parenchyma. (b) Transverse B-scan over a portion of the mass in the superior breast (arrow) shows acoustic attenuation distal to the mass.

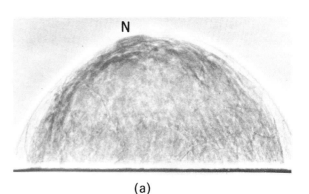

(a)

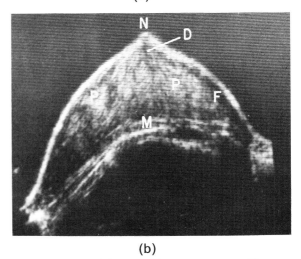

(b)

Fig. 8.19 Dysplastic breast on X-ray mammogram, 35-year-old woman. (a) X-ray mammogram, cephalocaudal view shows dense glandular tissue. N, nipple. (b) Transverse B-scan at the nipple level (N) shows echogenic breast parenchyma (P) with ducts (D) extending from below the nipple into the parenchyma in a radial fashion. Subcutaneous fat (F); pectoralis major muscle, (M).

breast during this period. Thus, when pregnant patients develop palpable breast masses or questionable abnormalities, ultrasound is the imaging modality of choice. The hyperplastic glandular tissue changes are imaged on an ultrasound mammogram as uniform fine echoes throughout the gland, and there is a decrease in the amount of fatty tissue. In late pregnancy and during the lactating period, the breast duct lumens increase in diameter and are fluid-filled. These ducts image as tubular, branching echo-free structures radiating from the nipple into the breast parenchyma (Fig. 8.20).

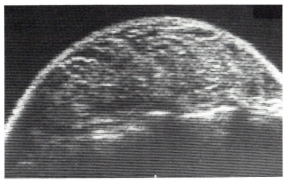

Fig. 8.20 Lactating breast, 32-year-old woman. Sagittal B-scan medial to the right nipple under compression demonstrates echogenic glandular tissue with scattered echofree tubular spaces, the milk-filled ducts.

Patients with clinically equivocal findings

Differentiation of normal breast tissue from a true abnormality is a valid indication for an ultrasound mammogram. Frequently, clinicians palpate areas of breast parenchyma thickening or nodularity, and a clear distinction between a palpable mass and an area of normal breast parenchyma cannot be made on the basis of a physical examination. Ultrasound images of this area can be compared to the surrounding breast parenchyma, and to a similar area in the contralateral breast, for evaluation of a true abnormality.

Recurrent masses in patients with fibrocystic disease

Frequently, patients with a diagnosis of fibrocystic disease develop breast masses in the third and fourth decade. Ultrasound imaging of the mass quickly determines if the patient has developed an additional or recurrent cyst. A potential use of ultrasound breast parenchymal imaging in this group of patients is for the documentation of the effects of therapy for fibrocystic disease. Patients may have an initial ultrasound mammogram prior to the institution of medication for this condition. The progress of the disease can be monitored at intervals to obtain an objective evaluation of the effects of specific medications or dietary control. In a patient with a documented fluid-filled cyst, change in size and volume of the cyst can be accurately assessed. The cyclic changes that frequently occur in these patients with the hormonal stimulation of the menstrual cycle causes variations in the intensity of the symptoms encountered in this disease. It is difficult to differentiate the normal cyclic changes occurring with the menstrual cycle from those encountered with specific therapies.

Patients with inflammatory conditions of the breast

'Mastitis' is a term applied to diffuse inflammation of the breast. This is occasionally seen in a patient in the postpartum period or during lactation. The inflammation of the breast may be diffuse or localised. When localised, an abscess may occur in the subareolar area. X-ray mammography is not indicated, as it is impossible to apply compression to an inflamed breast without considerable discomfort. Ultrasound mammography images the breast parenchyma and localised fluid collections within abscesses appear as irregularly margined, strongly echogenic regions containing weak internal echoes. During lactation, or in the older woman, diffuse inflammation of the breast can occur as a result of rapid growth of an infiltrating duct carcinoma. The resultant inflammatory response within the breast parenchyma is termed an 'inflammatory carcinoma'. Skin thickening frequently occurs in conjunction with this entity. On an ultrasound B-scan, the volume of the breast parenchyma is generally increased compared to the opposite breast. The supporting structures of the breast, Cooper's ligaments, and the breast parenchyma may appear altered and spread apart. Thickening of the fibrous connective tissue strands can often be imaged as brightly echogenic curvilinear lines extending to the skin surface.

The normal skin of the breast has a thickness of less than 2 mm. When thickened, the skin images as an increased echodense line measuring up to 6 mm. Occasionally, the thickened skin demonstrates a narrow echo-lucent line within the centre giving the appearance of a relatively echo-dense outer and inner surface and a narrow echo-lucent centre (Fig. 8.21).[28] X-ray mammography, of course, is also capable of imaging thickened skin. Frequently, the site of the inflammatory carcinoma is not seen. Such indirect signs as thickening of Cooper's ligaments, and straightening of the ligaments in a direction towards the tumour site, however, can sometimes be seen. An area of acoustic attenuation frequently indicates the site of the bulk of the malignant tissue.

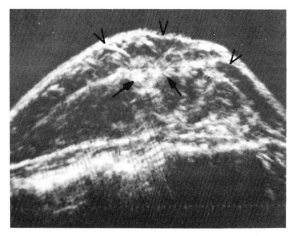

Fig. 8.21 Inflammatory carcinoma in a post menopausal breast with thickened Cooper's ligaments. Sagittal B-scan lateral to the nipple. The skin of the breast is irregular, puckered (arrowheads) and thickened. Cooper's ligaments are thickened and converge towards an echogenic focus (arrows).

Patients with breast trauma

Trauma to the breast can occur as a result of a direct blow, an automobile accident or surgical intervention. Haemorrhage within the breast may present clinically as a bruise on the skin surface. When imaged ultrasonically, the area immediately below the skin surface shows an increase in echogenicity distributed in a uniform fashion. Serial imaging over three to six months demonstrates a gradual disappearance of the bright echoes, and usually thickening of the fibrous connective tissue ligaments in the area. Fat necro-

sis, also occurring as a result of trauma, can be recorded as an area of increased echogenicity with variable degrees of distal acoustic attenuation. It is usually difficult to differentiate the appearance of fat necrosis from infiltrating malignancy. A history of breast trauma or a recent surgical procedure is needed before such a specific diagnosis can be suggested.

Patients with augmented breast

Augmentation of the breast is performed for cosmetic reasons or following post-subcutaneous mastectomy for minimal or in-situ breast carcinoma. Silicone gel prostheses are inserted in a retromammary space for enlargement of a small breast, and the glandular tissue is located over the surface of the conical prosthesis. Clinical evaluation of the glandular tissue over the prosthesis is frequently difficult due to fibrous changes, which occasionally occur in the tissue immediately adjacent to the envelope of the silicone gel. Nodularity along the incision site may represent scar tissue formation or mass lesions. With X-ray mammography, the glandular tissue overlying the prosthesis is incompletely imaged due to the extreme radiodensity of the silicone. Ultrasound is able easily to image both the overlying glandular tissue and the prosthesis. Since the speed of sound in silicone is slower than the speed of sound in the breast parenchyma, the image of the posterior surface of the silicone prosthesis may project posteriorly into that of the chest wall, but this presents no diagnostic problems. Of course, if the prosthesis is saline-filled, there is no such posterior margin displacement. A small fraction of patients develop a hard fibrous capsule around the silicone bag prosthesis.[21] When this occurs, the prosthesis may assume a distorted shape and image on a B-scan as a dense echogenic line over the flattened or distorted silicone bag (Fig. 8.22).

Male breast

Enlargement of the soft tissues in the region of the nipple in the male may represent deposition of adipose tissue on the anterior chest wall or proliferation of the breast ducts in the subareolar area. This latter condition is termed 'gynaecomas-

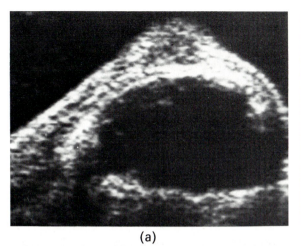

(a)

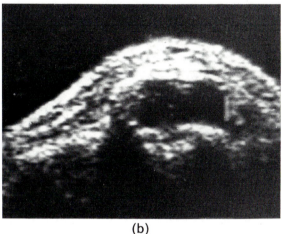

(b)

Fig. 8.22 Augmented breast. 29-year-old woman with silicone bag breast augmentation, duration one year. Gradual hardening of the left prosthesis. (a) Sagittal B-scan, right breast. Rounded echo-free silicone bag behind the breast tissue. (b) Sagittal B-scan, left breast. Distorted angular-shaped silicone bag with fibrous capsule formation.

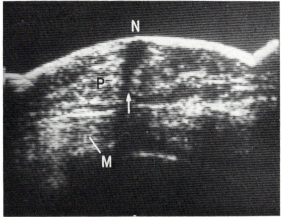

Fig. 8.23 Diffuse gynaecomastia. 51-year-old man with left breast enlargement. Sagittal B-scan through the left nipple (N). Echogenic parenchyma (P) between the skin and the chest wall muscles (M). Note shadow (arrow) in the subareolar area from the nipple.

tia'. Localised duct proliferation in the subareolar area can clinically present as a palpable mass. Carcinoma of the breast in the male also presents as a palpable mass in the subareolar area, and therefore it is difficult to differentiate from true gynaecomastia. X-ray imaging of the soft tissues of the anterior chest wall in the male can be technically difficult.[23, 34] Ultrasound imaging of the enlarged male breast using water path techniques allows for evaluation of the tissues of the anterior chest wall. Both localised and diffuse duct proliferation can be imaged in the subareolar area (Fig. 8.23). Imaged masses are evaluated according to the criteria for mass lesions in the female breast.

SCROTUM AND TESTIS

Recently, ultrasound examination of the scrotum has assumed an increasingly important role in the diagnostic evaluation of scrotal swelling. It is often difficult on physical examination to determine whether a palpable abnormality is testicular in origin or arises from adjacent structures within the scrotum, the epididymis, spermatic cord or scrotal sac wall. It is possible to image the scrotum and its contents using automated water bath scanners, high resolution real-time scanners, as well as static B-scanners using high frequency transducers. With the water bath technique, a patient lies in a prone position or sits in a frog-leg position on a doughnut with the scrotum immersed in warmed sanitized water. Sagittal and transverse images of the scrotum are obtained in this position. Using a water bed, it is also possible to scan the scrotum of a newborn or paediatric patient for evaluation of descent of the testis. Dedicated small parts scanners with high frequency 7.5 or 10 MHz transducers sealed in a water-filled container may be placed directly on the skin of the scrotum for evaluation of its contents.[31, 32]

Scrotal anatomy

The scrotum contains the male reproductive organ, the paired testes and the accessory organs of the internal reproductive system, the epididymis, and the vasa deferentia.

The testes are ovoid glands normally 5 cm in length and 3 cm in diameter. Each is suspended by a spermatic cord within the cavity of the sac-like scrotum. During the last month of fetal life, the paired testes descend from behind the parietal peritoneum of the abdominal cavity near the kidneys to the lower abdominal cavity and then through the inguinal cavity into the scrotum. The testis carries with it the vas deferens, blood vessels and nerves which later form parts of the spermatic cord.

Each testis is enclosed by a fibrous tissue capsule, the tunica albuginea. Thin fibrous tissue septa extend from the inner surface of the capsule into the testis, dividing it into lobules. The septa join together along the posterior border of the testis to form the thick mediastinum testis. The lobules of the testis are filled with coiled seminiferous tubules which join posteriorly in a channel, the rete testis, located within the mediastinum testis. Ducts from these channels join the epididymis which is coiled on the outer surface of the testis.

The epididymis is a coiled tube that emerges from the superior aspect of the testis, descends along its posterior surface and then turns upward to become the vas deferens. Sperm cells, produced in the testis, migrate from the tubules of the testis into the epididymis where they are stored and undergo maturation. Peristaltic contractions occur that propel the sperm into the vas deferens.

The scrotal wall contains a layer of smooth muscle fibres, the dartos muscle. It is divided by a medial septum into two chambers, each containing a testis.

The normal scrotum

The testes occupy the majority of the scrotum. Each testicle images ultrasonically as an oval gland exhibiting fine homogeneous echogenic patterns similar to such other glandular structures as the thyroid (Fig. 8.24). Each testis has a smooth echogenic external surface and is usually surrounded by a small amount of fluid which remains echo-free. Along one aspect of the testes, an irregular tubular-like structure, the epididymis,

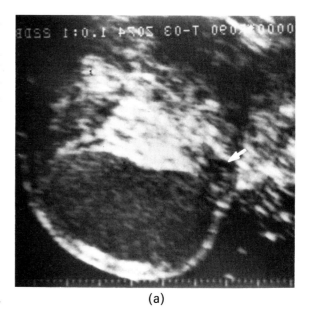

(a)

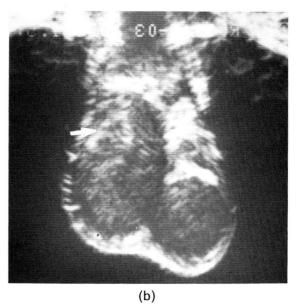

(b)

Fig. 8.24 Normal testis, 21-year-old student with no symptoms. (a) Sagittal magnified B-scan through the right testis. On the right note the coarser echotexture of the epidydimis (arrow). (b) Transverse B-scan of the mid-scrotum. Homogeneous echotexture of the testes and brighter, coarser echo-texture of the epidydimis (arrow).

can be seen which exhibits a much coarser echogenic texture. The spermatic cord is imaged along its anterior superior margin and it has a similar coarse echogenic pattern.

Testicular abnormalities

Patients who present with swellings or palpable mass lesions in the scrotum are candidates for ultrasonic imaging to determine if the mass lesion is located primarily within the testes or is extratesticular in location. Solid mass lesions within the testes usually are either primary or secondary testicular tumours. Seminoma, embryonal cell carcinoma, choriocarcinoma and lymphoma usually image as weakly echogenic irregular solid areas within the testes (Fig. 8.25). In many cases, it is poorly delineated, although some masses tend to be well circumscribed. These tumours must be differentiated from testicular inflammation (orchitis). In this condition, the testis is diffusely enlarged and shows a homogeneously weak echo pattern compared to the normal testes. Localised fluid-filled areas in these cases are due to focal abscesses. There is usually clinical evidence of inflammation in the scrotum. The testes of patients presenting with abdominal lymphadenopathy should be examined in search of a primary occult neoplasm (Fig. 8.26). Of course, the paraaortic region should always be surveyed whenever a testicular neoplasm is identified.

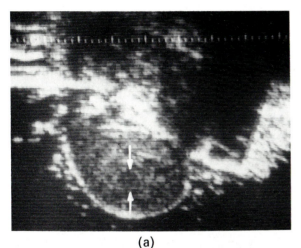

(a)

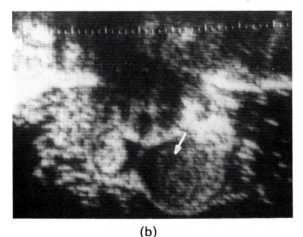

(b)

Fig. 8.25 Palpable scrotal mass — seminoma. (a) Sagittal B-scan using a water bath technique. 10 mm weakly echogenic mass within the testes centrally (arrows). Echogenic lines on the scrotal surface represent hair. (b) Transverse B-scan, mass within the left testis (arrow). At this scan plane, the epididymis on the right demonstrates a bright echogenic pattern.

Extratesticular abnormalities

Extratesticular mass lesions include fluid-containing areas and inflammatory changes. Fluid within the scrotum may be contained within the tunica albuginea, around the testes, in a hydrocele or enclosed within cysts of the spermatic cord or epididymis (Fig. 8.27). Dilated blood vessel varices may also be delineated ultrasonically.[41] Dilated veins (known as varioceles) frequently are imaged as tortuous tubular fluid-containing spaces (Fig. 8.28). Inflammatory conditions within the epididymis may be imaged as localised fluid-filled areas which represent focal abscesses. Less severe cases may just show diffuse enlargement with a weakly echogenic pattern.[41] Generally, inflammations are bilateral, whereas fluid collections may be either unilateral or bilateral. Hydroceles occur when fluid collects within the tunica vaginalis, a serosal-lined sac proximal to the testes (Figs. 8.29 and 8.30). Torsion of the testes may occur suddenly as a result of physical trauma. The resultant twisting of the spermatic cord may impair the venous drainage and arterial supply to the testes resulting in intense congestion. This is accompanied by sudden pain and oedema of the scrotal sac. Haemorrhage or ischaemic infarction of the testes may occur, and there may be extravasation of blood with the formation of a haematocele within the tunica vaginalis.

Complex mass lesions imaged within the scrotum can also represent herniation of bowel loops into the tunical vaginalis, a scrotal hernia.[41]

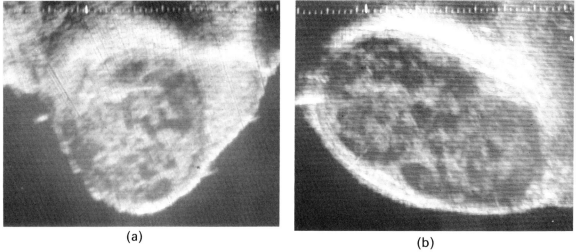

(a)   (b)

Fig. 8.26 Lymphoma of the testis. 55-year-old engineer who noticed a hard mass in the right scrotum. (a) Transverse B-scan demonstrates a normal appearing left testis and the enlarged, complex right testis. (b) Magnified sagittal B-scan of the right scrotum demonstrates an enlarged testis with gross architectural disruption. The predominantly solid area contains irregularly shaped echo-free regions representing degeneration.

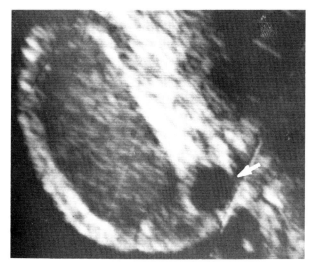

Fig. 8.27 Epididymal cyst. Small palpable hard nodule in the posterior right scrotum of a 26-year-old man. Sagittal B-scan demonstrates a normal homogeneous echo pattern of the testis. Note the 10 mm echo-free area (arrow) along the inferior posterior margin of the testis, an epididymal cyst.

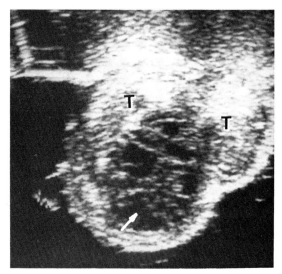

Fig. 8.28 Scrotal varices. Transverse B-scan shows tubular echo-free spaces (arrow) within the scrotum displacing the left testis laterally. The testes (T) have a normal echopattern.

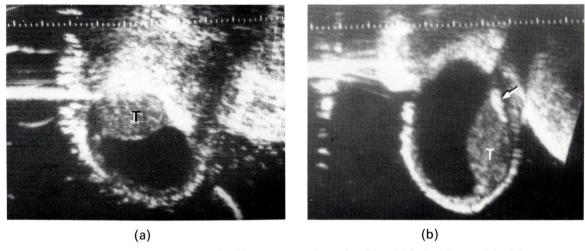

(a)   (b)

Fig. 8.29 Unilateral hydrocele. 38-year-old male with recent onset of scrotal swelling. (a) Sagittal B-scan of the right scrotum. Large echo-free space surrounds a normal appearing testis (T). (b) Transverse B-scan demonstrates displacement of the testis by a large echo-free space. Note the epididymis (arrow) imaged as a tubular, strongly echogenic structure along the surface of the testis.

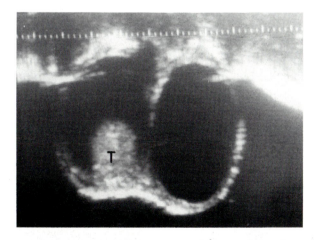

Fig. 8.30 Bilateral hydroceles. 58-year-old man with scrotal swelling. Transverse B-scan demonstrates large echo-free spaces within the scrotum. Note the normal-appearing right testis (T) at the inferior aspect of the scrotum.

OTHER SUPERFICIAL STRUCTURES

The use of specialised superficial small parts scanners has resulted in the sonic investigation of palpable abnormalities in the superficial layers of the body. In the extremities, mass lesions around joints and in the muscles have been evaluated. Around joints, lesions may develop over a period of time or become apparent rapidly due to communication with the synovial fluid-filled spaces. At the knee, a palpable mass lesion behind the joint may represent a Baker's cyst, a popliteal artery aneurysm, a bursae or an inflammatory mass lesion.[39, 47] The use of superficial scanners allows delineation of the mass lesion and with real-time imaging, intrinsic pulsations can be differentiated from transmitted pulsations to distinguish vascular structures from peri-articular and peri-vascular mass lesions. If a fluid-filled structure connects with the knee joint, a Baker's cyst is the most likely diagnosis.[18, 39] Occasionally, bursae and peri-articular granulomatous masses image as predominantly fluid-filled areas with internal echoes. These represent the hypertrophic cellular lining of the inflammatory processes associated with rheumatoid arthritis.[35]

Within the muscles, disruption of architecture may be noted with haematoma formation or neoplastic growths. Generally, the patient gives a history of trauma in the immediate or remote past. Initially, haematomas within the muscle cause disruption of the normal expected tissue planes with an ill-defined, weakly echogenic region. Over time, the size of the disruption diminishes and the area shows increased echogenicity as the blood resolves and the mass shrinks in size. Tumours within the muscles generally have well-defined margins and, if predominantly solid, show considerable echo attenuation. If adjacent to the bony cortex, disruption of the cortex may be imaged as in periosteal osteosarcomas. Calcification or bone formation within the soft tissues of the extremities results in a B-scan image of bright echoes along the anterior surface of the mass with considerable acoustic attenuation distally.

Patients on dialysis frequently have arteriovenous shunts in an extremity for vascular access. These superficial structures can be imaged with the high frequency scanners to evaluate the margins of the vessels for areas of stenosis, internal thrombus or pseudo-aneurysm formation.[44] Soft tissue haemorrhage with haematoma formation around the grafts can be distinguished from a true abnormality of the shunt (Fig. 8.31).

Superficial vessels can be evaluated in the femoral regions. Femoral artery aneurysms occur in peripheral vascular disease.[9] Following bypass vascular surgery to the iliac or femoral arteries, the graft and the anastomotic region can be imaged for evaluation of pseudo-aneurysm formation. Generally, real time imaging of the vessel with Doppler

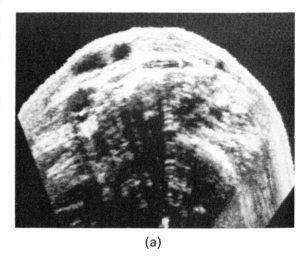

(a)

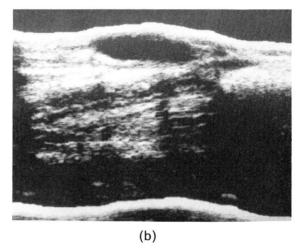

(b)

Fig. 8.31 7.5 MHz contact B-scan over the forearm in a patient with an arteriovenous shunt. (a) Transverse B-scan over two palpable masses reveals a round and an oval relatively echo-free region immediately below the skin. (b) Sagittal B-scan over one mass reveals an oval echo-free region corresponding to the site of a superficial haematoma.

blood flow recordings gives the maximum amount of information on the anatomy and flow through superficial vessels. Extensive investigations are underway to assess imaging and blood flow recording of the carotid arteries in the neck. Other potential superficial locations for analysis involve renal transplants in the more superficial area of the lower abdomen and pelvis. These topics are discussed further in Chapter 5.

In the newborn and infant, many of the organs are in a superficial location and the high frequency scanners have been utilised for evaluation of the neonatal kidney, superficial mass lesions in the abdomen, and the extremities.

Abdominal superficial structures

In the superficial muscles of the abdomen, mass lesions frequently occur in the postoperative period. Haematoma formation at the site of surgical scars or remote from the scar site can be evaluated.[22] A fusiform, weakly echogenic region within the muscle layers of the anterior abdominal wall may represent a haematoma or a superficial abscess. Generally, the abscess has an irregular margin and contains some scattered internal echoes from debris. The location of these superficial masses can be evaluated to determine if the region is extra-peritoneal or if it bridges the peritoneum to extend into the peritoneal cavity. Frequently, fluid collections in the superficial muscle layers of the abdomen can be drained under ultrasound guidance both for culture, antibiotic sensitivity and resolution.

The distal margin of shunt tubes can be evaluated when mass lesions or malfunction of the shunting procedure occurs. A fluid collection at the distal end of a peritoneal shunt for hydrocephalus demonstrates a well-defined, echo-free region within the superficial abdominal layers. The cerebrospinal fluid pseudocyst that forms at the distal end of the shunt causes blockage of the shunt tubing. It is possible to drain the pseudocyst superficially under ultrasound guidance and the shunt may not require surgical revision.

Follow-up scans of abscesses undergoing drainage can be used to document changes in size, location and configuration of these superficial collections for evaluation of their resolution.

Thoracic superficial structures

Intercostal scanning of the pleural space is an established procedure for the evaluation of the pleural fluid. Both free fluid and loculated fluid collections can be readily imaged when they are adjacent to the chest wall. Mobility of the fluid collections can be tested by changing the patient from recumbent to erect or into a decubitis position. Ultrasound guidance is frequently utilised for drainage of pleural fluid collections. The ability to image debris and septations within the collections greatly assists the clinician in planning drainage procedures.[33] Recently, solid nodules along the parietal pleura have been evaluated using ultrasound.[19] B-scan imaging of pleural thickening and evaluation of its characteristics can aid in the differentiation of benign or malignant processes.

REFERENCES

1. Bambach C P, Riley J W, Picker R H, Reeve T S, Middleton W R J 1978 Pre-operative parathyroid identification by ultrasonic scan. Med J Austral 2:227–229
2. Chatzkel S, Cole-Beuglet C, Breckenridge J W, Dubbins P A, Kurtz A B, Goldberg B B 1982 Ultrasound diagnosis of hypernephroma metastatic to thyroid and external jugular vein. Radiology 142:165–166
3. Clark O H, Okerlund M D, Cavalieri R R, Greenspan F 1979 Diagnosis and treatment of thyroid, parathyroid and thyroglossal duct cysts. J Clin Endocrinol Metab 48:983–988
4. Cole-Beuglet C 1980 Thyroid and neck ultrasound. In: Sarti D (ed) Diagnostic ultrasound text and cases. G. K. Hall, Boston, p 452–458
5. Cole-Beuglet C, Beique R A 1975 Continuous ultrasound

B-scanning of palpable breast masses. Radiology 117:123–128
6. Cole-Beuglet C, Goldberg B B, Kurtz A B, Rubin C S, Patchefsky A S, Shaber G S 1981 Ultrasound mammography: a comparison to X-ray mammography. Radiology 139:693–698
7. Crocker E F, Bautovich G J, Jellins J 1974 Grey-scale echographic appearance of thyroid malignancy. J Clin Ultrasound 2:305–306
8. Crocker E F, Bautovich G J, Jellins J 1978 Grey-scale echographic visualisation of parathyroid adenoma. Radiology 126:233–234
9. Davis R P, Neiman H L, Yao J S T, Bergan J J 1977 Ultrasound scan in diagnosis of peripheral aneurysms. Arch Surg 112:55–58

10. Doppman J L, Brennan M F, Koehler J O, Marx S J 1977 Computed tomography for parathyroid localization. J Comput Asst Tomogr 1:30–36

11. Duffy P, Picker R H, Duffield S, Reeve T, Hewlett S 1980 Parathyroid sonography: a useful aid to pre-operative localization. J Clin Ultrasound 8:113–116

12. Edis A J, Evans T C 1979 High-resolution, real-time ultrasonography in the preoperative location of parathyroid tumors. N Eng J Med 301:532–534

13. Gobien R 1979 Aspiration biopsy of the solitary thyroid nodule. Radiol Clin N Am 17:543–554

14. Green P S, Taenzer J C, Ramsey S D, Holzemer J F, Suarez J R, Marich K W, Evans T C, Sandok B A, Greenleaf J F 1977 A real-time ultrasonic imaging system for carotid arteriography. Ultrasound Med Biol 3:129–142

15. Harper P, Kelly-Fry E 1980 Ultrasound visualization of the breast in symptomatic patients. Radiology 137:465–470

16. Hassani S N, Bard R L 1977 Evaluation of solid thyroid neoplasms by grey-scale and real-time ultrasonography: the 'halo' sign. In: White D, Lyons E A (eds) Ultrasound in medicine, Vol 4, Plenum, New York, p 323–324

17. Heath H, Hodgson S F, Kennedy M A 1980 Primary hyperparathyroidism: incidence, morbidity and potential economic impact in a community. N Eng J Med 302:189–193

18. Hermann G, Yeh H-C, Lehr-Janus C, Berson B L 1981 Diagnosis of popliteal cyst: double contrast arthrography and sonography. Am J Roentg 137:369–372

19. Hirsch J H, Rogers J V, Mack L A 1981 Real-time sonography of pleural opacities. Am J Roentg 136:297–301

20. Jellins J, Kossoff G, Buddee F W Reeve T S 1971 Ultrasonic visualization of the breast. Med J Austral 1:305–307

21. Jellins J, Kossoff G, Reeve T S 1977 Detection and classification of liquid-filled masses in the breast by grey scale echography. Radiology 125:205–212

22. Kaftori J K, Rosenberger A, Pollack S, Fish J H 1977 Rectus sheath hematoma: ultrasonographic diagnosis. Am J Roentg 128:283–285

23. Kalisher L, Peyster R 1975 Xerographic manifestations of male breast disease. Am J Roentg 125:656–661

24. Karo J J, Maas L C, Kaine H, Gelzayd E A 1978 Ultrasonography and parathyroid adenoma. J Am Med Ass 239:2163–2164

25. Kelly-Fry E 1980 Breast imaging. In: Sabbagha R E (ed) Diagnostic ultrasound applied to obstetrics and gynecology. Harper & Row, Hagerstown, p327–350

26. Kobayashi T 1977 Grey-scale echography for brest cancer. Radiology 122:207–214

27. Kobayashi T 1978 Clinical ultrasound of the breast. Plenum, New York

28. Kopans D B, Meyer J E, Proppe K H 1981 The double line of skin thickening by scanning breast ultrasound. Radiology 141:485–487

29. Kossoff G, Carpenter D A, Robinson D E, Radovanovich G, Garrett W J 1976 Octoson — a new rapid general purpose echoscope. In: White D, Barnes R (eds) Ultrasound in medicine, Vol 2, Plenum, New York, p 333–339

30. Leopold G R 1980 Ultrasonography of superficially located structures. Radiol Clin N Am 18:161–173

31. Leopold G R 1981 Superficial organs. In: Goldberg B B (ed) Ultrasound in cancer. Churchill Livingstone, New York, p 123–135

32. Leopold G R, Woo V, Scheible W, Nachtsheim D, Gosink B B 1979 High resolution ultrasonography of scrotal pathology. Radiology 131:719–722

33. Lipscomb D J, Flower C D, Hadfield J W 1981 Ultrasound of the pleura: an assessment of its clinical value. Clin Radiol 32:289–290

34. Michels L, Gold R, Arndt R 1977 Radiography of gynecomastia and other disorders of the male breast. Radiology 122:117–122

35. Moore C P, Sarti D A, Louie J S 1975 Ultrasonographic demonstration of popliteal cysts in rheumatoid arthritis: a non-invasive technique. Arth Rheum 18:577–580

36. Neiman H L, Yao J S T, Silver T M 1979 Gray scale ultrasound diagnosis of peripheral arterial aneurysm. Radiology 130:413–416

37. Propper R A, Skolnick M, Weinstein B J, Dekker A 1980 The nonspecificity of the thyroid halo sign. J Clin Ultrasound 8:129–132

38. Rosen I B, Walfish P G, Miskin M 1979 The ultrasound of thyroid masses. Surg Clin N Am 59:19–33

39. Rudikoff J C, Lynch J J, Philips E, Clapp P R 1976 Ultrasound diagnosis of Baker's cyst. J Am Med Ass 235:1054–1055

40. Sacker J P, Passalaqua A M, Blum M, Amorocho L 1977 A spectrum of disease of the thyroid gland as imaged by gray-scale water bath sonography. Radiology 125:467–472

41. Sample W F, Gottesman J E, Skinner D G, Ehrlich R M. Grey-scale ultrasound of the scrotum. Radiology 127:225–228

42. Sample W F, Mitchell S P, Bledsoe R C 1978 Parathyroid ultrasonography. Radiology 127:485–490

43. Scheible W, Leopold G R, Woo V L 1979 High resolution real-time ultrasonography of thyroid nodules. Radiology 133:413–417

44. Scheible W, Skram C, Leopold G R 1980 High resolution real-time sonography of hemodialysis vascular access complications. Am J Roentg 134:1173–1176

45. Schottenfeld D, Gershman S T 1978 Epidemiology of thyroid cancer. CA-A Cancer J Clin 28:66–86

46. Shimstrak R R, Schoenrock G J, Taekman H P, Cianci P, Chambers R F 1979 Preoperative localization of a parathyroid adenoma using computed tomography and thyroid scanning. J Comput Asst Tomogr 3:117–119

47. Silver T M, Washburn R L, Stanley J C, Gross W S 1977 Gray scale ultrasound evaluation of popliteal artery aneurysms. Am J Roentg 129:1003–1006

48. Teixidor H S 1980 The use of ultrasonography in the management of masses of the breast. Surg Obstet Gynec 150:486–490

49. Wagai T, Takahashi S, Ohasi H, Ichikawa H 1967 A trial for quantitative diagnosis of breast tumor by ultrasono-tomography. Jap Med Ultrasonics 5:39

50. Walfish P G, Hazani E, Strawbridge H T, Miskin M, Rosen I B 1977 A prospective study of combined ultrasonography and needle aspiration biopsy in the assessment of hypofunctioning thyroid nodule. Surgery 82:474–482

51. Wells P N T, Evans K T 1968 An immersion scanner for two-dimensional ultrasonic examination of the human breast. Ultrasonics 6:220–228

9

Miscellaneous organs and structures

B. B. Goldberg and P. N. T. Wells

INTRODUCTION

The major clinical applications of diagnostic ultrasound are reviewed in Chapters 2–8. In this chapter, clinical applications which do not fall conveniently under the main headings of these chapters are drawn together and briefly discussed. Appropriate references are given so that readers can conveniently follow up any particular specialised topics in which they may be interested.

ALIMENTARY TRACT

Teeth and mouth

In the developed world, one of the main sources of exposure of the population to ionising radiations is due to dental examinations. Moreover, for the individual patient, the time required to process the films may cause delays and return visits. Therefore, important advantages would be gained by the development of safe and rapid ultrasonic diagnostic measurement techniques in dentistry, provided that the clinical results were as reliable as radiography and the apparatus was economical to purchase and to operate.

Both reflection and transmission methods have been shown to be capable of detecting degenerative pulpitis.[26] Probes of 1 mm diameter were used, operating at 18 MHz. The transmission method depends on the attenuation due to gas associated with pulpitis. It is more reliable than the reflection method, because it is not affected by the angulations of surfaces, but it does require access to both sides of the tooth.

The A-scope has been used to measure the positions of the dentinoenamel junction and the pulp chamber wall.[30] A stand-off probe was used, with an aluminium rod (which has a similar characteristic impedance to that of dental enamel) positioned between the transducer and the tooth. The results of preliminary experiments showed that it may be possible to estimate the thickness of the water film between an amalgam filling and the dentine. It may also be possible to assess changes in the mineralisation of the enamel.[31]

Although these experimental results are interesting, there is no doubt that formidable problems remain to be solved before ultrasonic examination of the teeth could become a routine procedure.

It has been reported that it is possible to distinguish between solid and cystic lesions of the salivary glands by transcutaneous two-dimensional scanning through the upper neck.[34]

Stomach and intestine

Generally, ultrasonic scanning of some of the contents of the abdomen may be made difficult (or impossible) by the gas which tends to exist in the alimentary tract.

In suitable patients, the gas problem can be largely eliminated by utilising special techniques. Thus, an ultrasonic window to the upper left abdomen can be created by filling the fasting stomach with water or such thicker solutions as methylcellulose suspension, giving intravenous glucagon, or examining the patient in the prone position[50] (Fig. 9.1). Glucagon relaxes the gastric musculature and the prone position allows any stomach gas to rise to the fundus. This makes it possible to assess the relationships of the stomach to adjacent structures and permits the display of structures forming the stomach bed. In patients

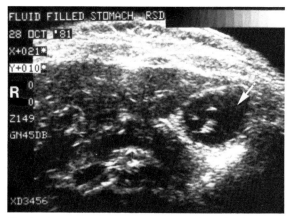

Fig. 9.1 Supine transverse ultrasonogram shows liquid within the stomach (arrow) allowing for improved visualisation of distally located structures such as the tail of the pancreas. (R, right.)

Within the limitations imposed by the presence of intraabdominal gas, it is possible to visualise space occupying lesions in the stomach and the intestine. Although ultrasonic imaging is a weak competitor with conventional X-ray contrast studies, there are occasions on which a quick search with a real-time scanner may be quite illuminating. The characteristic ultrasonic appearance of a neoplasm in these parts of the alimentary tract is that of strong echoes in the centre with a virtually anechoic thickened mantle[32, 35](Fig. 9.2). In addition to suggesting the diagnosis, the ultrasonic scan may provide a method of monitoring therapy, and it may also be helpful in guiding biopsy.

who have had a gastroenterostomy or vagotomy, the methylcellulose suspensions is not retained in the stomach; it rapidly enters the small bowel, demonstrating the jejunal loops and outlining the bowel walls. Even in patients who have not been prepared in this way, the possibility that a dilated jejunal loop may be wrongly identified as a pseudocyst, abscess or haematoma should not be forgotten.[4] Real-time ultrasound usually eliminates this problem by demonstrating peristalsis. Another possible source of error is due to omentum, which may mimic cystic masses in the pelvis.[12, 38]

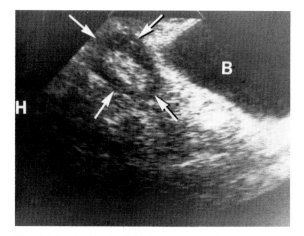

Fig. 9.2 Supine longitudinal ultrasonogram demonstrates typical pattern of abnormal bowel (arrows). In this case, the thickening (weakly echogenic area around the brightly echogenic lumen) is due to tumour. (B, bladder; H, superior.)

INTERVENTIVE ULTRASONIC IMAGING

Guided puncture

Ultrasonically-guided percutaneous puncture of organs and pathological lesions is an important technique, particularly in internal medicine and in obstetrics. The method makes it possible to obtain material for cytological, chemical, bacteriological, histological and perhaps even immunological studies, from almost anywhere in the abdomen.[17] Any region which can be ultrasonically visualised can be subjected to puncture or fine needle biopsy in this way.

Although in some situations satisfactory results can be obtained without the use of specialised equipment, it is generally better to employ a transducer drilled to allow the needle to be introduced in a controlled way. The transducer may, for example, be the single element device of a static B-scanner, or an electronically switched linear array for real-time imaging.[13] This method gives an important, but rather unexpected, advantage: an echo emanates from the region of the tip of the needle, and so it is possible to guide the tip so that it is precisely positioned within the area of interest (Figs. 9.3a and b). Moreover, the inadvertent puncture of normal structures, such as lung, heart wall, blood vessel and placenta, are easily avoided. Real-time sector imaging can also be used effectively as a guide (Fig. 9.3c).

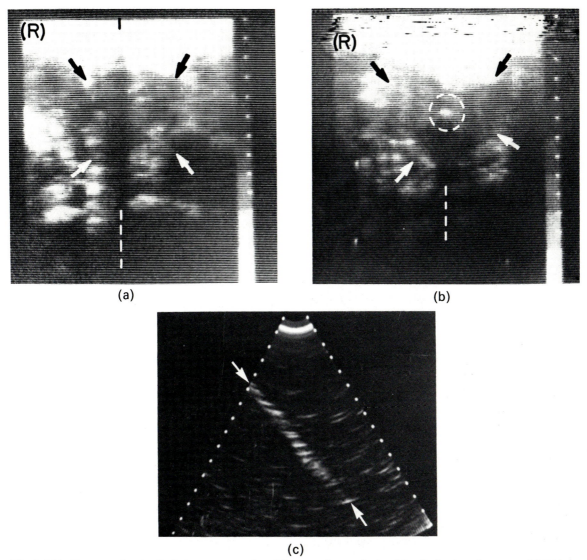

(a)

(b)

(c)

Fig. 9.3 (a) Transverse longitudinal ultrasonogram of a solid pancreatic mass (arrows) localised using a linear array (dotted line indicates pathway for insertion of needle). (R, right.) (b) After insertion of needle, bright echo (dotted circle) seen within the solid mass (arrows). (R, right.) (c) B-scan real-time sector image shows how needle path can be followed (arrows).

Intraoperative ultrasonic imaging

There are several surgical procedures which may be expected to be improved significantly by the use of intraoperative ultrasonic imaging.

The first requirement for this purpose is a suitable ultrasonic scanner. At present, the mechanically operated real-time scanner, of the type originally designed for ophthalmic visualisation, is really the only type of scanner which is at all suitable. The transducer, of course, must be sterilised or covered by a sterile sleeve.

In renal surgery, ultrasonic real-time imaging may be invaluable for the localisation of calculi. Before the kidney has been cut (this may allow air to enter), stones of 2 mm in diameter may easily be detected.[33] The positions of the stones can be marked by inserting needles to make contact with them.

In arterial surgery, the principal value of

ultrasonic imaging is in the detection of technical errors, which can be corrected immediately, thus avoiding re-arteriotomy[29] (Fig. 9.4). Intraoperative Doppler studies of blood flow can also give supplementary information.[21]

In biliary surgery, ultrasonic imaging is useful for locating the common bile duct in the presence of acute inflammation and other abnormal anatomy, detecting small calculi in the gallbladder,

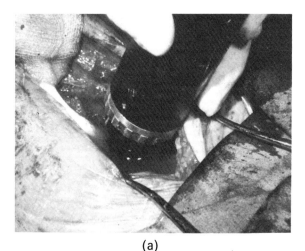

(a)

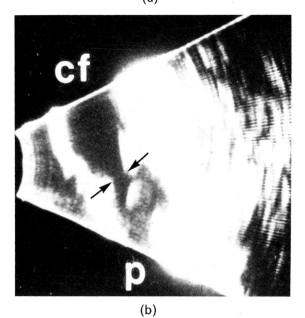

(b)

Fig. 9.4 (a) Real-time scanner being used to examine the adequacy of the anastomosis between a graft and the femoral artery. (b) Real-time intraoperative image showing the common femoral (cf) and profunda femoris (p) arteries. The profunda femoris artery is stenosed (arrows).

especially when the wall is thickened, measuring the diameter of the common bile duct, and identifying calculi within it.[46] Moreover, the search for calculi can be extended to intrahepatic ducts.[29]

In pancreatic surgery, ultrasonic imaging is particularly useful in locating the gland where the anatomy has been distorted by previous surgery and for establishing whether or not the head of the gland is involved in the neoplastic process.[29] This latter procedure may save hours of operating time to assess the degree of infiltration of the portal vein, and to design the optimal palliative drainage scheme.

In neurosurgery, preliminary results with the scanner applied directly to the surface of the brain have been encouraging in locating tumours.[49] In newborn up to one year of age, ultrasound has been used successfully to guide shunts into the ventricles. The transducer is placed on the skin surface in the region of the open anterior fontanelle. The position of the catheter can be easily identified within the ventricular system.

ONCOLOGY

Radiotherapy planning

Traditionally, radiotherapy planning is based on a sketch, made by the clinician, showing an approximation of the cross-sectional body outline and the position and shape of the lesion to be treated in that plane. The clinician states the radiation dose which he wishes to be delivered to the tumour, and indicates radiosensitive regions which he wishes to be spared. The planner then uses skill and judgement to position radiation field plots over the clinician's drawing, to give the optimal dose distribution. In doing this, the calculations may be speeded up by computer.

The accuracy of this planning procedure — and it is only by having a sufficiently high accuracy that the treatment can be successful — depends on the accuracy of the clinician's original sketch. In planning the treatment of tumours in ultrasonically accessible sites, ultrasonic two-dimensional scanning can in principle greatly improve on the accuracy obtainable by clinical judgement.[2,6,27] Unfortunately, not all important sites are accessible because of the limitation imposed by gas and bone. Accessible sites include neoplasms of the retroperitoneum, kidneys, pancreas, adrenals, uterus, ovaries, bladder, prostate and lymph nodes.

The information required for radiotherapy planning differs in certain aspects from that needed for diagnosis.[3] In the planning application, accurate spatial relationships must be determined (whereas, in diagnosis these do not usually need to be known). Therefore, as little pressure as possible is applied to the scanning probe, to avoid distorting the body outline and displacing internal structures. The patient is scanned in the proposed treatment position, with normal respiration, and the physiological conditions, such as the fullness of the bladder, are also the same.

Although ultrasonic tumour localisation is extremely attractive as an aid in radiotherapy planning, it has not been widely adopted. Recently, however, great enthusiasm has developed for the use of X-ray computed tomography for this purpose.[42,45] CT scanning is excellent because it does not have limitations due to bone or gas and, at least in principle, it allows the radiation dose calculations to take into account the estimated X-ray attenuations in the different tissues and materials in the treatment field. CT is relatively expensive, however, and in this respect ultrasound has a substantial advantage.

A rather specialised, but nevertheless important role of ultrasound in radiotherapy planning is the measurement of chest wall thickness. This information is essential in electron beam radiotherapy, because it allows the treatment to be optimised so that the entire chest wall is irradiated while the lungs are spared.[20]

Tumour staging

The treatment strategy for many types of malignancy is critically dependent on the stage which the disease has reached. The technique of ultrasonic staging seems to be worked out best in the management of bladder tumours.[39] The tumour can be staged according to depth of infiltration, but assessment of the extension of tumour into the pelvis is better done with CT. Other conditions in which ultrasonic staging may be useful include cancer of the breast, prostate and lymph glands.

Monitoring treatment

Direct observation by ultrasonic imaging of tumours treated by chemotherapy or radiotherapy is helpful in assessing regression and progression.[5,23,24,25,27] If the regression rate is disappointing, the oncologist can change the treatment schedule, or test the effect of different drugs.

Detecting recurrences and metastases

Patients who have been successfully treated for malignant disease, whether by surgery, radiotherapy or chemotherapy, should be periodically scanned to detect recurrences or metastases. This is especially important for patients who have had Hodgkin's disease, or tumours of the genital tract.[27] Early detection of these complications allows the most effective therapy to be instituted.

ORTHOPAEDICS AND RHEUMATOLOGY

Fat thickness

Subcutaneous fat thickness can be measured with an A-scope, either with the transducer in contact with the skin, or separated by a water-filled or solid plastic delay path. For example, the correlation between needle puncture and ultrasonic measurements in 100 individuals at a site 20 mm below and to the right of the umbilicus was 0.96.[9] It is considered that the ultrasonic method is more accurate than Harpenden calipers.[4] Recent work using automated B-scan equipment has shown promise in being able to measure, without deformity, subcutaneous and muscular layers (Fig. 9.5).

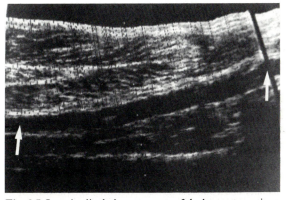

Fig. 9.5 Longitudinal ultrasonogram of the lower extremity about the region of the knee clearly shows the subcutaneous and muscular layers (arrows denote the femoral artery).

Muscular dystrophy

A comparative study of ultrasonic images of the thigh[16] showed consistent differences in 10 children with muscular dystrophy compared with 40 healthy controls. This noninvasive technique could be useful in assessing the extent of pathological change, and could also prove to be a valuable diagnostic aid.

Fluid collections

Ultrasonic scanning can reliably detect popliteal cysts of clinically significant size[14] (Fig. 9.6). Although arthrography does demonstrate some small cysts which may be missed by ultrasound, ultrasound can show cysts which do not fill on arthrography. Moreover, cyst rupture may be apparent on the ultrasonic image if the distal margin of the cyst is diffuse.

Intra-articular fluid collections can be ultrasonically visualised in joint disorders of the hips, shoulders and elbows.[44] Ultrasound images clearly delineate the contours of bony surfaces, normal muscles and other soft-tissue structures around joints. In addition to fluid collections of as little as 10 ml, intra-articular loose bodies can be identified.

Ultrasonic imaging is a useful adjunct in the evaluation and follow-up of rheumatoid arthritis of the knee,[10] since it allows the degree of suprapatellar effusion and synovial thickening to be estimated.

The relatively anechoic spaces representing rectus sheath haematomas can be visualised by

ultrasound[47] (Fig. 9.7) and so also can those associated with venous thrombi, and here Doppler methods can provide confirmatory data.[11]

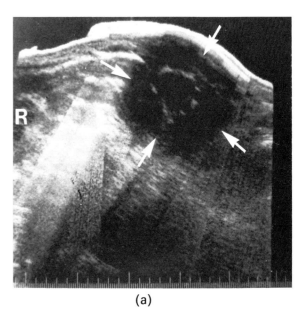

(a)

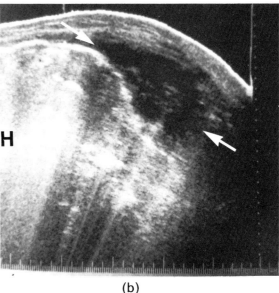

(b)

Fig. 9.7 (a) Transverse and (b) Longitudinal supine ultrasonograms show evidence of a complex predominantly cystic area (arrows) collecting in the region of the left rectus sheath which proved to be a haematoma. Note that the complex area does not cross the midline on the transverse ultrasonogram which is quite typical for rectus sheath collections. (R, right; H, superior.)

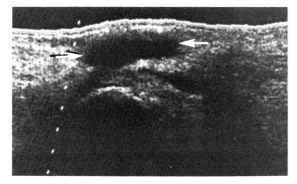

Fig. 9.6 Longitudinal scan of the popliteal region shows evidence of a predominantly cystic area (arrows) extending distally along the posterior aspect of the knee consistent with a diagnosis of popliteal cyst.

Bony tumours

Ultrasonic imaging allows the size of accessible bony tumours to be assessed.[28] The relationship between the bone and the tumour can almost always be established; this is important information in planning surgery. It is not possible, however, to examine the bone itself by ultrasound, or to establish whether the tumour is adherent to the periosteum or originates in the bone. The echoes from within the tumour are usually inhomogeneous, probably because of degeneration and haemorrhage. Ewing tumours and osteosarcomas generally have quite well-defined boundaries, but liposarcomas and chondrosarcomas tend to be irregular and nodular.

Spinal canal

With the patient lying prone, it is possible by ultrasonic contact scanning to visualise the lumbar vertebrae and laminae. By inclining the ultrasonic beam at 15° to the sagittal plane, echoes from the laminae and the posterior surfaces of the vertebral bodies can be displayed on an A-scope.[40] Thus, it is possible to measure the oblique sagittal depth of the spinal canal, and it is fortuitous that this particular dimension is that which is most affected in stenosis when laminar hypertrophy exaggerates the trefoil shape. This bony encroachment may not be detected by myelography when the mid-sagittal and coronal dimensions are adequate. Moreover, ultrasonic measurements at each lumbar level helps the surgeon to decide on the segmental extent of necessary decompression. Recently, using high-resolution small parts scanners, images of the entire spinal canal have been obtained in the newborn and young paediatric patient (Fig. 9.8).

Assessment of fracture healing

The results have been reported of some preliminary experiments designed to test the possibility of assessing fracture healing in limbs in terms of the attenuation of shear waves transmitted across the fractures.[19] The rationale of the method is that shear waves are attenuated rapidly in fluids, so that the transmission of a shear wave across the region of a fracture is controlled largely by the extent of

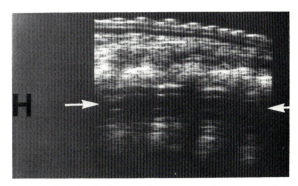

Fig. 9.8 Longitudinal prone ultrasonogram demonstrates the spinal canal (arrows) in a newborn. The areas of distal acoustic shadowing are due to the sound being blocked by the vertebral bodies. Sufficient sound, however, is transmitted through the interspaces to allow for adequate imaging of the canal. (H, superior.)

bone fusion, since liquid fills the space where fusion has not occurred. Another proposed method is based on the measurement of the change in transit time along the bone as the fracture unites.[1] This approach may also have an application in monitoring callus formation.[7]

OTORHINOLARYNGOLOGY

Pharynx and vocal fold

M-mode recording has been used to study the motion of the lateral pharyngeal wall during phonation.[22] The transducer is placed on the neck just below the angle of the mandible. The technique may be useful in the evaluation of cleft palate patients prior to therapy, and in the rehabilitation of patients following laryngectomy.

Vocal fold motion during phonation has been measured by the Doppler method.[37] Positioning of the probe is not easy, however, and a pulse technique involving both transmission to align the beam and pulse-echo to measure movement has been developed to solve this problem.[18] Work in these areas is continuing.

Liquid collections

The A-mode ultrasonic equipment has been used to detect the presence of liquid both in the maxillary sinus and in the middle ear.[8] If the cavity contains liquid, an echo is received from its posterior wall; if it contains air, no such echo is received. These ultrasonic approaches have the potential to reduce the necessity to X-ray individuals, particularly children, suspected of having sinus problems or otitis media.

THE THORAX

Lung

In patients suspected of having pulmonary embolism, the diagnosis can usually be made clinically. It can generally be substantiated by scintigraphy. If this fails, pulmonary arteriography may very occasionally be justified

The interpretation of ultrasonic echoes obtained subcutaneously from lung is difficult because transmission is attenuated rapidly, and reverberation gives rise to artifactual echoes which appear on the display. It has been reported that the length of the A-scope timebase occupied by these echoes is much greater when the transducer is over an area of ischaemia than when it is over normal lung.[36] This may be due to decreased ventilation following interruption by pulmonary embolism in the perfusion of a lung segment.[43] A simple instrument has been developed with a 'yes-no' display and an audible signal output the pitch of which increases with increasing 'penetration'.[15] Ultrasound, however, has not replaced scintigraphy in the routine evaluation of pulmonary embolism.

Pleural effusion

The differential diagnosis of large opacities in a chest radiograph can on occasion be very difficult.

In this situation, the question may be resolved using ultrasound since a liquid collection is ultrasonically anechoic and transonic. It is possible to delineate the anatomical structure, demonstrate the movement of the diaphragm, visualise liquid, and determine the best site for pleural tap.[48] Thus, the presence of pleural effusion may be confirmed, and doubts about subphrenic collections and pleural thickening may be resolved (Fig. 9.9).

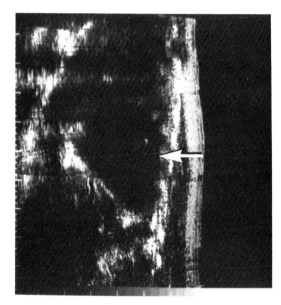

Fig. 9.9 Erect longitudinal ultrasonogram of the left hemithorax confirms the presence of a moderate pleural effusion shown by the echo-free area (arrow) located just above the diaphragm.

ACKNOWLEDGEMENT

The illustration and scan in Fig. 9.4 were kindly provided by R. J. Lane.

REFERENCES

1. Abendschwein W, Hyatt G W 1972 Ultrasonics and the physical properties of healing bone. J Trauma 12:297–301
2. Badcock P C 1977 Ultrasound scanning in the radiotherapy department. Clin Radiol 28:287–293
3. Banjavic R A 1980 Ultrasound in the radiotherapy department: past, present and future. In: Fullerton G D, Zagzebski J A (eds) Medical physics of CT and ultrasound. American Institute of Physics, New York, p 469–487
4. Booth R A, Goddard B A, Paton A 1966 Measurement of fat thickness in man: a comparison of ultrasound, Harpenden calipers and electrical conductivity. Br J Nutr 20:719–725
5. Brascho D J 1972 Clinical applications of diagnostic

ultrasound in abdominal malignancy. S Med J 65:1331–1339
6. Brascho D J 1974 Computerized radiation treatment planning with ultrasound. Am J Roentg 120:213–223
7. Brown S A, Major M B 1976 Ultrasonic assessment of early callus formation. Biomed Eng 11:124–127
8. Brzezinska H 1972 Application of the ultrasonic echo method for laryngeal diagnostics in children. In: Filipczynski L (ed) Ultrasonics in biology and medicine. Polish Scientific, Warsaw, p 29–33
9. Bullen B A, Quaade F, Olesen E, Lund S A 1965 Ultrasonic reflexion used for measuring subcutaneous fat in humans. Hum Biol 37:375–384
10. Cooperberg P L, Tsang I, Truelove L, Knickerbocker W

J 1978 Gray scale ultrasound in the evaluation of rheumatoid arthritis of the knee. Radiology 126 : 759–763

11. Day T K, Fish P J, Kakkar V V 1976 Detection of deep vein thrombosis by Doppler angiotherapy. Br Med J 1 : 618–620

12. Engel J M, Deitch E A 1980 Omentum mimicking cystic masses in the pelvis. J Clin Ultrasound 8 : 31–33

13. Goldberg B B, Pollack H M 1976 Ultrasonic aspiration biopsy techniques. J Clin Ultrasound 4 : 141–151

14. Gompels B M, Darlington L G 1979 Grey scale ultrasonography and arthrography in evaluation of popliteal cysts. Clin Radiol 30 : 539–545

15. Gordon D 1974 A new ultrasonic technique for lung diagnosis. In : Vlieger M de, White D N, McCready VR (eds) Ultrasonics in medicine. Excerpta Medica, Amsterdam, p 207–211

16. Hechmatt J Z, Dubowitz V, Leeman S 1980 Detection of pathological change in dystrophic muscle with B-scan ultrasound imaging. Lancet 1 : 1389–1390

17. Holm H H, Als O, Gammelgaard J 1979 Percutaneous aspiration and biopsy procedures under ultrasound visualization. In : Taylor K J W (ed) Diagnostic ultrasound in gastrointestinal disease. Churchill Livingstone, New York, p. 137–149

18. Holmer N-G, Kitzling P 1975 Localization of the vocal folds and registration of their movements by ultrasound. In : Kazner E, Vlieger M de, Muller H R, McCready V R (eds) Ultrasonics in medicine. Excerpta Medica, Amsterdam, p 349–354.

19. Horn C A, Robinson D 1965 Assessment of fracture healing by ultrasonics. J Coll Radiol Austral 9 : 165–167

20. Jackson S M, Naylor G F, Kerby I J 1970 Ultrasonic measurement of post-mastectomy chest wall thickness. Br J Radiol 43 : 458–461

21. Keitzer W F, Lichi E L, Brossart F A, Weese M S de 1972 Use of the Doppler ultrasonic flowmeter during arterial vascular surgery. Arch Surg 105 : 308–312

22. Kelsey C A, Hixon T J, Minifie F D 1968 Ultrasonic measurement of lateral pharyngeal wall displacement. IEEE Trans Biomed Eng BME-16 : 143–147

23. Kobayashi T, Takatani O, Hattori N, Kimura K 1972 Clinical investigation of ultrasonographic patterns of malignant abdominal tumor in special reference to changes in its pattern after irradiation or chemotherapy (preliminary report). Med Ultrasonics 10 : 18–22

24. Kobayashi T, Takatani O, Hattori N, Kimura K 1972 Clinical investigation of ultrasonographic patterns of malignant abdominal tumor in special reference to changes in its pattern after irradiation or chemotherapy (II). Med Ultrasonics 10 : 132–135

25. Kobayashi T, Takatani O, Hattori N, Kimura K 1974 Echographic evaluation of abdominal tumor regression during antineoplastic treatment. J Clin Ultrasound 2 : 131–141

26. Kossoff G, Sharpe C J 1966 Examination of the contents of the pulp cavity in teeth. Ultrasonics 4 : 77–83

27. Kratochwil A 1981 Treatment planning. In : Goldberg B B (ed) Ultrasound in cancer. Churchill Livingstone, New York, p 167–185

28. Kratochwil A, Zweymuller K 1975 Ultrasonic examination in orthopaedic surgery. In : Kazner E, Vlieger M de, Muller H R, McCready V R (eds) Ultrasonics in medicine. Excerpta Medica, Amsterdam, p 343–348

29. Lane R J 1980 Intraoperative B-mode scanning. J Clin Ultrasound 8 : 427–434

30. Lees S, Barber F E 1971 Looking into the tooth and its surfaces with ultrasound. Ultrasonics 9 : 95–100

31. Lees S, Gerhard F B, Oppenheim F G 1973 Ultrasonic measurement of dental demineralization. Ultrasonics 11 : 269–273

32. Lutz H Th, Petzold R 1976 Ultrasonic patterns of space occupying lesions of the stomach and the intestine. Ultrasound Med Biol 2 : 129–132

33. Lytton B, Cook J III 1979 Intraoperative ultrasound. In : Rosenfield AT (ed) Genitourinary ultrasonography. Churchill Livingstone, New York, p 175–181

34. Macridis C A, Kouloulas A, Koutsimbelas B, Yannoulis G 1975 Diagnosis of tumours of the salivary gland by ultrasonography. Electromedica 43 : 130–134

35. Mascatello V J, Carrera G F, Telle R L, Berger M, Holm H H, Smith E H 1977 The ultrasonic demonstration of gastric lesions. J Clin Ultrasound 5 : 383–387

36. Miller L D, Joyner C R, Dudrick S J, Eskin D J 1967 Clinical use of ultrasound in the early diagnosis of pulmonary embolism. Ann Surg 166 : 381–392

37. Minifie F D, Kelsey C A, Hixon T J 1969 Measurement of vocal fold motion using an ultrasonic Doppler velocity meter. J. Acoust Soc Am 43 : 1165–1169

38. Mittelstaedt C 1975 Ultrasonic diagnosis of omental cysts. J Clin Ultrasound 3 : 673–676

39. Morley P 1978 Clinical staging of epithelial bladder tumours by echotomography. In : Hill C R, McCready V R, Cosgrove D O (eds) Ultrasound in tumour diagnosis. Pitman Medical, Tunbridge Wells, p 145–161

40. Porter R W, Wicks M, Ottewell D 1978 Measurement of the spinal canal by diagnostic ultrasound. J Bone Joint Surg 60–B : 481–484

41. Pozderac R V, Doust B D 1978 Confusing appearance of a dilated jejunal loop. J Clin Ultrasound 6 : 165–166

42. Purdy J A, Prasad S C 1980 Computed tomography applied to radiation therapy treatment planning. In : Fullerton G D, Zagzebski J A (eds) Medical physics of CT and ultrasound. American Institute of Physics, New York, p 221–250

43. Ross A M, Genton E, Holmes J H 1968 Ultrasonic examination of the lung. J Lab Clin Med 72 : 556–64.

44. Seltzer S E, Finberg H J, Weissman B N 1980 Arthrosonography technique, sonographic anatomy and pathology. Invest Radiol 15 : 19–28

45. Sharpe W P, Jenkins D 1980 Flexible interaction with C T scan data for radiotherapy planning and other purposes. Br J Radiol 53 : 897–900

46. Sigel B, Coelho J C U, Spigos D G, Donahue P E, Renigers S A, Capek V, Nyhus L M, Popky G L 1980 Real-time ultrasonography during biliary surgery. Radiology 137 : 531–533

47. Spitz H B, Wyatt G M 1977 Rectus sheath hematoma. J Clin Ultrasound 5 : 413–416

48. Taylor K J W 1974 Use of ultrasound in opaque hemithorax. Br J Radiol 47 : 199–200

49. Voorhies R M, Patterson R H 1980 Preliminary experience with intraoperative ultrasonographic localization of brain tumors. Radiol Nucl Med Mag 10 : 8–9

50. Warren P S, Garrett W J, Kossoff G 1978 The liquid-filled stomach — an ultrasonic window to the upper abdomen. J Clin Ultrasound 6 : 315–320

10

Choosing equipment and developing a department

P. N. T. Wells and B. B. Goldberg

INTRODUCTION

Ultrasound is an essential component of modern diagnostic medicine, and most large hospitals now have established ultrasound departments. Opportunities for establishing totally new departments are infrequent and the practical day-to-day problems discussed in this chapter are related to the choice of replacement equipment and of equipment to provide new capabilities, with the provision of appropriate physical facilities, and with the role and training of personnel. The problems of the small specialised unit and of the clinician working in his own office are also addressed.

One topic which is not discussed further in this chapter is that of the hierarchical management of ultrasonic departments. Different professional groups may have varying views and interests, and there may be a variety of situations in different countries. Therefore, we confine ourselves to one comment on this subject. This is that it must not be forgotten that the management arrangements, whatever they may be, exist primarily to allow the department to provide the best possible care for the patient, and that there may be other vital roles in training students and in research. All other considerations should be secondary.

EQUIPMENT

Once the decision has been made to obtain equipment for ultrasonic diagnosis, it becomes necessary to choose from the bewilderingly wide range that is now available. For a very few general clinical applications, almost any instrument of the appropriate type is suitable, but, much more often,

success depends on the use of an instrument designed for application in a specific diagnostic area. The novice finds the selection of equipment particularly difficult. This is unfortunate as making the correct initial decision may be crucial in subsequent clinical practice.

Ultrasonic equipment can be purchased or leased.[1] The choice of the best method of equipment acquisition depends on local rules and circumstances, and so varies from institution to institution and from country to country. Consequently, general guidelines cannot be given.

Equipment for specific clinical applications

At the present time, ultrasound can provide useful information in the following clinical specialities, listed roughly in order of frequency of referral in a large hospital:

Obstetrics
Cardiology
Internal medicine (liver and biliary system, pancreas, reticuloendothelial system and spleen)
Urology (kidney, urinary bladder, prostate)
Gynaecology
Vascular surgery
General surgery
Paediatrics
Oncology
Ophthalmology
Accident and emergency

The type of facility needed, whether a generalised clinical-diagnostic ultrasound facility or a specialised diagnostic unit, depends upon the interests of the clinicians involved, the type of hospital or office, the patient population, and the demography of the area to be served.[1]

The principles of the different types of ultrasonic equipment are described in Chapter 1. In deciding what equipment is necessary, the following questions need to be considered:

1. Which clinical specialities does the department wish to serve?

2. What equipment is necessary to provide these services?

3. Which services require specialised equipment?

4. How many services can be satisfied by a single piece of equipment?

5. Which services have a patient load that economically justifies an independent piece of equipment, or dedicated staff?

6. If a specialised unit is justified from a clinical viewpoint, will it be advantageous economically in terms of cost-effectiveness of the hospital's operation?

Equipment performance and specification

Once the type of equipment which is required has been determined, the next step is to decide which commercially available instrument is most suitable.[2]

The following considerations apply to the selection of pulse-echo imaging systems:

Display. What size is the display? How bright is it? Is the spot size acceptably small? What facilities are there for photography? Can the display be viewed and photographed simultaneously? Is there an image storage (frame freeze) capability of adequate performance? Is the grey scale performance satisfactory? Can the image be either white-on-black or black-on-white?

Transmitter. Is the output variable, so that the ultrasonic intensity can be kept as low as possible?

Transducer. What size is the transducer? How heavy is it? At what frequency does it operate? Is the resolution adequate? Can the transducer be used for or adapted for puncture guidance? Can it be sterilised?

Receiver. Is the gain adequate? Can time gain control be applied? Are the controls clearly calibrated? What are the signal processing characteristics and capabilities?

Special considerations apply to each particular type of imaging instrument, as follows:

A-mode instruments. Is the timebase variable? Is it calibrated in terms of distance? Can distance markers be displayed? Can the timebase be delayed relative to the transmission time or the image scale factor changed to allow echoes from deep structures to be studied in detail? Can the trace be inverted? Is the instrument suitable for special studies, such as brain midline localisation?

Two-dimensional static B-scanners. Selection of the static B-scanner represents a major financial investment for an institution and is one of the more difficult choices which has to be made, due to the large number of such instruments available on the market today. While no one instrument meets all of the desirable criteria for the ideal static B-scanner, the following are points which should be evaluated when considering purchase of such a device.

The instrument should offer the capability of utilisation of a wide range of transducer frequencies from 1 MHz up to or beyond 7.5 MHz. Ideally, the manufacturer should provide, as part of the initial purchase price, three transducers in the 2.25 MHz, 3.5 MHz, and 5 MHz range frequency. Transducers should be easily and rapidly interchangeable and some automatic indication of the transducer frequency selected should appear on the displayed image. Transducers should be of such size and shape as to be comfortable to hold and easy to manipulate. The scanning arm should be durable, stable and well counterbalanced so that no resistance to motion of the transducer is evidenced within the scan plane. It should be easy to tilt the scan arm and a visual indication of the degree of tilt should be provided either on the arm or within the displayed image.

Instrument controls should be arranged in a logical fashion and the function of each control should be clearly indicated on the control panel. Gain control should be incremental and clearly calibrated so that gain settings may be reproduced at some later date if so desired. A graphic swept gain curve display should be continuously visible on the instrument and that display should immediately reflect any changes in the settings of gain controls. Ideally, all relevant gain, transducer and scale factor data should be visibly displayed as part of the ultrasound image.

The instrument should be easy to update and should have the capacity to accept a real-time linear array scanner module or a sector scanner module, or both. Biopsy transducers should be available for the instrument. Pre- and post-signal processing capability should be available and white-on-black or black-on-white image should be switch-selectable for both monitor display and photographic purposes.

Power output of the instrument should be variable and accessible to the user so that ultrasonic intensity delivered to the patient can be kept as low as is consistent with adequate imaging.

M-mode systems. Is the recording system adequate? Is it calibrated in terms of distance and time? Are there facilities for recording ECG and PCG? Can the M-mode operate simultaneously with real-time display of the two-dimensional image, if this is available?

The consideration of Doppler systems may need to take into account other factors as follows:

All Doppler systems. What size is the transducer? What is its operating frequency and is it a pulsed or continuous wave system? Is the ultrasonic field geometry and sample volume appropriate to the clinical application? Is the ultrasonic intensity as low as possible? Are special transducers available for applications such as blood flow studies? Is there a satisfactory audio monitor? Is there a suitable recording system? Does the unit provide, or can it be upgraded with, a spectral analysis module?

Blood flow detectors. Is the detector sensitive to flow direction? Is the output derived from a zero-crossing detector or a frequency spectrum analyser? Is it real-time? Is the system range-gated? What recording system is provided?

Duplex Doppler/real-time instruments. Is the blood flow detection performance adequate? Can arterial and venous flow be separated? Is the resolution satisfactory? Is the system pulsed? Is the display satisfactory? Can blood flow volume rate be computed? Can the real-time image be photographed with an indication of the position of the sample volume accurately displayed?

Recording methods

A-scans. Generally, A-scans are most conveniently recorded on self-processing film (e.g., Polaroid).

Two-dimensional B-scans. Two-dimensional B-scans made with static B-scanners can be recorded either as individual images on self-processing film (e.g., Polaroid), or on 35 mm, 70 mm or 90 mm negative film, or, by means of a multiformat camera in groups of images (typically 2 × 3) on 200 mm × 250 mm X-ray film. Several factors must be taken into account in deciding on the best method in any particular situation. The quality of the recorded image is, of course, the primary consideration. Although it is sometimes said that Polaroid film lacks the necessary dynamic range, this is untrue. The adjustment of the brightness and contrast of the display certainly has to be carried out with more care for Polaroid than for negative film photography; but, provided that this is done properly, the total dynamic range of even the best cathode ray tube display is easily accommodated. In fact, the only relevant factors in deciding on the recording method are cost and convenience of operation, viewing, and storage. In arguments about cost, Polaroid is often dismissed because of expense except for small workloads or where film processing facilities are not available. This kind of reasoning needs careful scrutiny, however, because the true costs of alternative film methods are often concealed by failure to take into account factors such as staff time, unnecessary usage, and equipment depreciation.

Another method makes use of the dry silver paper recorder. The cost of the instrument is quite high, but the paper is relatively inexpensive and the grey-scale performance is good. Archival qualities should, however, be carefully evaluated.

Electronic storage of two-dimensional B-scans can be provided by video magnetic tape recording or by video magnetic disk. The tape method poses problems of access and of deterioration during extended viewing, but the disk method has considerable potential.

Real-time scans are usually stored on video magnetic tape. The only practical alternative is ciné film, but this is inconvenient and unpopular except for illustrating lectures.

M-mode recordings. Nowadays, M-mode recordings are almost always made on paper strip charts. Paper sensitive to the ultraviolet light of a cathode ray tube with a fibre optic faceplate is commonly used, but the higher quality recording obtainable with dry silver paper makes this an attractive alternative.

Accessories

Transducers. In addition to the standard transducers used for everyday investigations, it may be possible to justify the acquisition of special transducers, for example for paediatric applications, for puncture guidance, and for use during surgery.

Scanning beds. Many manufacturers supply suitable scanning tables with their equipment. Otherwise, an ordinary patient trolley is usually adequate, especially if it can be elevated at one end. It may sometimes be convenient if the bed has cut-out sections, for example to make it easier to gain access to the patient's flanks.

Test objects. The subject of quality assurance is discussed in Chapter 11. The department should be equipped with a suitable range of test objects and tissue phantoms.

Coupling agents. The choice of coupling agent is a matter of the personal preference of the clinician or the sonographer. Mineral oil, liquid paraffin, and olive oil are all used, and there are some excellent proprietary coupling gels. Cost and availability have made mineral oil the most popular couplant for static B-scanning, although it may have deleterious effects on cables and plastics. The proprietary gels, however, are most widely used for real-time scanning.

Potential pitfalls

Because of the variety and complexity of ultrasonic systems, it is also not possible to give general rules about the choice of equipment. It may be helpful, however, to mention the factors which should be considered when deciding whether a particular instrument is suitable for specific clinical applications. In addition to assessing the specification of the equipment, it may also be wise to establish the following points:

1. The equipment meets the requirements of safety regulations.
2. The equipment is reliable, and the servicing arrangements are satisfactory.
3. Single-handed operation of the equipment is possible in clinical practice.
4. The performance of the equipment is predictable and reliable.
5. Training of operators does not pose insoluble problems.
6. The equipment is suitable for the proposed accommodation.
7. If the equipment is to be used in several locations, it is easily mobile.
8. The delivery time is acceptable.
9. The price does not exclude necessary accessories, and does include import duty and taxes if liable.

It is vital to remember that ultrasonic diagnostic equipment is constantly being developed and improved. The current state of the market must be fully reviewed before deciding which equipment to buy, and the dangers of choosing new and perhaps unproved systems must not be forgotten. Several publications exist which are invaluable as aids to keeping up-to-date.[3, 4]

WORKLOAD

Considerations concerning the types and numbers of instruments, the size of the department, and the staffing levels, must take the present and projected workloads into account.

In a review of experience in North America,[1] it has emerged that the caseload of a comprehensive ultrasound service may be expected to undergo a rapid growth of 35–50 per cent per year over a three year period, at which stage the demand should level out to approximately one examination per year for every fifty people in the population served in established ultrasound laboratories. Development of new types of ultrasound examinations and growth of the population presently creates a sustained caseload growth rate of approximately 15 per cent per year. A typical distribution of the workload in a large teaching hospital is such that 40–60 per cent of the studies are in obstetrics, 15–35 per cent in abdominal specialities, 10—15 per cent in cardiology, with the remaining 10–15 per cent in neurology, ophthalmology, and so on. As a rough guide, a busy department examines approximately five inpatients per year for every acute hospital bed.

Having estimated the workload in each clinical specialty to be served, the next step is to use this information to determine the numbers of instruments required. The capacity of one static B-scanner operated by one sonographer is 6–15 patients per day, depending on the experience of the sonographer and the clinician, the adequacy of the supporting services, and the types of examinations. Similarly, the capacity of an M-mode echocardiographic system is 10–20 patients per day. Real-time scanning can be up to three times faster than static B-scanning in experienced hands, but it is not reasonable to expect that even the most enthusiastic can keep up this rate of working for long periods without a rest.

Doppler studies require patience, and Doppler imaging is a lengthy process. A maximum of 4–18 patients per day is probably as many as either the machine or the operator can manage.

These estimates of maximum workloads assume that the operators — sonographers and clinicians — are highly trained and fully competent, and that they are able to work without distraction. In departments with heavy teaching loads, the number of patients examined may be reduced by a factor of two.

Having established the numbers and types of instruments needed, it is then possible to estimate the minimum number of personnel required to provide the service. At this point, it is worth mentioning that the constraint which is usually most seriously considered is the minimum staffing level. The provision of an efficient service, however, also depends on the staff maintaining adequate competence. Thus, in a general ultrasound service, a clinician needs to examine at least four patients per day, although not less than two patients per day may be adequate in a specialist service. A sonographer, who generally should be technically more competent than his or her medical colleagues, needs to scan perhaps twice as many patients. Staff carrying out lengthier procedures such as Doppler examinations can remain competent although examining smaller numbers of

patients, but the actual time during which they are involved with patient care needs to be about the same.

ACCOMMODATION

Although, in ideal circumstances, the ultrasonic department should be in a self-contained suite of rooms, such as that shown in Figure 10.1, it is

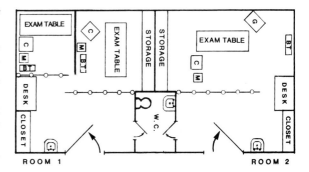

Fig. 10.1 Idealised layout for a two-room, three-instrument ultrasound examination area. Each room is equipped with a desk and closet, a washbasin and curtains which provide privacy without reducing accessibility. Room 1 is equipped with two real-time examination stations and room 2 is equipped for static examinations. Each examination station has a bedside table for temporary storage of film, couplant and other supplies. A lavatory is accessible from each room. Both room doors are wide enough to permit access for a typical hospital bed. Patient reception and waiting areas should be conveniently nearby. G = gantry-scan arm for static imager; C = control console; M = multiformat or fixed format camera; BT = bedside table.

extremely desirable for it to be in close proximity to other imaging departments such as radiography, computed tomography, and radionuclide imaging. Good working relationships between the staff providing these complementary services ensures that the patients are investigated by the most effective modality, and many of the supporting services can be shared.

Room layout

A principal room for ultrasonic scanning should have an area of about 10 m². The door should be wide enough to allow the patients to be brought in on a bed if necessary. The floor should be safe for the weight of the equipment and the people, and there should be good ceiling clearance.

It must be possible to lower the brightness of the lights in the room, and air conditioning should be provided if necessary to maintain a comfortable environment for patient and staff. A washbasin is essential.

There must be conveniently placed electrical power points of adequate capacity with at least one supply line for each instrument. There should also be a telephone and a worktable. Adequate shelf and cupboard space should be provided.

Privacy is an important factor in ensuring the comfort of the patient and the concentration of the sonographer. If several patients are to be scanned at the same time, each procedure should be carried out in a separate room, or at least within a cubicle separated by curtains.

Support space

The scanning rooms must have convenient access to the patient waiting area, and to lavatories. Emergency resuscitation equipment must be located nearby. If film is used, the automatic processor must be conveniently situated. A steriliser may be required.

There must be an adequate number of offices and viewing areas for the clinicians and there should be a common room for the staff.

In departments with students, there should be convenient access to a lecture room, and to teaching carrels.

Last but not least, the necessity for storing the results of many years of work must not be forgotten. It is remarkable how quickly the volume of scans builds up.

Provision for the future

In the short term, it is economically sensible to plan a department so that its size is exactly matched to the contemporary application of ultrasound. This is a short-sighted policy, however, because substantial changes are inevitable in the future as new capabilities are developed, both in ultrasound and in complementary and competing technologies. It is sterile to try to predict long term trends; but the luxury of unused space today will inevitably become the salvation of tomorrow!

PERSONNEL

The provision of a successful service depends on the support of engineers, physicists, instrument technicians, clerks, aids, and housekeepers. Some of these individuals may be shared with neighbouring departments. Engineers and physicists often provide their services to several departments in neighbouring hospitals.

Whatever the administrative arrangements of the department each member of the staff has a distinct professional role.

The clinician is personally responsible for the medical care of the patient. The advice given to the patient or to the clinicians attending the patient must either come directly from the clinician, or from paramedical staff acting under his or her supervision.

The clinician may actually scan the patient, or he may delegate this task to the sonographer. In the latter case, the sonographer must behave according to the established professional standards. Indeed, although often the sonographer is technically more competent than the clinician to whom the sonographer is accountable, the chain of clinical legal responsibility must not be disturbed.

A physicist may be responsible for quality control. This individual should be involved in the choice of equipment, and often is involved in research. Like the engineer and instrument technician or serviceman, the physicist may look after the maintenance of the equipment.

All the professional staff have important roles in teaching. Teaching the art and science of ultrasonic diagnosis requires both formal lectures and hands-on experience. Various professional societies test the competence of students, and issue certificates of proficiency. The largest such organisation devoted entirely to ultrasound is the American Registry of Diagnostic Medical Ultrasound,* which provides comprehensive testing and certification in the various speciality areas of ultrasound. In the United Kingdom, the College of Radiographers† is the examining body for the Diploma in Medical Ultrasound. Certificates and diplomas of

* American Registry of Diagnostic Medical Ultrasound, 2810 Burnet Avenue, Suite N, Cincinnati, OH 45229, USA.
† College of Radiographers, 14 Upper Wimpole Street, London W1M 8BN, UK.

this kind are valuable indicators of the standings of the individuals who held them.

Although frequently overlooked, the contributions made by the lay members of the staff are essential to the wellbeing of the patients and to the success of the department.

REFERENCES

1. Alliance for Engineering in Medicine and Biology 1977 System design of a clinical facility for diagnostic ultrasound. Technical report N-1977-2. AEMB, Bethesda
2. Carlsen E N 1980 Instrumentation considerations in establishing a clinical ultrasound facility. In: Wells P N T, Ziskin M C (eds) New techniques and instrumentation in ultrasonography. Churchill Livingstone, New York, p 1–19
3. Clinical ultrasound purchaser's catalogue. Published annually by McGraphics, 371 S Emerson, Denver, CO 80209, USA
4. Diagnostic imaging. Published monthly by Miller Freeman Publications Ltd, 500 Howard Street, San Francisco, CA 94105, USA

11

Quality assurance

Albert Goldstein

INTRODUCTION

The function of quality assurance in clinical ultrasound is to keep the diagnostic quality of the ultrasound image at a consistently high level. An adequate quality assurance programme includes proper equipment operation, routine preventive maintenance, and acceptance and routine performance tests. It is suggested that routine preventive maintenance procedures and performance tests be performed at least once a week. The individuals whose involvement is essential to quality assurance include the physician, the sonographer, the equipment service engineer and, if available, a physicist or engineer who is familiar with clinical ultrasound images and the operation of ultrasound equipment.

Quality assurance in a clinical subspeciality should exist on many different levels. Physicists or engineers can perform both acceptance tests, and the more technical routine performance tests, and be available for consultations. The equipment service engineers can repair or replace faulty components, calibrate the equipment, and perform the manufacturer's own quality assurance procedures. With the latest equipment the clinician has more information present in the ultrasound image. Besides improving the patient diagnosis, the increased information can aid the clinician in diagnosing operator error or equipment malfunction. The sonographer can learn to use and maintain the equipment better, as well as to know when to call in one of the technical experts.

In this chapter, a system description of image formation is considered. The areas where quality assurance procedures are most needed are pointed out, and some of the procedures to be followed are outlined.

A SYSTEM APPROACH TO IMAGE FORMATION

The data flow in ultrasound image formation is shown diagrammatically in Figure 11.1. The ultrasound data (static image or real-time) are acquired from the patient, processed and then stored in the scan converter which drives the TV display. Hard copy obtained from the TV display is viewed and the diagnosis is made. Note that the diagnosis relies on the quality of the perceived image in the viewer's brain. Consideration of the properties of the hardcopy-eye and eye-brain interfaces is vital in preserving the information obtained from the patient.[1]

The dotted lines in Figure 11.1 outline the clinical equipment. The weakest links in the

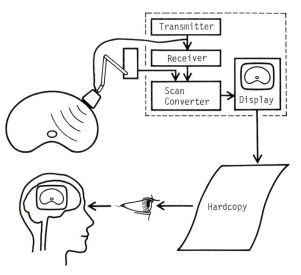

Fig. 11.1 Diagram of the information flow in ultrasound image generation and interpretation.

imaging chain occur outside of the equipment, in the patient scanning and display-hardcopy-view steps. Since the operator has direct control over them, adequate attention should be placed on transducer selection and use and in obtaining high quality hardcopy. The equipment controlled process and store functions are largely inaccessible to the operator. Thus, the necessary equipment quality assurance is properly to choose the equipment variables for each scan and routinely to ascertain that the equipment is working well and consistently.

TESTING EQUIPMENT PERFORMANCE

A routine performance test is a qualitative measurement performed at a fixed interval in order to verify consistent equipment performance. An acceptance test is a test performed by technically competent personnel to make quantitative measurements of equipment performance for comparison with the recommended performance parameters or with the manufacturer's stated performance specifications. The recommended procedures for both routine performance and acceptance tests are still being developed.[2-5, 7, 9, 10]

DATA ACQUISITION

Transducer selection

Image resolution

In order to obtain the best image resolution (least blurring), the proper transducer should be chosen for each particular examination. Usually the lateral resolution (determined by the beam width) is the most crucial factor. Choose a transducer with the highest frequency that penetrates to the depth of interest and whose focal zone coincides with the depth of interest. Measure the transducer beam width and axial resolution with the appropriate test objects. Keep these data for comparison in case transducer damage is suspected.

Changing transducers

When changing the ultrasound transducer, the electrical transmission pulse should be inhibited or rerouted to prevent shock to the operator. Consult the service engineer for the proper technique for the particular equipment.

Transducer case

The transducer is pulsed at rates up to several thousand times per second with voltages as high as 900 V. Although the energy in each pulse is not great, there is a hazard of shock to the patient or the operator. The transducer case is usually made of nonconductive materials to prevent electrical connection between the voltage inside the case and the patient or operator. A transducer that develops any openings, cracks, or other voids in the nonconductive case is not suitable for patient use and should be immediately removed from clinical service.

Ultrasound equipment should be checked initially and then once a year for excessive leakage currents. This is a safety procedure to protect the patient from potential shocks. Invasive ultrasound procedures, such as needle biopsy or cyst puncture, or ultrasound examinations of cardiac patients, pose an electrical shock hazard to the patient. The primary danger is electrical shock to the heart muscle. Applications of external electrical currents to myocardium can disorganise its normal rhythm and put the heart into fibrillation. Even low currents through a needle or catheter piercing the skin near the heart can produce at the heart current densities sufficient to cause fibrillation.

If any ultrasound equipment has high leakage potentials (electrical voltages 'leaking' through nonconductive insulation) it might transmit these voltages to the biopsy needle or catheter and introduce microampere currents directly to the heart. Since the threshold of perception for unbroken skin (the sonographer) is in the milliampere range, this current can go unnoticed.

The danger to the patient can come not only from cracks in the transducer case but from any leakage potential or bad insulation (leakage paths) in the equipment. Leakage tests should be performed at the recommended intervals, and proper records should be kept. The transducer manufacturer can probably recertify the transducer for a nominal cost, and this should include a leakage test.

Besides leakage testing the prudent ultrasound user tries to maintain the patient electrically isolated from both the equipment and electrical ground. The commercial coupling gels which are highly electrically conductive should be used with caution. This is especially important in any needle puncture procedure or on scans of a patient with an external cardiac catheter or external cardiac pacemaker leads.

Sterilisation

For biopsy, aspiration and intraoperative procedures, the transducer must be sterile. An autoclave or any other high temperature technique must *not* be used. High temperature, even for a brief period of time, can corrupt the adhesives and plastics used in the fabrication of the transducer. Cold gas sterilisation should be used.

Scanning technique

Scanning motion

Proper scanning technique is important when dealing with the potential generation of image artifacts. One such case is the acoustic shadow caused by trapped air in the digestive system. The acoustic shadow is always distal to the transducer along the 'line of sight' of the ultrasound beam. Angulation of the beam causes the acoustic shadow to change position in the image. For example, a preferred scan plane orientation for scans of the pancreas is one which contains the gallbladder, duodenum, and the head of the pancreas. If the duodenum contains an air bubble and the transducer is positioned too far laterally, then when it is aimed medially the accoustic shadow of the air in the duodenum can obscure the head of the pancreas. In this case it is better to position the transducer medially and to orient the beam to cause the acoustic shadow to fall lateral to the head of the pancreas.

Coupling medium

An adequate amount of ultrasound coupling gel should be used in the examination. If the amount of coupling gel on the patient's skin diminishes, the amplitude of the received echoes can decrease. Sometimes this effect is very slight and goes unnoticed by the operator. The ultrasound operator should also make sure that the transducer is moving over the coupling gel rather than acting as a snowplough and pushing the gel aside.

Placing the coupling gel on the display TV in order to warm it up is not a recommended practice; if the coupling gel gets on the ultrasound equipment it could lead to future serious difficulties. The coupling gel should be stored off the equipment in available heaters designed to keep it warm. Hot plates should never be used.

Distance measurements

In the generation of an ultrasound image to be used for making a distance measurement, it is recommended that the 1 cm depth marker dots should be used. During the scan it is essential that the transducer beam should be parallel to the measurement direction and that the 1 cm depth marker dots should be placed along the line of measurement. Because the depth marker dots are often obscured due to tissue echo information, it is good practice, when necessary, to put another set of adjacent parallel depth marker dots in an area where fewer echoes are present.

On digital equipment it is convenient to use the digital caliper feature to make distance measurements. Since the accuracy of the digital caliper measurement depends upon proper B-mode registration, this should be periodically checked. After the B-mode registration has been verified, the calibration of the digital caliper itself should be checked by projecting 1 cm depth marker dots on the image and measuring 10 cm along the depth marker dots with the digital calipers. If there is a measured difference, the inaccuracy is most likely in the digital caliper circuit.

EQUIPMENT OPERATION

Logic lockup

The latest ultrasound equipment contains digital circuitry. In digital circuits numbers (not voltages or currents as in analogue circuits) are used to represent signal (echo) amplitudes and for control purposes. On rare occasions this equipment refuses to operate properly or is totally unresponsive to the operator controls. This condition is known as 'logic lockup'.

The cause of logic lockup lies in the fact that digital circuitry functions by storage and transfer of numbers. If a stored or transferred number used for control purposes changes spontaneously (due to heating or component malfunction), then the logical sequence of events necessary for equipment function is interrupted. Since this is an 'unplanned' change in a number the equipment cannot, by itself, get out of this logic lockup state.

The way to get out of logic lockup is completely to turn off the equipment, wait one minute and then turn it back on. This procedure takes advantage of the digital circuits whose function is to put the correct initial numbers in key locations when the equipment is turned on. Thus, the turn-on procedure ('initialisation', in jargon) automatically corrects a logic lockup state unless the incorrect number is stored in the mass storage (computer tape, disc, or floppy disks). In this case it is necessary to replace the mass storage device.

If repeated applications of this remedy do not work, the advice of the service engineer should be sought. If logic lockup occurs frequently, then some aspect of the environment (high temperature, static electricity, and so on) is causing spontaneous changes in a susceptible portion of the equipment.

Air filters

The solid-state circuit components in the equipment are very sensitive to temperature. It is necessary to keep a supply of cool air circulating through the equipment in order to prevent excessive temperatures. This is usually accomplished by mounting a fan on the back of the unit which forces air out of the equipment and creates an internal vacuum. Input openings located at several strategic spots on the equipment to take in outside air sucked in by the vacuum are each covered by an air filter. If the air filters are not clean, they lose their efficiency in passing air and the fan instead sucks the cooling air through any hole in the shell of the equipment — usually the small holes from which the control knobs protrude. These holes are a bad source of cooling air because dust trapped on oil spills or oily knobs can be sucked into the machine and get on critical elements. This layer of dust acts as an insulator and allows temperature rises which can lead to premature component failure. The air filters should be removed periodically and cleaned by hand, compressed air, or by putting them under a stream of water. The service engineer should be consulted about the recommended procedure for each particular instrument.

Integrity of enclosures

All doors and other openings on the equipment must be closed during the routine operation of the equipment. If certain doors or drawers are left open, the cooling fan may suck in air and dust through that opening on to the electronics. This is detrimental to consistent long-term operation of the equipment.

Worn or frayed cables

The electrical coaxial cables which connect the various components of the system suffer mechanical stress from daily use. A stress-related occurrence is an internal break in the centre conductor of a cable. This is frustrating to the operator because the machine 'works' or 'doesn't work' depending upon various gyrations performed. If equipment begins to function in an intermittent fashion, check the cables for internal breaks by wiggling them. The most vulnerable portion of a cable is the joint between the cable and the connector. The service engineer should be told if it is felt that the cable is broken.

Scanning arm lateral stability

The scanning arm of static B-scan equipment performs two vital functions. The first is to track the position and angulation of the ultrasound transducer beam. The second is to confine the ultrasound beam motion to the plane of the scanning arm (the scan plane in patient). If there is excessive lateral mechanical play or movement in the scanning arm, echoes from adjacent tissue structures may be present in the scan plane image. There is disagreement as to exactly how 'stiff' the ultrasound arm should be. Whatever the initial choice of the user, it is possible for the arm to become looser and consequently to introduce improper echoes in the image.

Image spatial registration

In static imaging (B-scan) the proper tracking of the transducer position and angulation leads to good image spatial B-mode registration, i.e., each reflector is placed in its correct anatomical location in the image independent of transducer position or angulation. Improper B-mode registration leads to image distortion in single pass scans and image distortion and blurring in compound scans. B-mode registration can be adequately checked with the AIUM 100 mm Test Object.[2, 3, 7, 9]

Receiver

The combination of amplifier, rectifier and demodulating circuits is called the 'receiver'. Excessive electrical noise and variations in receiver sensitivity (gain) affect image quality.

Electrical noise

The electrical noise present on the power lines in the hospital affects the maximum penetration in a clinical scan. With high gain this noise presents on the image as background low amplitude shades of grey. It is undesirable to have these background low amplitude shades of grey present on the image because they do not represent true echoes coming from inside the patient and the electrical noise may change character from scan to scan and in repeat ultrasound clinical examinations.

When installing ultrasound equipment, the engineer turns the receiver gain to maximum and looks for electrical noise on the A-mode baseline. He then adjusts an internal gain control to reduce the effects of the electrical noise. It is important to note that he is not cheating the user. He is not removing any clinical information from the scan because at the highest levels of gain the electrical noise overrides or 'swamps' the low amplitude echoes.

It is recommended that the clinical ultrasound user should reduce the electrical noise on the power lines to the minimum possible. This may be accomplished by putting the ultrasound laboratory on a separate transformer, by removing other pieces of electrical equipment, such as elevators or centrifuges, from the same power line, or by removing conduit grounds. When the electrical noise has been reduced, the service engineer should be asked to re-adjust the internal gain control commensurate with the new lower level of electrical noise. For maximum penetration, the gain should be set just below the level that gives excessive A-mode noise.

Sensitivity

The overall sensitivity of the receiver (amplifier) is important. The sensitivity of the receiver along with the sensitivity, frequency, and bandwidth of the transducer determines the maximum penetration in a given clinical examination. It is important that the receiver sensitivity be as high as possible and constant with time over the long term. New B-scanners are using more sensitive amplifiers and may require the use of heavier electrically shielded transducers to minimise low level electrical pickup.

The long term stability of receiver sensitivity may be monitored by repeat scans of a *stable* tissue equivalent phantom at maximum gain setting.

Digital circuitry

Modern ultrasound equipment utilises digital circuitry and techniques developed by the computer industry. Digital circuitry is inexpensive, reliable, flexible and readily available. Its use in ultrasound imaging equipment has led to many improvements in equipment performance.

Equipment 'self-checks'

One immediate advantage of the incorporation of microprocessors in ultrasonic scanners is the capability to perform simple tasks such as recording on the hard-copy the patient number, date, scan plane orientation, swept gain curve and transducer frequency. Another advantage is the ability of the equipment to perform 'self-checks' of its own digital circuitry. Manufacturers have built-in their own quality assurance programs. Some equipment goes through a comprehensive 'self-check' when turned on. Other equipment permits the service engineer to perform diagnostic test programs by using the patient data input keyboard as a control device. Even with this sophistication, however, some essential performance variables (such as receiver sensitivity) cannot be routinely checked.

Echo amplitude resolution

The digital scan converter in modern ultrasound equipment stores the acquired data prior to display. Since it contains digital computer memory circuits, it can only store data in the form of numbers. In order to retain good spatial resolution the scan plane is divided into a matrix of square pixels (picture elements) with the pixel size (spatial dimensions) small enough so that the image is aesthetically pleasing and the spatial resolution is determined by the transducer beam pattern. The analogue echo amplitude data acquired from each pixel are digitised (converted into a number) and stored in the pixel's assigned location (word) in the computer memory. This digitisation (analogue to digital conversion) is performed by one of a number of operator-selectable pre-processing schemes and is a potentially weak link in the imaging chain. For example, Figure 11.2 demon-

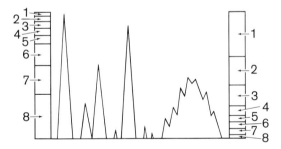

Fig. 11.2 Two pre-processing schemes for converting ranges of analogue echo amplitudes into the numbers stored in a digital memory.

strates two pre-processing schemes for converting ranges of analogue echo amplitude data into the numbers stored in the digital memory. The scheme on the right has the higher resolution for the low amplitude echoes, and vice versa. In the image, the ability to detect small amplitude differences in the low amplitude echoes depends on the exact number

and spacing of the digitisation ranges. Many digital scan converters offer the operator a choice of pre-processing schemes. Each has its own inherent echo amplitude resolution, which permanently determines the echo amplitude resolution in the image. Poor operator choice of pre-processing can degrade the information in the image. It is important to learn to select the proper pre-processing for each clinical examination.

IMAGE VIEWING

Proper grey bar

Photographic hardcopy technique should be properly set up and routinely checked for each hardcopy camera. The image display and hardcopy camera in present ultrasound equipment are among the weakest links in the imaging system, so they should be checked daily. When testing the photographic technique, a grey bar generated by the equipment should be used. The spatial relationships between the shades of grey in a clinical scan are different from those in the grey bar. This difference affects the viewer's perception of the grey levels. The proper technique for establishing a good grey bar is to adjust the levels until a good clinical scan (or, alternatively, a good standing liver or tissue phantom scan) is obtained and then to document these settings with a grey bar image.

Film processor quality assurance

With the addition of a multi-image camera it is essential that the automatic film processor used for development of the X-ray film should be maintained at the correct temperatures and have the correct chemistry. If the institution does not have a processor quality assurance programme,[6, 8] then one should be started. In the meantime, some simple tests can be carried out. Sensitometry, densitometry and the like need not be employed, but just some commonsense checks during the clinical day. The image grey bars should be checked for consistency. The replenisher tank in the processor should be inspected at least once a day to make certain there is an adequate supply of replenisher. Nothing degrades the chemistry in the processor faster than the replenisher pump sucking air into the developer! The level of liquid in the replenisher tank must be high enough to prevent this from happening.

Polaroid images

There are three essential points to remember when using Polaroid film. The first is that it is a high-contrast film. It is essential when using Polaroid film to maintain proper photographic technique. Secondly, if Polaroid film is exposed to average room humidity for longer than several hours, the shades of grey in the image begin to be 'washed out'. It is strongly recommended not to remove the film from its protective foil wrapper until immediately prior to use. Thirdly, Polaroid film must be developed within one minute, and never for more than three to five minutes. Development of a Polaroid image begins when the film is pulled from the camera and terminates when the Polaroid print is peeled off the negative. Excessive development time leads to a Polaroid image that degrades in the film file.

Proper film handling, just discussed, is the first area of preventive maintenance associated with Polaroid images. The other is routinely cleaning of the Polaroid camera. The rollers inside the Polaroid camera become contaminated with the development chemicals. It is necessary to clean these rollers

with an alcohol wiper at least every other day, and preferably once a day. Also dust off the back of the camera on a routine basis.

Multi-image cameras

Due to image quality and cost consideration, multi-image cameras (usually utilising 200 × 250 mm X-ray film) are becoming standard in ultrasound clinics. Such a camera is shown schematically in Figure 11.3a. Above the high quality black-and-white TV cathode ray tube is a lens and shutter assembly. At the top, accessible to the user, is an X-ray sheet film cassette. Since the lens is positioned midway between the TV monitor (object) and the film (image), the optical magnification is unity and a single image is produced on the film. This is the so-called '1-on-1' format.

If the lens is positioned closer to the film, then a smaller image is obtained. When the magnification of the optical system is chosen such that the image is one-fourth the size of the film, then four images are obtained on one sheet of film ('4-on-1') as shown in Figure 11.3b. Similarly, a 6-on-1 format can be obtained by positioning the lens even closer to the film (Fig. 11.3c). The lens is mechanically translated between exposures to the proper positions for these multi-image formats. The light rays from the TV to the film are not

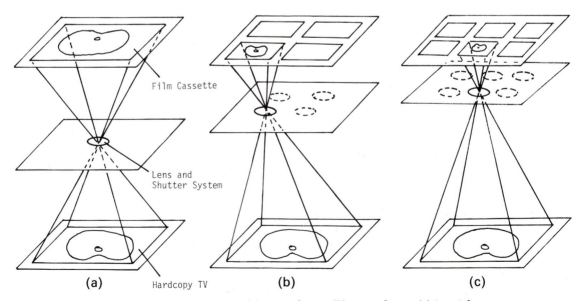

Film Cassette

Lens and Shutter System

(a)   Hardcopy TV   (b)   (c)

Fig. 11.3 Multi-image formatting. (a) 1-on-1 format. (b) 4-on-1 format. (c) 6-on-1 format.

cylindrically symmetrical around the optical axis of the lens. This optical geometry is called 'off-axis', and it is more susceptible than 'on-axis' geometry (Fig. 11.3a) to image geometrical aberrations. The use of flat face TV monitors and high numerical lens aperture (f-numbers) minimises geometrical distortions on the image. Usually, the multi-image camera is properly designed to be relatively free from geometrical off-axis distortions.

Another way to obtain multi-image hardcopy is permanently to mount multiple lenses at the proper positions between the TV and film. In sequence the shutter of each lens is opened to expose the film. This arrangement, although more expensive optically, has fewer mechanical moving parts. Other multi-image cameras keep the single lens and TV fixed and simply translate the film to the proper spatial positions for obtaining the multi-image hardcopy. Still other manufacturers keep the film stationary but translate the single lens TV as a unit. These last two schemes have the advantage of on-axis geometry for all images.

A multi-format camera is one in which the operator can choose between several multi-image formats. Popular cameras in ultrasound (and CT) clinics are those providing '4- and 9-on-1' and '4-, 6- and 9-on-1' formats. Note that a multi-format camera is much more mechanically complex than a multi-image camera. In some multi-format cameras, when changing the multi-image format, both the TV monitor and the lens must be changed in a vertical position to obtain properly focused images of the correct magnification. This additional mechanical complexity (and the potentially greater 'down-time') should be considered when deciding on the purchase of a multi-image or multi-format camera.

Dust and dirt accumulating inside the multi-image camera affect the resulting hardcopy. In some cameras excessive dust build-up can be prevented by keeping a cassette in the camera at all times, closing the film cassette opening at the camera top. Dirt on the lens is not apparent in the images but, in extreme cases, changes the photographic technique required for proper film exposure. Dust on the face of the TV monitor produces an 'in-focus' shadow on the film. Dust on the film or on a protective sheet of glass adjacent to the film also produces an 'in-focus' shadow on the film.

These causes of image artifacts are illustrated in Figure 11.4. The insides of the camera should be

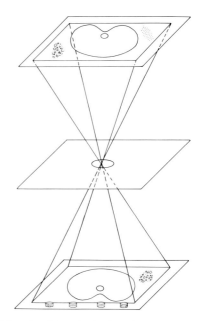

Fig. 11.4 'Dust shadow' artifact generation.

cleaned daily. Since the X-ray film produces a negative image of the TV display, the so-called 'black-on-white' hardcopy display is more susceptible to dust shadow artifacts.

Image interpretation

In B-scanners the real-space horizontal and vertical orientation is usually recorded on the image. For example, the force of gravity points vertically down and this fact may be used in the differentiation of real or false gallbladder 'debris'. In real-time equipment the image is oriented with respect to the transducer. The transducer face is at the top of the image and lower image positions correspond merely to further distances from the transducer face. The real-space orientation of real-time images must be ascertained by consistent image labelling and good communication between the person doing the scanning and the person interpreting the images.

In order properly to adjust equipment operating controls and to help in identifying image artifacts a firm understanding of the fundamentals of ultrasound propagation in tissues is required. The person who performs the scan may well be the best qualified to interpret the resulting image.

DOCUMENTING EQUIPMENT PERFORMANCE AND REPAIRS

It is essential that the performance of the ultrasound equipment should be documented each time tests are made. These tests frequently have value at a later time when knowledge of the baseline performance of a particular component in the ultrasound equipment is required.

There is a great potential risk of miscommunication between the equipment operator and the service personnel. Often the danger lies in incorrectly or inadequately explaining the problem to the service engineer. Often the service engineer repairs a known problem on the equipment and also fixes conditions discovered in the course of his work without telling the equipment operators what he has done to the machine. Certain adjustments made by the service engineer affect clinical performance or the results of the next routine performance test. In order to facilitate the communication between the machine operators and the equipment service personnel, it is essential to maintain a complete log of equipment repairs.

REFERENCES

1. Brendel K, Filipczynski L S, Gerstner R, Hill C R, Kossoff G, Quentin G, Reid J M, Saneyoshi J, Somer J C, Tchevnenko A A, Wells P N T 1976 Methods of measuring the performance of ultrasonic pulse-echo diagnostic equipment. Ultrasound Med Biol 2:343–350
2. Carson P L, Dubuque G I 1978 Ultrasound instrument quality control procedures. Centers for Radiological Physics, Chevy Chase, Report III
3. Carson P L, Zagzebski J 1980 AAPM Report 8: Pulse-echo ultrasound imaging systems: performance tests and criteria. American Institute of Physics, New York
4. Chesters M S 1982 Perception and evaluation of images. In: Wells P N T (ed) Scientific basis of medical imaging. Churchill Livingstone, Edinburgh, p 237–279
5. Department of Health and Social Security, Scottish Home and Health Department, Welsh Office and Department of Health and Social Services (Northern Ireland) (UK) 1981 Code of practice for acceptance

testing of medical electrical equipment. Health Equipment Information, No. 95
6. Goldman L W 1981 Radiographic film processing quality assurance: a self-teaching workbook. DHEW publication No. (FDA) 76-8146
7. Goldstein A 1980 Quality assurance in diagnostic ultrasound. American Institute of Ultrasound in Medicine, Washington
8. Gray J E Photographic quality assurance in diagnostic radiology, nuclear medicine and radiation therapy. DHEW publications No. (FDA) 76-8043, 1976: Vol 1; and No. (FDA) 77-8018, 1977: Vol 2
9. Hospital Physicists' Association (UK) 1978 Methods of monitoring ultrasonic scanning equipment. Topic Group Report, No. 23
10. Lopez H, Smith S W 1979 Implementation of a quality assurance program for ultrasound B-scanners. DHEW publication No. (FDA) 80-8100

12

Biological effects of ultrasound

C. R. Hill

INTRODUCTION

The generation of ultrasound in solids and liquids was first achieved in 1917 by the French physicist Langevin, who had the idea of using the piezoelectric effect to excite electrically a quartz crystal into mechanical oscillation at one of its resonant frequencies. The technique was rapidly developed in connection with naval interest in echo-sounding and, quite early in this work, it was observed that small fish were being killed by the action of the sound beams. This observation was later confirmed by the results of work carried out on fish, mice, unicellular organisms and blood[52] and, subsequently, by many other workers. It thus led, in the 1930s, to a considerable interest in the possibilities for using ultrasound for tissue modification or destruction in a wide range of disorders, from cancer to *Violinspieler Krampf*.[1] In retrospect, much of the work of this period may now appear naive and partially to justify the subsequent reaction, that seems to have occurred in some quarters, against the use of ultrasound in medicine. From a comprehensive and critical survey of this period,[1] however, it is clear that the subject of ultrasound biology had already been shown to be of great interest although it undoubtedly suffered from the absence of a rigorous physical and quantitative approach.

A new phase in the medical and biological application of ultrasound commenced following the Second World War. On the one hand the military development of radar and sonar had provided the technology necessary for the introduction of pulse-echo techniques in diagnosis and, at the same time, the foundations were being laid for a quantitative, biophysical approach to the study of the biological effect of ultrasound. The latter owes much to the pioneering work at the University of Illinois.[14] In spite of these developments, serious interest in the safety of medical ultrasonic procedures, and specific experimental investigation of the subject, date only from about 1964.

The period since 1970 has seen a tremendous growth in the use of ultrasound for diagnosis and this development has brought with it a concern about the existence of any possible hazards that might be associated with in vivo ultrasonic irradiation. Such concern does not, of course, necessarily reflect a belief that any such hazards exist; it largely arises from the analogies that can be seen between ultrasound and ionising radiations, used as diagnostic tools, and from the general need to balance benefit and risk in forming clinical decisions on all aspects of patient management. Against this background there has been a steady flow of experimental investigation of the question, and corresponding publication of research results. Much of this has proved difficult to assess and interpret and has been the subject of review by a number of expert groups, and particularly one established by the World Health Organization.[25] Such reviews are continuing and it is likely that concensus reports will be issued from time to time by (either separately or in conjunction) the World, European, American and Japanese organisations for ultrasound in medicine and biology. Supplementary to these, individual annual reviews of the literature are starting to appear.[15, 16] In the following sections of this chapter the phenomena of the biological action of ultrasound are presented in a manner that should help the reader to form a reasoned judgement on questions of radiation safety.

ULTRASOUND BIOPHYSICS

In order adequately to understand and interpret the large amount of experimental evidence now available relating to the 'biological effects of ultrasound', it is necessary to appreciate, at least in outline, the physical nature of the processes that occur when ultrasonic energy is absorbed by tissues and by experimental systems of interest to biologists. What follows is an attempt to summarise this subject, and necessarily omits much detail that can be found in the references. In this discussion it is implicit that the media of interest are liquids and soft tissues (which behave acoustically like liquids); in particular, except where specifically indicated, only longitudinal (i.e., compressional) wave motion is considered.

Ultrasonic absorption

Attenuation of an ultrasonic beam by a medium occurs as a result of several processes, which may be classified under the headings of true absorption, scattering and cavitation.

True absorption is a process that occurs at the level of inter- or intra-molecular organisation and corresponds to the irreversible transfer of coherent mechanical energy to molecular or structural energy levels. Most, or all, of this energy ultimately appears as heat.

Ultrasonic absorption is a complex subject that is far from being fully understood even in simple media,[2] and the further complexities that arise with biological media are very considerable.[6, 8, 49] In a few very simple monatomic liquids absorption is due primarily to the viscous forces opposing mechanical movement of the medium but in molecular liquids, and particularly for large biological molecules, the predominant absorption mechanisms are related to inter- and intra-molecular structure. In living tissue it appears that absorption is due almost entirely to macro-molecules. For example, in a study on protein, it has been shown that a major part of the absorption is due to the rearrangement that occurs, under the action of oscillatory pressure, in the hydration shell surrounding the protein molecule.[29] Since such rearrangement requires a finite time it tends to lag behind, and so to absorb energy from the pressure wave. It is not yet known whether so-called *relaxation processes* such as this can lead to any permanent modification of molecular structure other than that resulting from the thermal energy release to which they ultimately lead.

Scattering

Ultrasonic scattering by tissues[24] is a subject that has been given very little attention hitherto, although evidence for its occurrence comes from observations of the return of diffuse echoes from macroscopically homogeneous tissue volumes in the process of ultrasonic pulse-echo diagnosis. Such scattering may arise from the existence in the medium of acoustic inhomogeneities (e.g., density differences) and particularly those on a scale comparable with the ultrasonic wavelength (0.1–1 mm in tissue in the medical ultrasonic region). When an ultrasonic wave traverses such an inhomogeneous medium the different components vibrate with different amplitudes and thus move relatively to each other. It is this relative movement that is partly responsible for the scattering process, but it should be noted that not all the incident energy is re-radiated; a significant fraction is absorbed by viscous forces.

Reliable information is not yet available as to the relative importance of this process of *inelastic scattering* as a contributor to overall attenuation but, for a number of soft tissues, it appears to be responsible for some 10 per cent of the total.[24] A further interesting aspect of this phenomenon is that relative oscillatory motion of tissue components can be expected, on theoretical grounds, to give rise to steady fluid streaming patterns within the tissue and, under certain experimental conditions, it has been shown that such streaming movement can lead to a variety of structural changes in cells and tissues.[39]

Cavitation

The third attenuation process to be considered, cavitation, may be thought of as a form of scattering in which the scattering objects are themselves created by the action of the ultrasonic field. All normal liquids contain a considerable population of sub-microscopic gas bubbles and, under the action of mechanical vibration, these tend to grow by a process of *rectified diffusion* (net transfer of gas to the bubble from solution in the surrounding liquid). When such bubbles grow to a certain size in relation to the ultrasonic wavelength (about 6 μm diameter at 1 MHz) they behave as resonant cavities and their vibration amplitudes can become very large (several orders of magnitude greater than the vibration amplitude of the incident ultrasonic wave).

Such large amplitude vibration of bubbles within a liquid can lead, in two rather different ways, to modification of biological structures. In the first place, as in the case just described, of relative oscillatory motion occurring within tissue structures, the bubble vibration tends to set up a steady streaming pattern in the surrounding liquid, with consequent occurrence of localised high velocity gradients sufficient to shear cell membranes and large biological molecules.[37] In addition, however, as a result of the extremely rapid oscillatory compression of gas within the micro-bubbles, a phenomenon akin to ionisation occurs in the gas volume and chemically highly reactive free radical species are produced and released into the surrounding liquid in high local concentrations. In this respect cavitation bears an interesting resemblance to ionising radiation, and specifically to radiation of high linear energy transfer.[48] It has been shown that the rate of release of free radicals resulting from ultrasonic cavitation in water can be as high as that due to an absorbed dose rate of the order of 100 Gy min^{-1}.

The phenomena just described are sometimes referred to in the literature by the term *stable cavitation*, in distinction from *transient* or *collapse* cavitation, which is a phenomenon that generally only occurs at very high ultrasonic intensities and in which the liquid structure breaks down, giving rise to a cavity which collapses at the end of the negative half-cycle of the pressure wave.[11] There is some doubt about the extent to which cavitation in any form may take place in organised living tissues. Prolonged exposure to high intensities, however, has been reported to give rise, in mammalian tissues, to localised lesions that are suggestive of cavitation damage, and stable cavitation has been observed in association with the small gas bubbles that are naturally present in

certain plant tissues. More recently, exposure of guinea pigs to physical therapy ultrasound conditions (0.5 W cm^{-2} at 0.75 MHz for 1–5 min) has been shown to result in formation of intra-tissue cavities similar to those occurring as a result of decompression.[17] It is not yet clear whether such cavities, which presumably act as foci of mechanical vibration, are associated with significant tissue damage, but it is relevant to point out here that the process of stimulation of stable cavitation, as just described, requires a finite period of time for its full attainment and thus, even in liquids, it does not occur in irradiations involving pulses of a few microseconds duration, relatively widely spaced, such as are used in pulse-echo diagnosis.[23]

ULTRASOUND BIOLOGY

A very considerable volume of experimental work on the chemical and biological effects of ultrasound has been published in the past fifty years and it can, perhaps, best be classified in ascending order of complexity of the experimental system involved.

Chemical systems

The action of ultrasonically stimulated cavitation on simple chemical substances in aqueous solution has been studied by a number of authors and is reviewed elsewhere.[10] The effects here seem to be consistent with the action of the free radicals H and OH formed from water, but the precise nature of the reactions that occur is found to be dependent on the identity and concentration of gases dissolved in the water and thus capable of diffusing into the active cavities. A reaction that has been well studied and is commonly used as an indicator of cavitation activity is the release of free iodine from potassium iodide solution.

In addition to specific *sonochemical* action of this type, ultrasound is also capable of accelerating existing chemical reactions, such as the development of photographic emulsions,[30] apparently as a result of the fluid stirring action that it induces.

Macromolecules

Much attention has been given to the action of ultrasound on biological macromolecules in aqueous solution and, in particular, to effects on DNA.[38] Here it is found that the primary effect is of a degradation process in which double strand breaks are induced and arise preferentially at the midpoint of the molecule, which suggests fluid shear as a causative mechanism. Although much of the work on this phenomenon has been done at low ultrasonic frequencies, in the region of 20 kHz, it has also been shown to be effective in the 0.25–4 MHz range at intensities of the order of 1 W cm^{-2}.[23] Evidence for free radical attack, based on observed changes in ultraviolet absorption and melting temperature in ultrasonically treated DNA solutions, has also been reported but this appears to be a relatively minor effect even in irradiations leading to very considerable degradation (e.g., to a mean molecular weight of 0.25 million from an initial ten million).

In general the mechanical and chemical effects are clearly attributable to cavitation, since they can be inhibited entirely by degassing the solution or by increasing the effective ambient pressure. It has, however, been reported that, at moderately high intensities (25 W cm^{-2} at 1 MHz), DNA can be degraded by a mechanism other than cavitation, [20] although it has not been possible to detect degradation in proteins in the absence of cavitation.[7]

Cells

Much of the early work on the biological effects of ultrasound was carried out on aqueous suspensions of micro-organisms or other single cells, and great interest in this type of system continues because of the compromise it appears to provide between biology and simplicity. The most commonly observed type of effect here is a catastrophic rupture of the cell membrane,[27] and this in fact provides the basis for the action of the ultrasonic cell disintegrator.[26] The mechanism responsible for this effect is again cavitation and it has been shown that, although sonochemical action is normally present, its effects are likely to be buffered by the nutrient medium in which the cells are irradiated and it is thus, as with DNA degradation, the purely mechanical action of cavitation that is predominant.[4] Non-lethal damage is isolated cells, generally acting at the cell membrane, but possibly

also at other sites, has also been demonstrated. Several experimenters, for example, have reported changes in the electrophoretic mobility of various mammalian cells, apparently reflecting a reduction in the net surface electrical charge on the cell membrane, and this has been shown to be associated with the cavitational action of ultrasound, and also to be reversible.[28] A possibly related phenomenon is the observed increase in the permeability of certain cells to specific cytotoxic drugs following ultrasonic irradiation of an in vitro suspension culture.[31]

As previously noted, cavitation is primarily a phenomenon of the liquid state and there is some question as to the extent to which it occurs in organised tissues. Thus, in attempting to understand the effects of ultrasound occurring at the cellular level in organised tissue, evidence from experiments on liquid suspensions of cells is of somewhat limited value unless effective steps have been taken to inhibit cavitational action. Surprisingly little work has been attempted in this direction although it has been shown to be experimentally practicable, even for mammalian cells, for example by irradiating in gel suspension.[5] Another method of inhibiting cavitation is to use very short ultrasonic pulses and, by this means, it has been possible to study the response of the proliferation pattern of mammalian cells to ultrasonic irradiation and to show that no detectable change in this pattern is caused by irradiations for 5 h at a peak intensity of 15 W cm^{-2}, with pulses of 1 ms duration and duty factor 0.1.[3]

An interesting demonstration that ultrasound can act on living cells by a nonthermal, noncavitational mechanism is provided by experimental evidence that mammalian cells under temperature stress (at about 43°C) are destroyed much more rapidly in the presence of ultrasound than they are by the effects of temperature alone.[18]

Microscopic observation of single cells undergoing ultrasonic irradiation has been carried out on large plant and marine egg cells which are brought in close proximity to the tip of a needle vibrating at a frequency of about 20 kHz. Under these conditions acoustic streaming of nucleoplasm and cytoplasm is seen to occur, with movement, deformation and eventual fragmentation of some intracellular bodies.[50] The irradiation fields employed in this work, however, differ from those typical of an ultrasonic beam in that they are very nonuniform, and it is uncertain to what extent similar effects may occur under uniform beam conditions and at the considerably higher ultrasonic frequencies used in medical diagnostic applications.

Tissues

A central interest of ultrasound biology is in effects that take place in organised tissue but it is in just this area that, until now, interpretation of experimental data has been particularly difficult.

Two characteristics in which organised tissues differ appreciably from the simpler systems considered above are a relatively high ultrasonic absorption coefficient and a relatively low thermal mobility (i.e., low conductivity, zero convection, and variable heat transport by blood flow). Thus temperature rise can become an important consideration and, in practice, is a mode of action that can readily lead to modification or destruction of tissue function. It seems likely that this provides part of the basis for the therapeutic effectiveness that is claimed for ultrasound in physical medicine and it has been applied specifically in techniques, based on focused ultrasonic beams, for the destruction of small deep localised regions of tissue.[13]

Another situation in which localised heating can become significant, and inadvertently so, occurs when an ultrasonic beam is incident on an interface between soft tissue and bone. In this case it is possible for an appreciable fraction of the beam energy to be radiated into soft tissue in the form of transverse vibrations, for which the absorption coefficient in soft tissue (and thus the local rate of heat deposition) may be several orders of magnitude greater than that applicable to the more common longitudinal vibrations.

Whilst such temperature effects can evidently be of great importance in some situations, a potentially much more interesting field of study has arisen from the demonstration that other, nonthermal and noncavitational mechanisms of action of ultrasound can be effective in organised tissue. The first clear evidence for such a phenomenon

was obtained in an experimental series of irradiations of mouse spinal cord, in which hind leg paralysis was used as an end-point.[13] It was shown that paralysis could be obtained in conditions where neither temperature rise nor cavitation were significant factors; and a potentiating effect was also demonstrated, in which paralysis followed from two exposures well separated in time, neither of which would be effective separately. Unfortunately these early results were not thoroughly followed up at the time but some later work[43, 45] appears to confirm the general pattern of a threshold of accumulated exposure time, again with evidence against a thermal mechanism. Further support for this latter finding comes from a study on rat liver subjected to pulsed irradiation, in which marked qualitative differences in the lesions produced were found for different pulsing regimes but constant average exposure, and thus constant thermal effectiveness.[44]

The effects contributing to these observations all appear to be biologically destructive in their nature, and are thus in contrast with the demonstration of a significant increase in the rate of wound healing following low-intensity irradiation under conditions in which a thermal explanation can be discounted and cavitation is most improbable.[9] An enhanced rate of synthesis of DNA, and possibly also of protein, has been demonstrated in the regenerating tissue following treatment with ultrasound; it has been suggested that induced microstreaming movements in tissue fluid may be the physical mechanism responsible for these effects, but another interesting possibility is that they may result from the activity of enzymes released following ultrasonically induced rupture of lysosomal membranes.[19]

Another reported finding, which the original authors believed might also have a nonthermal basis, is that ultrasound, applied as an exposure that is in itself ineffective, is capable of enhancing by a factor of 1.7 the tumour-therapeutic effectiveness of a given dose of X-rays.[51] The possibility of a synergistic effect of this type is supported by some other reports, but in a later attempt to repeat the original experiments, using both in vivo and in vitro techniques, it was not possible to demonstrate significant evidence for such synergism under conditions in which temperature was constant.[5]

INVESTIGATIONS OF HAZARD

It is possible to distinguish three rather different but complementary approaches to the elucidation of the hazards that might be involved in medical application of ultrasound.[22] The general study of the biology and biophysics of ultrasound, as already described in this chapter, should eventually provide a fundamental understanding of all the processes that could lead to clinically harmful effects. This, however, is clearly a long term approach and it is necessary, in the shorter term, to carry out specific empirical investigations in which certain biological end-points, such as fetal abnormalities in rats or mice, are examined in relation to exposure parameters typical of particular medical applications. The third approach is that of epidemiology — the study of groups of individuals who have been exposed to ultrasound for medical or other reasons.

Experimental investigations

A practical difficulty encountered in planning specific experimental investigations of ultrasound hazard is that of knowing what effect to seek. Experience from the field of X-ray hazard is not necessarily of direct relevance here although it is indirectly of great value to the extent that it suggests useful criteria and high standards of judgement for such work. Viewed in this light the corresponding work on ultrasound appears to be at an early stage of development.

The first study directed specifically towards the problem of hazard of diagnostic ultrasound was designed to look for histological changes in animal brain tissue,[12] but subsequent work has been concerned mainly with the possibility of genetic changes or somatic mutation. A considerable, and somewhat perplexing, body of work has been reported on this subject. A major critical review, however, has concluded that, with the possible exception of certain extreme exposure conditions where significant heat shock may be induced, present evidence does not point to a high risk of genetic effects from medical ultrasound, and that current diagnostic procedures in particular are very unlikely to result in a genetic hazard.[46] This conclusion is supported by the results of systematic searches for genetic changes in yeast systems[47] and

in mice[33] following ultrasonic exposures greatly in excess of those in normal medical use.

Whilst a number of other similar studies have been equally reassuring,[25] it is important to note that some isolated, apparently contradictory findings have been reported particularly in relation to induction of cytogenetic[35] and teratogenic[42] changes. Although these reports were preliminary in nature, because of their potential significance they have both stimulated rather widespread series of follow-up studies, including subsequent investigations by one of the original authors.[34] These have produced uniformly negative results for the mammalian cell systems that have been studied.[25, 40] Similarly, no evidence has been found of increase in the 'sister chromatid exchange' phenomenon when human leukocytes were irradiated at intensities up to $36\ W\ cm^{-2}$,[32, 36] although there is some suggestion that cells irradiated ultrasonically in vitro could be transformed to have malignant potential.[32] The practical significance of this finding, if it can be upheld, is at present unclear.

Epidemiology

As is again evident from experience in the related field of radiological safety, the epidemiological approach to hazard of ultrasound is likely to be difficult and beset with pitfalls, and few investigations in this category have yet been reported. The first serious study[21] followed up 1114 apparently normal pregnant women examined by ultrasound in three different centres and at various stages of pregnancy. A 2.7 per cent incidence of fetal abnormalities was found in the group, which compared with a figure of 4.8 per cent reported elsewhere in a separate and unmatched survey of women who had not had ultrasonic diagnosis; neither the time in gestation at which the first examination was made, nor the number of examinations, seemed to increase the risk of fetal abnormality. More recently, the results of a study have been reported in which no apparent sequellae, at one year of age, could be found due to diagnostic ultrasound irradiation in the course of placental localisation prior to amniocentesis.[41] Valuable as such studies are, in the absence of better data the quantitative significance of their findings must be somewhat open to question and, in the long term, it is highly desirable to conduct more extensive and statistically controlled studies, preferably on a prospective rather than retrospective basis. Plans to carry out at least one major study of this kind are already well advanced.

CRITERIA FOR 'SAFE' EXPOSURES TO DIAGNOSTIC ULTRASOUND

It should have become clear from the discussion so far that the evidence on biological effects of ultrasound is still somewhat incoherent. In particular it is to be noted that an intellectually satisfactory approach to 'dosimetry' — the correlation of physical measures of radiation exposure with magnitude of consequent biological change — is still lacking. Whereas for ionising radiations there is an identifiable quantity (absorbed dose) that is usefully predictive of biological change, no comparable quantity has yet been satisfactorily demonstrated for ultrasound and it now seems that several different quantities may need to be employed in different circumstances.

Notwithstanding this lack of adequate foundation there are increasing practical pressures to provide simple and preferably quantitative indicators of the limits of exposure that may properly be considered 'safe'. This subject has been considered in the World Health Organization report already referred to,[25] in the following terms:

There is at present no clearly established evidence to indicate that the ultrasonic exposures involved (in diagnostic medicine) constitute a hazard to the patient. The suggestion has nevertheless been made that it would be desirable to set some upper limit to the physical exposure of a patient undergoing diagnostic examination. There is certainly no general agreement on the merits of this suggestion and there is a serious opposing point of view based on the fact that, other factors (e.g., system noise level, bandwidth) being equal, there is a direct relationship between the amount of diagnostic information obtainable and the level of primary ultrasonic energy directed into a patient.

The WHO report goes on to point out, however, that, in spite of the above difficulties, at least one systematic attempt to establish exposure criteria has achieved some degree of recognition. This

analysis,* due to the American Institute of Ultrasound in Medicine (AIUM), is made from both theoretical and experimental points of view. In the first place, making a series of fairly major assumptions, it calculates the maximum local ultrasonic intensity which, over a given period of time, could be applied without inducing a damaging degree of temperature rise in living tissue. Secondly, it documents the minimum ultrasonic intensities which have been reported to produce a 'biological effect' (regardless of whether the 'effect' was necessarily damaging). This analysis led to the conclusion that no 'bio-effects' occur in tissues at spatial peak, temporal average intensities less than 0.1 W cm^{-2} for long exposures or less than $50/t$ W cm^{-2} for exposures of less than 500 s (where t is the irradiation time in s).

Such a statement is superficially attractive as it seems to provide some practical guideline. In the view of the present author, however, it is dangerous to the extent that it is likely to be used uncritically and out of context, and this could be particularly undesirable if, as has already been seriously suggested, it were used as the basis for official regulations. As already indicated, the scientific basis of ultrasound dosimetry has not yet progressed to the point of confirming that time averaged intensity (W cm^{-2}) is a particularly appropriate measure of biological hazard. The AIUM guideline may be the best available for the time being but its inadequacies should always be borne in mind.

CONCLUSIONS

The study of the biological effects of ultrasound is still, after fifty years, in a rather primitive state. Of the several distinct mechanisms of action that can be identified, the thermal effect is the simplest and

* The AIUM statement (published in Ultrasound Med Biol 1976, 2:351, and subsequently amended) is as follows: 'In the low megahertz frequency range there have been, as of this date, no independently confirmed significant biological effects in tissues exposed to intensities (spatial peak, temporal average (SPTA) as measured in a free field in water) below 100 mW cm^{-2}. Furthermore, for ultrasonic exposure times (total time: this includes off-time as well as on-time for a repeated-pulse régime) less than 500 s and more than 1 s, such effects have not been demonstrated at higher intensities when the product of the intensity and exposure time (as defined above) is less than 50 J cm^{-2}.'

best documented although the actual processes underlying ultrasonic absorption in mammalian tissues are poorly understood. Cavitation constitutes a very effective mechanism of action in liquid systems under the conditions in which much of the published experimental work on ultrasound biology has been carried out, but its occurrence under conditions experienced in diagnostic medical applications is very doubtful. Evidence is accumulating for the existence of effects that are neither thermal nor cavitational: their actual nature is unclear, and there are no substantial data to indicate that they could constitute a hazard in the medical use of ultrasound. This latter conclusion has been uniformly supported, whenever the work has been carried out under conditions that have been shown to be repeatable and quantitatively significant, by the evidence of toxicological studies on genetic and related phenomena. Finally, the epidemiological evidence is sparse and satisfactory studies will almost certainly prove expensive and technically difficult to implement. Nevertheless, in view of the very widespread use of diagnostic ultrasound that seems likely in the future, and not least in obstetrics, it is of great importance that good follow-up information should be soon obtained from a proportion of the current large number of human irradiations.

REFERENCES

1. Bergman L 1954 Der Ultraschall. Hirzel Verlag, Stuttgart
2. Bhatia A B 1967 Ultrasonic absorption. Clarendon Press, Oxford
3. Clarke P R, Hill C R 1969 Biological action of ultrasound in relation to the cell cycle. Exp Cell Res 58: 443–444
4. Clarke P R, Hill C R 1970 Physical and chemical aspects of ultrasonic disruption of cells. J Acoust Soc Am 47: 649–653
5. Clarke P R, Hill C R, Adams K 1970 Synergism between ultrasound and X-rays in tumour therapy. Br J Radiol 43: 97–99
6. Dunn F, Edmonds P D, Fry W J 1969 Absorption and dispersion of ultrasound in biological media. In: Schwann H P (ed) Biological engineering. McGraw Hill, New York, p 205–332
7. Dunn F, Macleod R M 1968 Effects of intense noncavitating ultrasound on selected enzymes. J Acoust Soc Am 44: 932–940
8. Dunn F, O'Brien W D 1978 Ultrasonic absorption and dispersion. In: Fry F J (ed) Ultrasound, its application in medicine and biology. Vol 1. Elsevier, Amsterdam, p 393–440
9. Dyson M, Pond J B, Joseph J, Warwick R 1970 Stimulation of tissue regeneration by pulsed plane-wave ultrasound. IEEE Trans Sonics Ultrason SU-17: 133–140
10. El'piner I E 1964 Ultrasound: physical, chemical and biological effects. Consultants' Bureau, New York
11. Flynn H G 1964 Physics of acoustic cavitation in liquids. In: Mason W P (ed) Physical acoustics. Vol IB, Academic Press, New York, p 57–214
12. French L A, Wild J J, Neal D 1951 Attempts to determine harmful effects of pulsed ultrasonic vibrations. Cancer NY, 4: 342–344
13. Fry W J 1958 Intense ultrasound in investigation of the central nervous system. Adv Biol Med Phys 6: 281–348
14. Fry W J, Dunn F 1962 Ultrasound: analysis and experimental methods in biological research. In: Nastuk W L (ed) Physical techniques in biological research. Vol IV, Academic Press, New York, p 261–394
15. Haar G ter 1980 Safety of medical ultrasound. In: Kurjak A (ed) Progress in medical ultrasound: reviews and comments. Vol 1, Excerpta Medica, Amsterdam, p 313–318
16. Haar G ter 1981 Safety of medical ultrasound. In: Kurjak A (ed) Progress in medical ultrasound: reviews and comments. Vol 2, Excerpta Medica, Amsterdam, p 271–275. See also Vol 3 (1982) p 349–355
17. Haar G ter, Daniels S, Eastaugh K C, Hill C R 1982 Ultrasonically induced cavitation in vivo Br J Cancer 45, Suppl 5: 151–155
18. Haar G R ter, Stratford I J, Hill C R 1980 Ultrasonic irradiation of mammalian cells in vitro at hyperthermic temperatures. Br J Radiol 53: 784–789
19. Harvey W, Dyson M, Pond J B, Grahame R 1975 The 'in vitro' stimulation of protein synthesis in human fibroblasts by therapeutic levels of ultrasound. In: Kazner E et al (eds) Ultrasonics in medicine. Excerpta Medica, Amsterdam, p 10–21
20. Hawley S A, Macleod R M, Dunn F 1963 Degradation of DNA by intense, noncavitating ultrasound. J Acoust Soc Am 35: 1285–1287
21. Hellman L M, Duffus G M, Donald I, Sunden B 1970 Safety of diagnostic ultrasound in obstetrics. Lancet 1: 1133–1135
22. Hill C R 1968 The possibility of hazard in medical and industrial applications of ultrasound. Br J Radiol 41: 561–569
23. Hill C R 1972 Ultrasound exposure thresholds for changes in cells and tissues. J Acoust Soc Am 52: 667–672
24. Hill C R, Chivers R C, Huggins R W, Nicholas D 1978 Scattering of ultrasound by human tissues. In: Fry F J (ed) Ultrasound: its application in medicine and biology. Vol 1, Elsevier, Amsterdam, p 441–494
25. Hill C R, Haar G R ter 1982 Ultrasound. In: Suess M J, (ed) Nonionizing radiation protection. Geneva, World Health Organization, p 199–223
26. Hughes D E 1961 The disintegration of bacteria and other micro-organisms by the MSE-Mullard ultrasonic disintegrator. J Biochem Microbiol Technol Engng 3: 405–433
27. Hughes D E, Nyborg W L 1962 Cell disruption by ultrasound. Science NY 138: 108–114
28. Joshi G P, Hill C R, Forrester J A 1973 Mode of action of ultrasound on the surface charge of mammalian cells in vitro. Ultrasound Med Biol 1: 45–48

29. Kessler L W, Dunn F 1969 Ultrasonic investigation of the conformal changes of bovine serum albumin in aqueous solution. J Phys Chem 73:4256–4263
30. Kossoff G 1962 Calibration of ultrasonic therapeutic equipment. Acustica 12:84–90
31. Kremkau F W, Kaufmann J S, Walker M M, Busch P G, Spurr C L 1976 Ultrasonic enhancement of nitrogen mustard cytotoxicity in mouse leukemia. Cancer NY 37:1647
32. Liebeskind D, Bases R, Elequin F, Neubort S, Leifer R, Goldberg R, Koenigsberg M 1979 Diagnostic ultrasound: effects on the DNA and growth patterns in animal cells. Radiology 131:177–184
33. Lyon M F, Simpson G M 1974 An investigation into the possible genetic hazards of ultrasound. Br J Radiol 47:712–722
34. Macintosh I J C, Brown R C, Coakley W T 1975 Ultrasound and 'in vitro' chromosome aberrations. Br J Radiol 48:230–232
35. Macintosh I J C, Davey D A 1972 Relationship between intensity of ultrasound and induction of chromosome aberrations. Br J Radiol 45:320–327
36. Morris S M, Palmer C G, Fry F J, Johnson L K 1978 Effect of ultrasound on human leukocytes: sister chromatid exchange analysis. Ultrasound Med Biol 4:253–258
37. Nyborg W L 1965 Acoustic streaming. In: Mason W P (ed) Physical acoustics. Vol IIB, Academic Press, New York, p 265–331
38. Peacocke A R, Pritchard N J 1968 Some biological aspects of ultrasound. Prog Biophys Chem 18:186–208
39. Ravitz M J, Schnitzler R M 1970 Morphological changes induced in frog semitendinosus muscle fiber by localized ultrasound. Exp Cell Res 60:78–85
40. Roseboro J A, Buchanan P, Norman A 1978 Effect of ultrasonic irradiation on mammalian cells and chromosomes in vitro. Phys Med Biol 23:324–331
41. Scheidt P C, Stanley F, Bryla D A 1978 Investigations for effects of intrauterine ultrasound in humans. Am J Obstet Gynec 131:743–748
42. Shoji R 1975 Influence of low-intensity ultrasonic irradiation on prenatal development of the inbred mouse strains. Teratology 12:227–231
43. Taylor K J W 1970 Ultrasonic damage to the spinal cord and the synergistic effect of hypoxia. J Path 102:41–47
44. Taylor K J W, Connolly C C 1969 Differing hepatic lesions caused by the same dose of ultrasound. J Path 98:291–293
45. Taylor K J W, Pond J B 1972 A study of the production of haemorrhagic injury and paraplegia in rat spinal cord by pulsed ultrasound of low megahertz frequencies in the context of the safety for clinical usage. Br J Radiol 45:343–353
46. Thacker J 1973 The possibility of genetic hazard from ultrasonic irradiation. Curr Top Radiat Res Quart 8:235–258
47. Thacker J 1974 An assessment of ultrasonic radiation hazard using yeast genetic systems. Br J Radiol 47:130–138
48. Weissler A 1959 Formation of H_2O_2 by ultrasonic waves: free radicals. J Am Chem Soc 81:1077–1081
49. Wells P N T 1975 Absorption and dispersion of ultrasound in biological tissue. Ultrasound Med Biol 1:369–376
50. Wilson W L, Wiercinski F J, Nyborg W L, Schnitzler R M, Sichel F J 1966 Deformation and motion produced in isolated single cells by localized ultrasonic vibration. J Acoust Soc Am 40:1363–1370
51. Woeber K 1965 The effect of ultrasound in the treatment of cancer. In: Kelly E (ed) Ultrasonic energy. University of Urbana Press, Urbana, p 137–149
52. Wood R W, Loomis A L 1927 The physical and biological effects of high-frequency sound waves of great intensity. Phil Mag [7]4:417–436

Index